ABHANDLUNGEN DER DEUTSCHEN AKADEMIE
DER WISSENSCHAFTEN ZU BERLIN

Klasse für Medizin

Jahrgang 1968 Nr. 2

SYMPOSIUM ÜBER BIOCHEMISCHE ASPEKTE DER STEROIDFORSCHUNG

Jena, 18. bis 21. September 1967

Herausgegeben von

KURT SCHUBERT

Mit 106 Abbildungen, 60 Tabellen und 4 Schemata

AKADEMIE-VERLAG · BERLIN

1969

Vorgelegt von Hrn. KNÖLL in der Klassensitzung vom 25. April 1968

Zum Druck genehmigt am gleichen Tage, ausgegeben am 17. Februar 1969

Erschienen im Akademie-Verlag GmbH, 108 Berlin, Leipziger Straße 3–4

Lizenznummer: 202 · 100/437/69

Gesamtherstellung: VEB Druckerei „Thomas Müntzer", 582 Bad Langensalza

Bestellnummer: 2001/68/IV/2 · ES 18 G 1/17 C 3

42,–

Contents

Introduction

KURT SCHUBERT

This symposium covers the biosynthesis, the metabolism, the degradation and the mechanism of action of steroids. The participants come from academic as well as industrial research institutions. This demonstrates the close interconnection between theory and practice.

The working with microorganisms has been widely applied in the field of steroid research. New insights have been gained and significant prodecures of practical value have been employed in industry. Investigations with microorganisms are extraordinarily serviceable to basic research.

By means of microorganisms model conceptions may be obtained which underlie biochemical investigations in higher organisms. Microbial enzymes have proved to be highly suitable for the investigations of special reactions, as for example the biosynthesis of sterols, the stereospecific hydrogenation of steroid hormones to metabolites and the oxidative degradation of the sterane skeleton. New products are being found.

Also steroids, which are difficult to produce chemically, are easily obtained in this way. A lot of exertions have yet to be made to elucidate the mechanism of action of steroids. In this respect, it will be necessary to look for further experimental starting-points.

Microbial investigations with steroids must not only be regarded as an end in itself, but they often form a section of a pathway towards attaining, in a rather short time, the object of clearing up the steroid metabolism and function of steroids. The results obtained with microorganisms may be a starting-point for further investigations in animals and men.

On the other hand, certain complex processes in mammals, by means of microorganisms, can be investigated more closely within a system that is rather easy to survey.

It is, therefore, necessary for the different lines of research to be coordinate and for the representatives of different departments to cooperate. Taking into account the present speedy development in the field of steroid research, the cooperation of biologists, chemists, and medical men is of decisive importance. Biochemistry in this connection takes a prominent position. Far-sighted experts in industry and at university level are agreed to point out the importance of basic biochemical research. It is the development in the field of steroids in particular which shows that the intensity of the centre of research has shifted from the chemistry of natural products to biochemistry.

In the present volume the lectures of the symposium have been compiled with the editorial assistance of Mrs. Dipl.-Chem. Karin Schade and Mrs. Dipl.-Chem. Gertrud Schumann.

On the Biosynthesis of Sterols and Pentacyclic Triterpenoids in *Tetrahymena pyriformis* [1, 2]

R. L. Conner

Department of Biology

F. B. Mallory

Department of Chemistry

Bryn Mawr College, Bryn Mawr, Pennsylvania, USA

Summary

Tetrahymena pyriformis synthesizes one and possibly two pentacyclic triterpenoid alcohols from acetate, mevalonate and squalene. The formation of these two alcohols, tetrahymanol and "diplopterol", is inhibited by the addition of cholesterol to the culture fluid of the ciliates. The synthesis of other compounds from acetate and mevalonate were not affected by the cholesterol addition which suggests a block occurs after "isoprene" formation and before squalene formation. Further, inhibition by cholesterol of squalene conversion to the pentacyclic alcohols has been established.

The mechanism for tetrahymanol formation from squalene may be a non-oxidative proton-induced cyclization of the triterpene rather than the intermediate formation of 2,3 oxido squalene as in lanosterol and β-amyrin synthesis.

Introduction

Investigations concerned with steroid metabolism range widely throughout physiological and biochemical circles as witnessed by the group meeting here today. The endocrinologist is concerned with the mode of action of the various steroidal hormones, with the degradative fate, excretion, and biosynthetic origin of these compounds. This latter facet included studies dealing with the conversion of cholesterol (or related compounds) to steroid hormones and these investigations, in turn, led directly to inquiries about the nature of cholesterol biosynthesis and degradation.

The mode of action of sterols and steroids in the metabolism of organisms, of tissues, and of cells in higher organisms has engendered considerable speculation based on an extensive series of investigations. The problem of establishing the relationship of *in vitro* experiments to *in vivo* studies remains our greatest difficulty in interpretation of the data available, due to the many types of interactions. A rigorous distinction between the primary and secondary effects of steroids has not been obtained, even though a given sequence of events can be described, at least in general terms, when such compounds are used experimentally.

A large number of sterols have been described in the plant and animal kingdoms. Early investigators such as Doreé, and later Bergman (1962), concluded that cholesterol has become the primary sterol in higher animals through the course of evolution and that several phytosterols, such as the sitosterols and stigmasterol, have become dominant in plants. Ergosterol is prevalent in lower plant forms. Most organisms have the ability to synthesize sterols from non-steroidal precursors making an investigation of the *in vivo* action of these compounds complex. Some organisms, however, have been found to require dietary sterols for growth; and hence, access to the investigation of the influence of these compounds on metabolism and membrane structure is readily available.

---

[1] This research was supported in part by Bryn Mawr College and by Grant GB-4605, National Science Foundation.

[2] The authors wish to thank Dr. L. J. Berry for his encouragement, Dr. D. Kritichevsky for the ^{14}C-squalene, and Mrs. J. R. Landrey and Miss K. Howard for their technical assistance.

A nutritional requirement for preformed sterols has been found in a number of lower biological forms including (1) mycoplasma (PPLO) (SMITH, 1964), (2) for all insects thus far investigated (CLAYTON, 1964), and (3) for a number of protozoa (HOLZ et al., 1962c). These organisms are unable to synthesize at least one necessary sterol or steroid. These organisms pose an interesting problem in that the same sterols cannot be used to satisfy all the known nutritional requirements, which indicates a range of stereochemical specificity and/or synthetic capacity.

Detailed studies relating the molecular structure of steroids to biological activity have been carried out by CAILLEAU (as cited in LWOFF, 1951) for the flagellate, *Trichomonas*; by VISHNIAC et al. (1953, 1955) for *Labyrinthula*; and by CONNER and VAN WAGTENDONK (1955) for the ciliated protozoal form, *Paramecium aurelia*. Similar sterol nutritional requirements have been noted and specificity studies have been carried out for some members of the genus *Tetrahymena* (HOLZ et al., 1961a, 1961b, 1962b).

The species *Tetrahymena pyriformis* has no sterol nutritional requirement (KIDDER et al., 1951) and at this time *de novo* sterol biosynthesis has not been demonstrated in these cells. The major solid alcohol of this ciliate was originally isolated by MCKEE et al. (1947) and identified as a compound isomeric with cholesterol; however, further examination of this material by MALLORY et al. (1963) established the alcohol to be a pentacyclic triterpenoid alcohol and an unambiguous synthesis by TSUDA et al. (1965) revealed the structure to be that represented in Figure 1. The occurrence of a pentacyclic triterpenoid alcohol in *Tetrahymena* is unusual since this type of compound has not been found previously in animal-like cells and further, pentacyclic triterpenoids have been found in no other case to be the dominant solid alcohol in an organism.

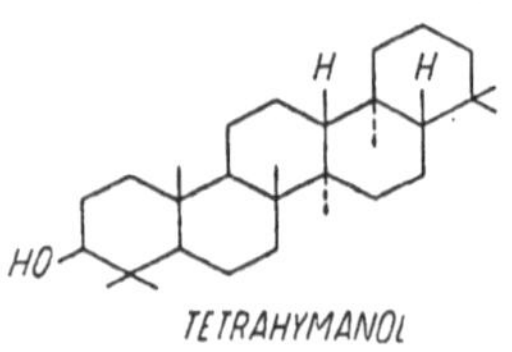

Fig. 1. Tetrahymanol, m. p. 315—320° (sublimed)

An absolute sterol nutritional requirement can be induced in *Tetrahymena pyriformis* by a number of growth inhibitors including compounds known to block cholesterol biosynthesis such as triparanol (JOHNSON et al., 1961; HOLZ et al., 1962a) as well as a number of substances whose mode of action is unknown, including certain substituted purines (DEWEY et al., 1959), colchicine (CONNER et al., 1958), steroids of the pregnane series (CONNER, 1959), the oxidative phosphorylation uncoupling agent, 2,4-dinitrophenol (CONNER, 1957), and 3-diakylaminoethoxy steroids (HOLMLUND et. al., 1966). This type of interaction between the growth inhibitors and sterols, whether primary or secondary, points out the importance of sterols or steroidal derivatives in the physiology of this organism.

SHORB et al. (1965) have demonstrated that triparanol blocks the synthesis of tetrahymanol with the resultant accumulation of squalene. CONNER et al. (1964b), as an aid to a further understanding of the mode of action of sterols in *Tetrahymena*, determined the fate of cholesterol when added to a cellular suspension of this protozoal form. Extensive accumulation of 4-^{14}C-cholesterol by cells suspended in a non-growth medium was observed. Fractionation by silicic acid column chromatography of the non-saponifiable lipids indicated the conversion of cholesterol to more polar materials. Subsequently, these metabolic products of cholesterol have been identified as 22-dehydrocholesterol, $\Delta^{5,7,22}$-cholestatrien-3β-ol, and possibly 7-dehydrocholesterol (CONNER et al., 1966b, 1968). It was noted, in the course of these studies, that the cellular content of tetrahymanol was markedly lowered when cholesterol was present in the growth medium, indicating a profound influence of sterols on the regulation of triterpene biosynthesis (CONNER, 1965). This observation led to an investigation as to where the cholesterol block occurs in tetrahymanol biosynthesis and to attempts to determine if the inhibition of the pentacyclic triterpene biosynthesis was a feedback mechanism similar to that noted for cholesterol by a number of investigators (GOULD et al., 1957; GOULD, 1959; BUCHER, 1959; BUCHER et al., 1959 and SIPERSTEIN et al., 1959) in mammalian tissues. To answer these questions a detailed knowledge of tetrahymanol biosynthesis was required in order to determine the site or sites of the inhibition induced by the addition of cholesterol to the growth medium. This problem had additional importance since TSUDA et al. (1965) proposed proton-induced cyclization of squalene to form tetrahymanol, thus differing from the biosynthetic scheme proposed for lanosterol formation (van TAMELEN et al., 1966; COREY et al., 1966a, 1966b).

This paper is a report of some of our recent investigations on the nature of tetrahymanol formation and on the possible sites of inhibition induced by exogenous cholesterol on the biosynthetic process.

Materials and methods

Axenic cultures of *Tetrahymena pyriformis* W were maintained in a proteose peptone-based culture fluid and the cells were harvested by methods described previously (CONNER et al., 1964a, 1966a). When the ciliate was grown with a cholesterol supplementation, an ethanolic solution of cholesterol was dispersed aseptically into culture fluid that had been previously autoclaved and reheated to 80°. After the mixture was cooled to 28°, the seed culture of ciliates was added. The cell population was claimed by centrifugation after a growth period of 36—40 hours (stationary phase) at 28 ± 1°. The cells were lyophilized and stored in the dark at —20° under nitrogen until the lipids were extracted.

Uninoculated flasks of culture fluid with and without emulsified cholesterol were incubated for an equivalent period of time to serve as controls. The water was sublimed by vacuum desiccation and the residue carried through the extraction procedure. No sterols were detected by gas-liquid chromatographic or thin-layer chromatographic analyses without cholesterol supplementation, and the cholesterol emulsified in the culture fluid did not give rise to discernable amounts of oxidation products in the time span employed.

The commercial samples of cholesterol were purified through the dibromide and recrystallized from ethanol (mp 150—151°, uncorrected, evacuated capillary). These preparations yielded a single spot, as visualized with iodine staining, on Silica Gel G (BRINKMAN) thin-layer chromatography plates with a variety of solvent systems, and gave only a single peak when analyzed by gas-liquid chromatography on the columns described below.

An F and M Model 400 Gas Chromatograph, equipped with a hydrogen flame ionization detector and a disc integrator, was used for gas-liquid chromatographic analyses and for quantitative determinations of tetrahymanol. Nitrogen was the carrier gas. The non-saponifiable lipids were analyzed using glass columns (3 mm I. D.) packed with 80—100 mesh Gas-Chrom S coated with 0.75% SE-52 Silicone Gum Rubber (6 ft.), or with 80—100 mesh Gas-Chrom S coated with 2% neopentyl glycol succinate (2 ft.).

Lipid Extraction

All organic solvents were redistilled before use. The lyophilized cellular residue was extracted three times with a chloroform–methanol mixture (2 : 1, v/v) at approximately 50°. The total volume of solvent was at least 20 ml/g solids. After the extract was filtered through sintered glass to remove cellular debris, the solvent was removed using a rotary evaporator. This total lipid fraction was saponified overnight in a nitrogen atmosphere at 37 ± 1° with 10% potassium hydroxide in 80% methanol. One volume of glass-distilled water was added to the reaction mixture followed by two volumes of diethyl ether. The two phases were separated, the aqueous layer was reextracted with one volume of ether, and the combined ethereal extracts were backwashed with water until neutral. The unsaponifiable fraction was dried over anhydrous magnesium sulfate and the ether was removed using a rotary evaporator.

Batches of cells were grown with 5 mg of one of the isotopically labeled precursors of tetrahymanol per liter of culture fluid in the presence and absence of 5 mg cholesterol per liter of culture fluid. The cells were grown with 2-^{14}C-acetate, DL-mevalonic-2-^{14}C-acid (DBED salt) and ^{14}C-squalene. The acetate and mevalonate were purchased from the New England Nuclear, Inc. and were used without further purification. Squalene was prepared by the incubation of rat liver slices with 2-^{14}C-acetate in a nitrogen atmosphere. The reaction was stopped by the addition of chloroform–methanol (2 : 1). The lipid fraction was saponified and the squalene was isolated by thin-layer chromatography, eluted from the plates, and purified as the thiourea adduct after the addition of carrier squalene.

Autoradiograms

When ^{14}C-precursors were employed, autoradiograms were used for the detection of the labeled materials on the thin-layer chromatographic plates. Eastman Kodak Royal-Blue X-ray No-Screen film was employed. After appropriate exposure times the film was developed in the usual manner.

The radiolabeled non-saponifiable lipid fraction obtained from *Tetrahymena* was resolved into a series of components by thin-layer chromatography using Silica Gel G plates developed with benzene-ethyl acetate (20 : 1). Squalene was identified using similar plates with petroleum ether (b.p. 40—60°) as the developing solvent. Figure 2 is a composite drawing from a number of autoradiograms. A series of graded amounts of each of the extracts was spotted on thin-layer plates in order to determine when overloading occurred, and also to detect

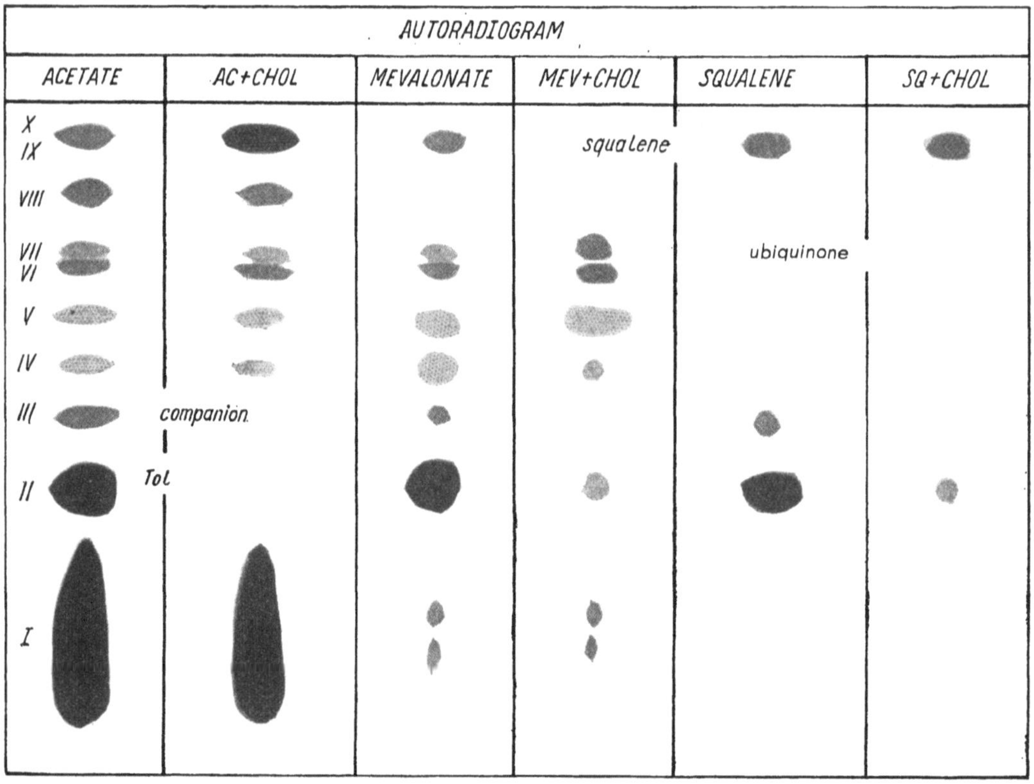

Fig. 2. A composite drawing of a number of autoradiograms prepared after the tlc separation of the non-saponifiable lipid fraction obtained from *Tetrahymena pyriformis*. The cells were grown in cultures supplemented with 2-^{14}C-acetate, 2-^{14}C-mevalonate or ^{14}C-squalene in the presence or absence of added cholesterol. Further details are described in the text

regions containing small amounts of radioactivity. In practice this was not a difficult technical problem since once it was established the plates were not overloaded, a series of exposure times of the film to the plates was used to ascertain when all regions of radioactivity capable of detection by the method had been located. A film was selected that gave optimal resolution of the components and showed all the regions of radioactivity with clarity. The density of each region was assigned by comparison to a graded series of exposures on the same film. All plates were stained with phosphomolybdic acid and heated at 125° for 15 minutes to visualize the materials found on the plates. The addition of the radiolabeled materials to the culture fluid did not alter the composition of the non-saponifiable lipid fraction in a qualitative fashion and the apparent quantitative changes in lipid composition are noted in the Results section.

Results and discussion

The lipid content of *Tetrahymena pyriformis* has been examined both chemically and cytochemically by a large number of investigators (Aaronson et al., 1961; Erwin et al., 1963; McCashland, 1963; McCashland et al., 1957; McKee et al., 1947; Taketomi, 1961; and Thompson, 1967). These cells contain lipid granules composed primarily of triglycerides which contain saturated or monounsaturated fatty acids, structural lipid components (tetrahymanol, phospholipids) and lipoids such as coenzyme Q. The lipid composition of these ciliates is dependent on the composition of the culture fluid, the age of the culture, the degree of aeration, and the strain. Triglycerides increase in amount as the culture enters the stationary phase of growth as do the number of lipid granules (Allison et al., 1967; Everhart et al., 1966). The lipids account for 14—24% of the dry weight. Erwin et al. (1963) reported 68% of the total lipids as phospholipids possessing a high percentage of polyenoic fatty acids and 28% neutral lipids for cultures of *Tetrahymena pyriformis*, variety I, mating type II, in the logarithmic phase of growth, while Thompson (1967) found *Tetrahymena pyriformis* W contained primarily phospholipids during the log-phase with increasing amounts of triglycerides as the culture entered the stationary phase until "accumulation of triglycerides may approach on a weight basis the phospholipid content". Taketomi (1961) noted that *Tetrahymena pyriformis* W in stationary phase contained 17.4% lipid by dry weight, of which 27% was phospholipid and 73% was neutral lipid. Table 1 lists the lipid composition of late log-phase cultures of *Tetrahymena pyriformis* W grown in our laboratory by the techniques as described in the Materials and Methods section.

The non-saponifiable fraction contains a number of components. The column labeled "acetate" in Figure 2 shows the various compounds, visualized by phosphomolybdic acid staining, of the non-saponifiable fraction spotted on a Silica Gel G thin-layer plate (tlc) and developed with petroleum ether–benzene (20 : 1). The large flame-shaped cone (I) observed at the

Table 1

Lipid composition of *Tetrahymena pyriformis*
(Late log phase culture — proteose peptone)

500 ml culture[1]	= ~ 4.0×10^8 cells
	= 1.27 g dry weight
	= 190 mg lipid (~15% dry weight)
	= 90 mg phospholipids
	= 140 mg "neutral" lipids
	= ~ 8.0 mg non-saponifiable material
	= 4,1 mg tetrahymanol

[1] 500 ml culture fluid in a 2600 ml low-form Erlenmeyer flask. The values are for cells in the late log phase of growth (28 ± 1°). The cultures were not aerated or shaken.

origin and extending well up the plate is an artifact of the hydrolysis procedure since this material does not appear in the total lipid fraction before saponification. In a recent paper, THOMPSON (1967) has confirmed and extended the investigations of TAKETOMI (1961) and HACK et al. (1962) showing that sizeable quantities of α-glyceryl ether derivatives occur in these ciliates. The ethers are found in the phospholipid fraction replacing the classical lecithin, to a significant degree as well as the phosphatydyl ethanolamine. Further, 2-aminoethylphosphonic acid is found to replace a portion of the usual 2-aminoethylphosphoric acid in the ethanolamine phosphatide (LIANG et al., 1966). The alcohol constituent of the ether seems to be exclusively chimyl alcohol (n-hexadecanol). THOMPSON (1967) has demonstrated that acetate is a metabolic precursor of the chimyl alcohol. The α-glyceryl ether diol, a product of alkaline hydrolysis of these unusual phospholipids, is included in the unsaponifiable lipid fraction obtained from *Tetrahymena* and seems to represent region I on the tlc plates.

Tetrahymanol (II) can be located on thin-layer plates by phosphomolybdic acid staining, but not by I_2 visualization, and the identity of this component can be readily confirmed by gas-liquid chromatographic analysis. Immediately above the tetrahymanol on the tlc plates is a compound which, at the present time, is termed the "companion" (III) since it also appears to be a pentacyclic triterpenoid alcohol. This compound has been tentatively identified as diplopterol on the basis of gas-liquid chromatographic comparison with an authentic sample. Six other components can be visualized on thin-layer plates, the identity of ubiquinone (CoQ_8, VI) and squalene have been confirmed. The squalene region (IX) contains at least two components which can be resolved on Silica Gel G tlc plates developed with 100% petroleum ether.

GLC analysis shows a number of the components in the non-saponifiable fraction. The sterol region of the chromatograms indicate at least six components with the dominant material being tetrahymanol (see Figure 3). The lower reproduction was obtained by increasing the attenuation of the instrument so that the tetrahymanol peak remained on the chart paper

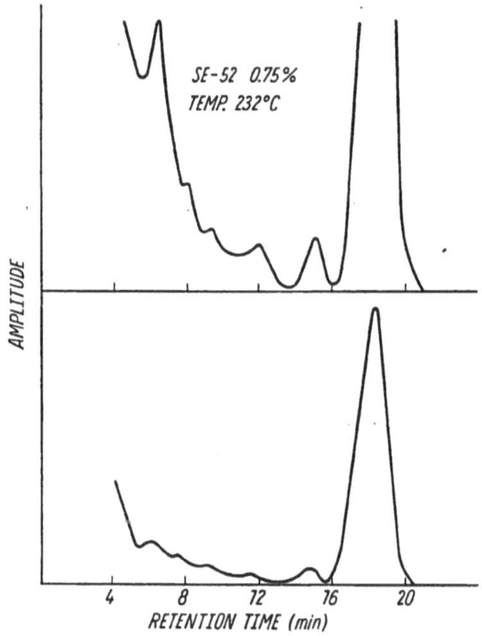

Fig. 3. GLC tracing of the non-saponifiable materials obtained from *Tetrahymena pyriformis*[1]

[1] SE-52 column and conditions as described in Materials and Methods. 4 mg of the non-saponifiable fraction dissolved in 0.2 ml benzene. Tracings from a 5 µl injection

and thus provides an indication of the relative amounts of the various compounds present in the sample. Authentic diplopterol dehydrates under the conditions used in the glc analysis; therefore, several of the peaks detected in the non-saponifiable fraction may be artifacts arising from the decomposition of this pentacyclic triterpenoid. The remaining components may either be synthesized by the cells or may originate from the culture fluid. We are attempting, at the present time, to isolate these compounds in sufficient quantities for more rigorous identification. Tetrahymanol and the companion are known to arise *de novo* from acetate, mevalonate, and squalene; however, the biosynthetic origin of the remaining components detected by glc has not been established. A quantitative glc analysis indicates that cells from approximately 400 liters of culture fluid would be required to isolate 10 mg of the companion and 1—2 mg of the other components present in this fraction. At the present time, all of these materials other than tetrahymanol and the companion can be catalogued solely on the basis of their chromatographic behaviors. It is clear that the sterol content, if indeed any sterol is present, of *Tetrahymena* is quite small and that tetrahymanol represents the major solid alcohol.

Table 2 shows data from a number of experiments where the cells were allowed to grow to stationary phase with a series of external concentrations of cholesterol. The amount of sterol present in the lipid extracts was estimated by glc where the detector response was recorded by the use of a disc integrator. Cholesterol is accumulated by the cells (CONNER et al., 1964b) and the percentage of the total sterol accumulated by the cells is approximately constant throughout all concentrations employed. These data suggest that the food vacuole may represent the major site of entry of cholesterol into the cells. When the amount of cholesterol accumulated by the ciliates is calculated on the basis of cell numbers there is some variation in amount at the various concentrations of sterol employed; however, this may be due to a change in cell size or volume and may not represent a change in the amount of sterol accumulated when calculated on a mass basis.

Table 2

The accumulation of cholesterol by *Tetrahymena* as a function of the external concentration

[External] (mg/500 ml culture fluid)	Recovery (mg)	%	mg/10^8 cells[1]
2.5	1.42	57	0.61
5.0	3.0	60	1.22
10.0	5.9	59	2.50

[1] Endogenous "sterol" = ~ 4—8 µg/10^8 cells.

% composition of the sterol recovered from the cells

mg	Cholesterol[1]	$\Delta^{5,7,22}$ Cholestratrien-3β-ol[1]	22-Dehydrocholesterol[1]
2.5	7	93	trace
5.0	14	79	5
10.0	25	52	23

[1] Amounts estimated by GLC detector response as recorded by a disc integrator. Colums: 0.75% SE-52, 6 ft., 235 °C and 2% neopentyl glycol succinate, 2 ft., 220 °C.

The amount of sterol accumulated by *Tetrahymena* is hundreds of times greater (see Table 4) than the amount estimated to be present in non-sterol supplemented cultures, if indeed any sterol is present in the latter. The composition of the sterol mixture recovered from the cells is listed in Table 4. Three major components have been detected including cholesterol, $\Delta^{5,7,22}$cholestatrien 3β-ol, and 22-dehydrocholesterol (CONNER et al., 1968). It is clear from these values regarding the sterol composition that *Tetrahymena* possess the ability to accumulate substantially greater quantities of cholesterol than can be metabolized during the time course of the experiment. For example, when cholesterol is supplied at 5 mg/500 ml culture fluid, approximately 80% of the cholesterol accumulated is converted to $\Delta^{5,7,22}$cholestatrien-3β-ol; however, at 10 mg of cholesterol/500 ml of culture fluid only 52% of the cholesterol accumulated by the cells is altered. Further, the amounts of 22-dehydrocholesterol found in the ciliates increases substantially with an elevation of the amount of cholesterol made available to the cells.

Table 3
A comparison of the GLC detector response of tetrahymanol and cholesterol

Amount	Integrator units[1]	Ratio
1.5 µg cholesterol	141.3	
1.5 µg tetrahymanol	127.3	1.11
6.0 µg cholesterol	566.6	
6.0 µg tetrahymanol	523.0	1.08

Tetrahymanol (M.W. = 428)/Cholesterol (M.W. = 386) = 1.11

[1] [Average of three determinations.] Range 10 Att. 32 Hydrogen Flame Ionization Detector. Column SE-52 (Silicone Rubber), 0.75% coating on Diatoport S (80–100 mesh), 6 ft., temp. 232 °C.

Table 4
Tetrahymanol content of cells when grown with various concentrations of cholesterol

Cholesterol Added (mg/500 ml)	Tetrahymanol[1] (µg/10⁸ cells)
none	880 ± 88 (16)
2.5 mg	320 ± 66 (5)
5.0 mg	12 ± 12 (6)
10.0 mg	8 ± 6 (3)

Initial inoculum = 1×10^6 cells = $\sim$ 8 µg tetrahymanol.

[1] Amounts estimated by GLC detector response as recorded by a disc integrator. Column: 0.75% SE-52 (Silicone Rubber) on 80–100 mesh Diatoport S.

The tetrahymanol content of the ciliates grown with exogenous cholesterol decreases markedly. The detector response of the glc unit for cholesterol and tetrahymanol is given in Table 3. Table 4 contains a listing of the results of a series of experiments where various concentrations of cholesterol were added to the culture fluid and the tetrahymanol content of the cells estimated. This unexpected finding led to a determination to see if the reduction in tetrahymanol content occurs by a mechanism similar to that found in mammalian liver for the cholesterol feedback inhibition of terpene biosynthesis. Before an attempt to compare the mechanisms involved in the feedback inhibition was possible, it was necessary to learn more about the biosynthesis of tetrahymanol and to establish the relationship of this synthesis to that of other lipids occurring in these protozoa. A condensed version of the lipid biosynthetic pattern found in most cells is presented in Figure 4. Acetate serves as a precursor for all fatty materials, mevalonate for the terpene derivatives, including sterols, pentacyclic triterpenes, ubiquinone with its polyisoprenoid side chain, and squalene. Squalene might be expected to give rise to tetrahymanol, a pentacyclic triterpenoid, as anticipated by ESCHENMOSER et al. (1955) and recently demonstrated by GOODWIN's group for β-amyrin (TREHARNE et al., 1966; REES et al., 1966; GOAD et al., 1966). A reproduction of an autoradiogram of the non-saponifiable materials obtained from *Tetrahymena* when then cells were incubated with radioacetate, radiomevalonate or radiosqualene is shown in Figure 2. All the lipids in the non-saponifiable and phospholipid fractions were found to be radioactive when 2-^{14}C-acetate was added to the culture fluid. Tetrahymanol and the suspected diplopterol clearly contained sizeable quantities of the label. Mevalonate and squalene proved to be highly effective precursors for the pentacyclic triterpenoid alcohols.

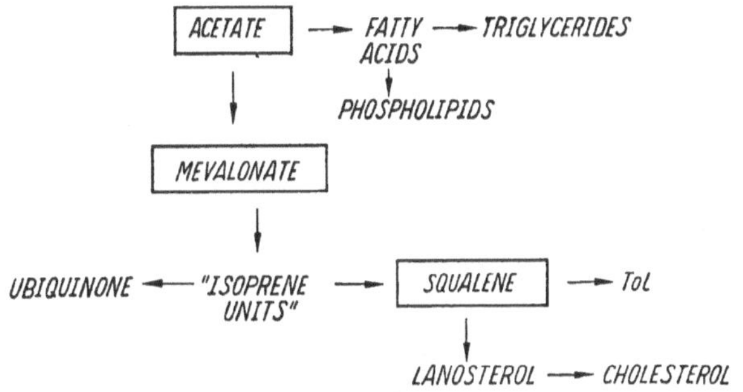

Fig. 4. An abbreviated schematic representation of the pathways of lipid biosynthesis

An examination of the autoradiograms revealed that the flame-shaped component (I), region VIII, and one component of IX were labeled only in acetate incubations. When radiosqualene was employed as a precursor only tetrahymanol (II) and the companion (III) contained detectable radioactivity. The intensity of the radioactivity in a given region showed a positive correlation to the color intensity obtained when the plates were stained and developed with phosphomolybdic acid. In summary, acetic acid, mevalonic acid and squalene can serve as precursors for the pentacyclic triterpene alcohols found in this ciliate. In these experiments, there was no detectable radioactivity in the sterol region of the tlc plates.

The addition of cholesterol to the culture fluid resulted in a reduction of the amount of label in tetrahymanol and the companion regions of the tlc plates. This loss is paralleled by the lack of phosphomolybdic acid staining in the appropriate regions for these two compounds. The reduction in amount of the two compounds was verified by glc analysis. When radioacetate was employed as a precursor in conjunction with cholesterol there was an accumulation of label in region IX. Thin-layer chromatography of the material from this region on Silica Gel G plates developed with 100% petroleum ether revealed two components. One component cochromatographed with squalene while the second unidentified compound remained at the origin. In extracts from cells grown with radioacetate alone both compounds were labeled; however, when cholesterol was added to the culture fluid there was no radioactivity in the squalene region, nor could squalene be detected by I_2 or phosphomolybdic acid visualization. The material remaining at the origin of these plates was labeled in the non-saponifiable extract from both cholesterol and from non-cholesterol supplemented cultures. Chromatograms obtained from the non-saponifiable lipids of cells grown with mevalonate showed radioactive squalene in the absence but not in the presence of exogenous cholesterol. Cells grown with radiosqualene, of course, showed the presence of this radioactive compound in both cases.

A summary of the data from the experiments employing the radioactive precursors of the pentacyclic triterpenoid tetrahymanol in the presence and absence of cholesterol indicates: (a) cholesterol or a metabolic derivative of cholesterol appears to reduce or block the incorporation of radiosqualene into tetrahymanol and "diplopterol", (b) mevalonate incorporation into squalene is blocked by cholesterol; however, mevalonate incorporation into other terpene derivatives is not reduced by a significant amount, (c) acetate incorporation into the pentacyclic triterpene alcohols is blocked by cholesterol; however, mevalonate formation from acetic acid appears to continue at a rate that allows the labeling of the metabolic derivatives of mevalonic acid.

From these observations it can be concluded that the major blocks induced by cholesterol in the metabolic sequence leading to tetrahymanol occur after isoprene formation and before squalene synthesis as well as at the squalene cyclization step. This pattern of inhibition would seem to differ considerably from that reported in mammalian liver and other tissues. Siperstein and Guest (1959) suggest that the primary point of feedback inhibition induced by exogenous cholesterol in liver is the reduction of a hydroxymethylglutaryl derivative to mevalonic acid. However, several groups (Gould et al., 1966; Dietschy et al., 1967) have also reported a lowering in the rate of incorporation of mevalonic acid and of squalene into sterols when rats are fed an exogenous source of cholesterol, but these blocks seem to be secondary in the inhibition pattern. These additional sites of inhibition on sterol synthesis in rat liver by dietary cholesterol have been confirmed and extended by Subba Rao et al. (1967) and by Krishnaiah et al. (1967), who have noted ubiquinone biosynthesis is not reduced by low levels of cholesterol feeding, and also found an inhibition of squalene cyclization to lanosterol. The situation in *Tetrahymena* would appear to be similar to these later reports with an additional control point occurring after isoprene formation, since cholesterol does not inhibit incorporation of the mevalonate into derivatives other than squalene, tetrahymanol and the "companion".

The rates of transformation of acetic acid to mevalonate, squalene, and tetrahymanol in *Tetrahymena* in the presence and absence of exogenous cholesterol have not been determined; therefore, it is not possible to argue that mevalonate biosynthesis from acetate occurs at the same level in the two experimental situations. Further work is necessary to determine if three sites of cholesterol inhibition are involved in the biosynthesis of tetrahymanol and to determine which specific enzymatic step or steps are affected by the addition of exogenous sterols to the culture fluid of these ciliates.

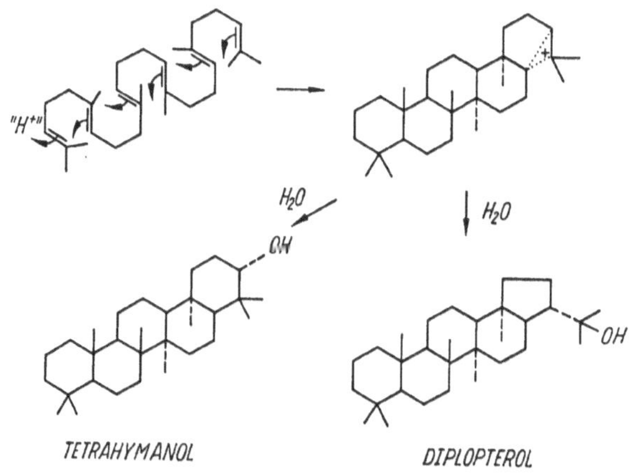

Fig. 5. Proposed scheme for the synthesis of tetrahymanol and diplopterol from squalene

One final point with regard to the mechanism of the formation of tetrahymanol from squalene in this ciliated protozoal form is important. β-amyrin biosynthesis in plants proceeds via the pathway: squalene → 2,3 oxido squaleve → β-amyrin (TREHARNE et al., 1966; REES et al., 1966; GOAD et al., 1966). Tetrahymanol biosynthesis, as suggested by TSUDA et al. (1965), may be a protoninduced cyclization of squalene. Figure 5 outlines a possible mechanism for tetrahymanol biosynthesis by the proton-induced cyclization of all chair-folded squalene. This cyclization may be viewed formally as proceeding by way of an intermediate carbonium ion that could react with water at either the secondary or the tertiary carbon to give tetrahymanol or diplopterol, respectively. Diplopterol has been isolated and identified in a Japanese fern (*Diplopterygium glaucum*) by KARIYONE and AGETA (1959). The tetrahymanol companion has recently been isolated and shown to have glc properties identical to an authentic sample of diplopterol that was kindly provided by Professor H. AGETA, Showa College of Pharmacy, Tokyo. This important observation on a possible second mechanism for the cyclization of squalene is currently being extended in this laboratory in collaboration with Dr. ELIAHU CASPI of the Worcester Foundation.*

While the *in vitro* synthesis of tetrahymanol has not yet been attempted, this type of study may provide the information required to determine if cholesterol exerts a direct influence on the enzymes involved in the formation of the pentacyclic alcohol or if a genetic repression mechanism is responsible for the inhibition.

References

[1] AARONSON, S. and H. BAKER. J. Protozool. 8, 274–277 (1961).
[2] ALLISON, B. M. and R. R. RONKIN. J. Protozool. **14**, 313–319 (1967).
[3] BERGMAN, W., In "Comparative Biochemistry", III A, eds. FLORKIN, M. and H. S. MASON. Academic Press, New York (1962).
[4] BUCHER, N. L. R., In "Ciba Symp. on Biosynthesis of Terpenes and Sterols", Little, Brown and Co., Boston, Mass. (1959).
[5] BUCHER, N. L. R., K. MCGARRAHAN, E. GOULD, and A. V. LOUD, J. Biol. Chem. **234**, 262 (1959).
[6] CAILLEAU, R. In LWOFF, M., ed. "Biochemistry and Physiology of the Protozoa", I, 235–250. Academic Press, New York. (1951).
[7] CLAYTON, R. B., J. Lipid Res. **5**, **3** (1964).
[8] CONNER, R. L., Science **126**, 698 (1957).
[9] CONNER, R. L., J. gen. Microbiol. **21**, 180–185 (1959).
[10] CONNER, R. L., Abst. 2nd International Conf. on Protozool. Excerpta Medica Found. Cong. Series **91**, 288 (1965).
[11] CONNER, R. L., and S. G. CLINE. J. Protozool. **11**, 486 (1964a).
[12] CONNER, R. L., S. G. CLINE, M. J. KOROLY, and B. HAMILTON., J. Protozool. **13**, 377 (1966a).
[13] CONNER, R. L., C. W. L. IYENGAR and J. R. LANDREY. J. Protozool. **13**, (Suppl. 31) (1966b).
[14] CONNER, R. L., F. B. MALLORY, C. W. L. IYENGAR., and J. R. LANDREY, J. Biol. Chem. submitted for publication (1968).
[15] CONNER, R. L., and M. NAKATANI, Arch. Biochem. Biophys. **74**, 175 (1958).
[16] CONNER, R. L., and F. UNGAR, Exptl. Cell Res. **36**, 134–144 (1964b).
[17] CONNER, R. L., and W. J. VAN WAGTENDONK, J. gen. Microbiol. **12**, 31 (1955).
[18] COREY, E. J., and W. E. RUSSEY, J. Am. Chem. Soc. 88, 4751 (1966a).
[19] COREY, E. J., W. E. RUSSEY, and P. R. O. DE MONTELLANO, J. Am. Chem. Soc. 88, 4750 (1966b).

*) Recent evidence, CASPI, E., J. M. ZANDER, J. B. GREIG, F. B. MALLORY, R. L. CONNER, J. R. LANDREY, J. Am. Chem. Soc. **90**, 3563–64 (1968); MALLORY, F. B., R. L. CONNER, J. R. LANDREY, J. M. ZANDER, J. B. GREIG, E. CASPI, J. Am. Chem. Soc. **90**, **3564**–66 (1968), lends strong support to the proposal of a non-oxidative proton-induced mechanism for cyclizing squalene directly to tetrahymanol and diplopterol. 2,3-oxidosqualene does not appear to serve as a precursor for the pentacyclic triterpenes synthesized by *Tetrahymena pyriformis*. A more complete description of the biosynthetic pattern of tetrahymanol biosynthesis as influenced by cholesterol can be found in *J. Protozool.* Dec. 1968.

[20] DEWEY, V. C., G. W. KIDDER, and D. G. MARKEES, Proc. Soc. Exptl. Biol. Med. **102**, 306 (1959).
[21] DIETSCHY, J. M., and M. D. SIPERSTEIN, J. Lipid Res. **8**, 97 (1967).
[22] ERWIN, J., and K. BLOCH, J. Biol. Chem. **238**, 1618 (1963).
[23] ESCHENMOSER, A., L. RUZICKA, O. JEGER, and D. ARIGONI, Helv. Chim. Acta **38**, 1890 (1955).
[24] EVERHART, L. P., Jr., and R. R. RONKIN, J. Protozool. **13**, 646 (1966).
[25] GOAD, L. J., and T. W. GOODWIN, Biochem. J. **99**, 735 (1966).
[26] GOULD, R. G., In "Hormones and Atherosclerosis", ed. G. PINCUS, Academic Press, New York (1959).
[27] GOULD, R. G., and G. POPJAK, Biochem. J. **66**, 51 P (1957).
[28] GOULD, R. G., and E. A. SWYRYD, J. Lipid Res. **7**, 698 (1966).
[29] HACK, M. H., R. G. YAEGER, and T. D. McCAFFERY, Comp. Biochem. Physiol. **6**, 247 (1962).
[30] HOLMLUND, C. E., and N. BOHONOS, Life Sciences **5**, 2133 (1966).
[31] HOLZ, G. G., Jr., J. A. ERWIN, N. ROSENBAUM, and S. AARONSON, Arch. Biochem. Biophys. **98**, 312 (1962a).
[32] HOLZ, G. G., Jr., J. A. ERWIN, and B. WAGNER, J. Protozool. **8**, 297 (1961a).
[33] HOLZ, G. G., Jr., J. A. ERWIN, B. WAGNER, and N. ROSENBAUM, J. Protozool. **9**, 359 (1962b).
[34] HOLZ, G. G. Jr., and S. H. HUTNER, Ann. Rev. Microbiol. **16**, 189 (1962c).
[35] HOLZ, G. G., Jr., B. WAGNER, and J. A. ERWIN, Comp. Biochem. Physiol. **2**, 202 (1961b).
[36] JOHNSON, W. J., and R. JASMIN, Fed. Proc. **20**, 281 (1961).
[37] KARIYONE, T., and H. AGETA, Pharm. Soc. Jap. J. **79**, 105–106 (1959).
[38] KIDDER, G.W., and V. C. DEWEY, In "Biochemistry and Physiology of Protozoa", I. Ed. A. LWOFF. Academic Press, New York (1951).
[39] KRISHNAIAH, R. V., V. C. JASHI, and T. KAMASARMA, Arch. Biochem. Biophys. **121**, 147 (1967).
[40] LIANG, C. R., and H. ROSENBERG, Biochim. Biophys. Acta **125**, 548 (1966).
[41] MALLORY, F. B., J. T. GORDON, and R. L. CONNER, J. Am. Chem. Soc. **85**, 1362 (1963).
[42] McCASHLAND, B. W., Growth **27**, 39–45 (1963).
[43] McCASHLAND, B. W., and V. A. JOHNSON, Growth **21**, 11–19 (1957).
[44] McKEE, C. M., J. D. DUTCHER, V. GROUPÉ, and M. MOORE. Proc. Soc. Exptl. Biol. Med. **65**, 326 (1947).
[45] REES, H. H., E. I. MERCER, and T. W. GOODWIN, Biochem. J. **99**, 726 (1966).
[46] SHORB, M. S., B. E. DUNLAP, and W. O. POLLARD, Proc. Soc. Exptl. Biol. Med. **118**, 1140 (1965).
[47] SIPERSTEIN, M. D., and M. J. GUEST, J. Clin. Invest. **38**, 1043 (1959).
[48] SMITH, P. F., Bacteriol. Revs. **28**, 97 (1964).
[49] SUBBA RAO, K., and R. E. OLSON, Biochim. Biophys. Res. Comm. **6**, 668 (1967).
[50] TAKETOMI, T., Z. Allgem. Mikrobiol. **1**, 331 (1961).
[51] THOMPSON, G. A., Jr. Biochem. **6**, 2015 (1967).
[52] TREHARNE, K. J., E. I. MERCER, and T. W. GOODWIN, Biochem. J. **99**, 230 (1966).
[53] TSUDA, Y., A. MORIMOTO, T. SANO, Y. INUBUSHI, F. B. MALLORY and J. T. GORDON, Tetrahedron Letters **19**, 1427–1431 (1965).
[54] VAN TAMELEN, E. E., J. D. WILLETT, R. B. CLAYTON and K. E. LORD, J. Am. Chem. Soc. **88**, 4750 (1966).
[55] VISHNIAC, H. S., J. gen. Microbiol. **12**, 464 (1955).
[56] VISHNIAC, H. S., and S. W. WATSON, J. gen. Microbiol. **8**, 248 (1953).

Biosynthesis in Plant sterols V. Formation of Cardenolides in Digitalis Plants[1,2]

E. Caspi

Worcester Foundation for Experimental Biology,
Shrewsbury, Massachusetts, USA

Summary

Evidence for the involvement of cholesterol in the biosynthesis of cardenolides is presented. The cholesterol is degraded to pregnenolone (3a) which in turn is oxidized to progesterone (5). The two C_{21}-steroids are incorporated into digitoxigenin (2), gitoxigenin (6) and digoxigenin (7). The saturation of the 3β-hydroxy-5-en of pregnenolone to a 3β-hydroxy-5β-H moiety prevailing in cardenolides proceeds through an obligatory formation of a 3-ketone. The mechanism of introduction of an hydroxyl at C-14 is discussed.

It is at present assumed that the biosyntheses of sterols in plants and in animals proceed by similar routes (Tschesche, 1955), (Heftman, 1967), (Frantz and Schroepfer, 1967). This supposition implies that the primary plant sterol would be cholesterol (1) or an analogous,

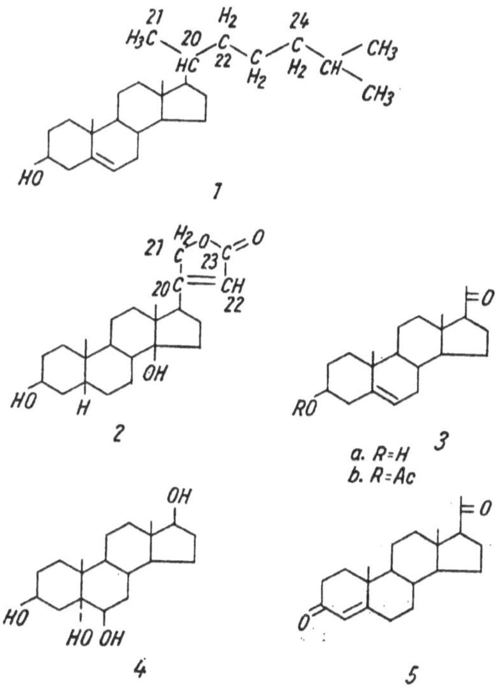

Fig. 1

[1] This work was supported by Grants HE 10566, IF3-HE-34477-01 and K3-16614 from United States Public Health Service.

[2] The permission of the Editors of Experientia, Phytochemistry and Science to use portions of previously published articles is gratefully acknowledged.

triterpene, which would then be metabolized to the various sterols found in the plant (HEFTMAN, 1967). Hence, for example in *Digitalis purpurea*, cholesterol or its analog would be the precursor of sterols with a butenolide moiety. It was previously demonstrated that digitoxigenin (2) can be biosynthesized from both 2-^{14}C mevalonic acid (MVA) and 1-^{14}C acetic acid (GREGORY and LEETE, 1960), (GROSS et al., 1965), (LEETE et al., 1965). However, carbons 22 and 23 in digitoxigenin biosynthesized from 2-^{14}C-MVA did not originate from the administered precursor, while the product obtained from 1-^{14}C-acetate contained ^{14}C at C-23. These observations support the view that the initial step in the conversion of cholesterol to butenolides could be cleavage between C-20 and C-22 to yield pregnenolone (3a) or a similar C_{21}-20 ketone. Subsequent addition of a two carbon unit, could give, after appropriate biological transformations, sterols with a butenolide moiety. For the theory to be tenable, we considered it important to determine whether indeed the plant has the capacity to metabolize cholesterol to pregnenolone (3a) and/or other C_{21}-20 ketones (CASPI et al., 1966).

With this in mind we undertook a study of the metabolism of cholesterol in *Digitalis purpurea* (CASPI et al., 1966). A sample of 4-^{14}C-cholesterol (ca 10 mg 8×10^8 dpm) in diglyme, was divided in half and each portion was fed through a cut leaf to a different plant. The choice of the solvent and the method of feeding proved rather unsatisfactory since the two leaves died after 48 hours. However, at that time the radioactivity became distributed in the plants particularly the root-crowns. Under the circumstances we did not expect a large conversion of cholesterol to C_{21} intermediates.

The experiment was terminated after 18 days and upon processing the plants a "steroidal" fraction containing 2.7×10^6 dpm was obtained. The extract was hydrolyzed with aqueous methanolic hydrochloric acid, and aliquots (10%) were chromatographed sequentially as described in Table 1. The material recovered from the last chromatography, which co-chromatographed with pregnenolone (3a), was further diluted with cold material and crystallized to constant specific activity as the free alcohol (3a) and then as the acetate (3b) (Table 2). The results of co-crystallization leave no doubt that the radioactivity remained associated with pregnenolone (3a) and its acetate (3b). Additional support for the structure of the metabolite was obtained by chemical degradation. A sample of pregnenolone (3a) (20.9×10^3 dpm/mM) from another isolation experiment was converted to the acetate (3b) (20.7×10^3 dpm/mM). The acetate upon Baeyer-Villiger oxidation with trifluoroperacetic acid and saponification gave 3β, 5α, 6β, 17β-tetrahydroxy-androstane (4) (20.9×10^3 dpm/mM) (CORNFORTH et al., 1965).

Table 1

Chromatographic isolation of radioactive pregnenolone from an aliquot (10%) of the metabolites recovered after the administration of 4-14C-cholesterol (4×10^8 dpm) to two *Digitalis purpurea* plants

Chromatographic system (temperature 25 °C)	Rf	Radioactivity[1] in pregnenolone zone 10^3 dpm
Propylene glycol-2% toluene/hexane	0,27	44.1
Hexane-methanol-water 100:70:30	0.68	7.0
Heptane-methanol-water 500:475:25	0.4	4.3
Propylene-glycol-4% toluene/hexane	0.37	2.9

[1] Corrections were made for the fractions removed for counting and color reactions. Whatman No. 1 paper was used. Solvents are expressed in volumes.

The presented evidence constitutes proof for the pregnenolone structure of the metabolite. The observation that pregnenolone (3a), its acetate (3b) and the tetrol (4) all have the same

Table 2
Co-crystallization of pregnenolone and pregnenolone acetate derived from 4-^{14}C-cholesterol (x 10^3 dpm)

Sample		Experiment I		Experiment II	
		per mg	per mM	per mg	per mM
1st crystallization	Crop 1	0.128	40.5	0.043	13.6
1st crystallization	Crop 2			0.042	13.3
2nd crystallization	Crop 1	0.128	40.5		
Acetate	Crop 1	0.108	39.5	0.038	13.6
Acetate	Crop 2	0.108	39.5	0.038	13.6

specific activity indicates that no randomization of the tracer occurred. Consequently, it may be inferred that the pregnenolone was formed from cholesterol via cleavage of its side chain. In view of the incorporation of pregnenolone into digitoxigenin (TSCHESCHE, 1964) gitoxigenin (6) and digoxigenin (7) (CASPI and LEWIS 1967) it can be concluded that cholesterol *may serve as a precursor* of cardenolides.

These observations are of interest because of the known importance of pregnenolone in the biosynthetic conversion of cholesterol to steroid hormones in animals (DORFMAN and UNGAR, 1965; p. 125). If it is assumed that pregnenolone plays a similar role in the formation of cardenolides the question arises as to the mechanism of reduction of its C-5 double bond. In animals the saturation of the homo-allylic alcohol (Δ^5-3β-OH) proceeds by way of oxidation of the 3β-hydroxyl to a 3-ketone and isomerization of the olefinic bond from C-5 to C-4 to yield a Δ^4-3-ketone (DORFMAN and UNGAR, 1965; p. 310) (TALALAY, 1957). Subsequently, the conjugated double bond is reduced to either a *cis*-(5β-H) or *trans*-(5α-H) junction between rings A and B, and this is followed by the reduction of the C-3 ketone (TOMKINS, 1956 and 1959), (DORFMAN and UNGAR, 1965; p. 406).

If a mechanism analogous to that described for animals operates in *Digitalis* plants, two conditions must first be satisfied. The plant should be able to convert pregnenolone (3) to progesterone (5) and to utilize the latter in the biosynthesis of cardenolides (2). To test the hypothesis we have investigated the metabolism of pregnenolone and progesterone in isolated *Digitalis lanata* leaves (CASPI and LEWIS, 1967).

One average-size leaf was cut with its stem from a *Digitalis lanata* plant, and an acetone solution of pregnenolone-7α-^{3}H (9) (9.0×10^8 dpm, disintegrations per minute) was then applied to the surface of the leaf. After evaporation of the acetone, the leaf was sprayed with a 10 per cent solution of silicon oil DC 200 in ligroin[1] (b.p. 60°C to 90°C) (BENNET and HEFTMANN, 1965). The leaf was then suspended in approximately 40 ml of a modified HOAGLAND's (MEYER et al., 1960) medium containing the requisite amounts of micrometabolic elements and continuously illuminated with light of 12,000 lu/m^2. The nutrient medium was changed at 48-hour intervals, and the experiment was terminated after 10 days. An analogous experiment was carried out with progesterone-7α-^{3}H (9.65×10^8 dpm).

Upon completion of the experiments the leaves were processed (EUW and REICHSTEIN, 1964), (CASPI and LEWIS, 1967) and the distribution of the tracers in the different extracts is summarized in Table 3. We found 19.2 per cent of the radioactivity of pregnenolone and 25.6 per cent of that of progesterone in the fractions containing "plant sterols." The bulk of cardenolides (2, 6, 7) is recovered with chloroform; consequently we undertook the processing of this extract (EUW and REICHSTEIN, 1966). The results of both experiments reported here refer to products from this fraction. The glycosides were hydrolyzed with aqueous methanolic sulfuric

[1] We thank the Dow Corning Co. for a gift of the silicon oil.

Table 3

Radioactivity in the "sterol fractions" recovered from isolated leaves of *Digitalis lanata* after the application of pregnenolone-7α-^{3}H (9.0×10^8 dpm) and progesterone-7α-^{3}H (9.65×10^8 dpm)

Extract	Radioactivity ($\times 10^7$ dpm)	
	Pregnenolone-7α-^{3}H	Progesterone-7α-^{3}H
Ligroin	2.98	0.48
Chloroform	8.6	14.43
Chloroform-ethanol (2:1)	3.7	0.62
Chloroform-ethanol (3:2)	0.62	0.85
Unabsorbed radioactivity[1]	4.8	35.0
Percentage recovery in "sterol fractions"[1]	19.2	25.6

[1] The percentage recovery is corrected for the unabsorbed radioactivity present in the ethyl acetate washings of the leaves at the termination of the experiment.

acid, and the aglycones were recovered with ethyl acetate. The hydrolyzate originating from progesterone (5) contained 8.6×10^7 dpm, and that from pregnenolone (3) 12.88×10^7 dpm. The isolation of digitoxigenin (2), gitoxigenin (6) and digoxigenin (7) was carried out on portions of the aglycones by chromatography in several systems (Caspi and Lewis, 1967).

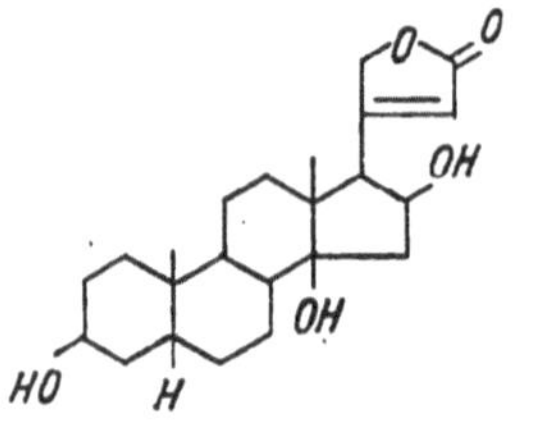

6

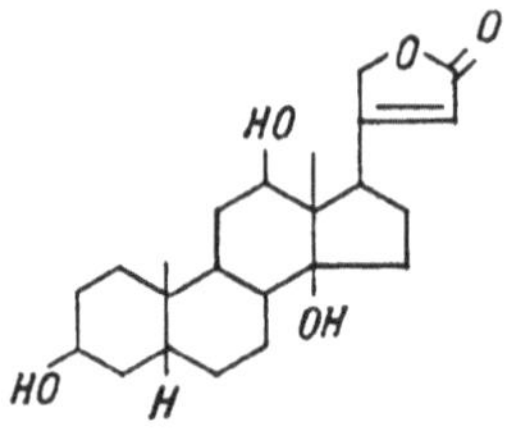

7

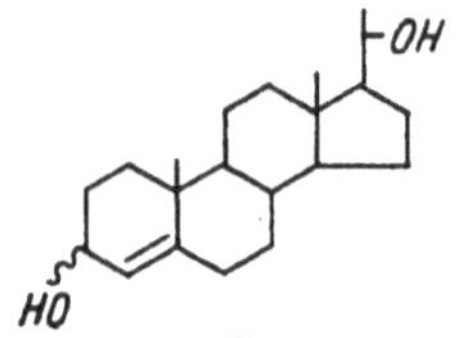

8

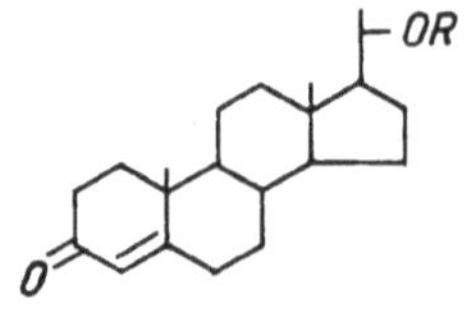

9 a. R=H
b. R=Ac

Fig. 2

The amounts of aglycones produced from the administered precursors are given in Table 4. The results represent the radioactivity recovered from the chromatogram in which the product

Table 4

Amounts of cardenolides formed from pregnenolone7α-^{3}H and progesterone-7α-^{3}H in isolated leaves of *Digitalis lanata*

Metabolite	Pregnenolone-7α-^{3}H (8.52×10^8 dpm)[1]		Progesterone-7α-^{3}H (6.15×10^8 dpm)[1]	
	($\times 10^6$ dpm)	(% conversion)	($\times 10^6$ dpm)	(% conversion)
Digitoxigenin (3)	10.57	1.24	12.40	2.02
Gitoxigenin (6)	3.18	0.37	1.87	0.30
Digoxigenin (7)	2.10	0.25	1.49	0.24

[1] This amount is corrected for the radioactivity recovered from the surface of the leaf at the termination of the experiment (Table 3).

appeared for the first time as a single symmetrical zone. A portion of the metabolite recovered from the last chromatography was then further diluted with nonradioactive material (20 to 30 mg) and crystallized to constant specific activity as the free alcohol and as the acetate or diacetate, as the case might be. At least two isolation and identification experiments were performed for each product.

Crystallization to constant specific activity was carried out for digitoxigenin (2), gitoxigenin (6), and digoxigenin (7) derived from progesterone as a precursor. In the experiment with pregnenolone (3a) digitoxigenin (2) was processed in the same way; but gitoxigenin (6) and digoxigenin (7) were identified by chromatography only. In most instances, some radioactive impurity accompanied the isolated metabolites and was removed by cocrystallization with cold material. Generally, the contamination was small (less than 5 per cent), and this amount does not significantly influence the calculated conversions shown in Table 4.

We then turned our attention to the conversion in the leaf of pregnenolone (3a) to progesterone (5). A portion (20 per cent) of the hydrolyzate of the chloroform extract from the pregnenolone experiment was fractionated by paper and thin-layer chromatography. The recovered progesterone "zone" (1.81×10^5 dpm) was diluted with 95 mg of unlabeled progesterone (5) and crystallized to constant specific activity (4.2×10^5 dpm/mmole). This material was then reduced with sodium borohydride and the mixture of allylic alcohols (Δ^4-3ξ-OH) (8) was reoxidized with dichlorodicyanoquinone to 9a (BURN et al. 1966). The 20β-alcohol (9a) was acetylated to yield 20β-acetoxypregn-4-en-3-one (9b), which showed the same specific activity (4.25×10^5 dpm/mmole) as the progesterone. The evidence presented constitutes proof that the exogenously supplied pregnenolone (3a) was metabolized to progesterone (5) in the surviving leaf of *Digitalis lanata*. In addition, the pregnenolone (3a) was also incorporated into digitoxigenin (2), gitoxigenin (6), and digoxigenin (7).

Similarly, externally fed progesterone was transformed into the *same three cardenolides under analogous experimental conditions*. Thus, both pregnenolone (3a) and progesterone (5) served as *precursors* of the cardenolides. Though it might be premature to draw conclusions concerning the relative merits of the two steroids as precursors of cardiac aglycones, results in Table 4 suggest a somewhat better incorporation of progesterone (5) than of pregnenolone (3a).

In view of the above results, it would seem likely that pregnenolone (3a) is first converted to progesterone (5) which is further utilized in the biosynthetic processes. Possibly the amount of progesterone in the plant (9.05×10^5 dpm, equivalent to 0.106 per cent of the pregnenolone) represents the steady-state concentration. However, the considerations should be viewed critically since the experimental conditions were not physiological.

Albeit these experiments proved the transformation of pregnenolone in the isolated surviving leaves of *Digitalis lanata* the actual role of the latter in the biosynthetic sequence remained obscure. We reasoned that if progesterone is an obligatory intermediate the 3α-hydrogen of the exogeneously supplied pregnenolone should be absent in the *de novo* formed cardenolides (2.6 and 7). Determination of the tritium content of labeled cardenolides resulting from the incorporation of doubly labeled 3α-^{3}H-4^{14}C-pregnenolone into *Digitalis lanata* would decide this point. (CASPI and HORNBY 1968).

For this reason we undertook the synthesis of the required 3α-^{3}H-pregnenolone. Deconjugation of progesterone-20-ethylene ketal (10) with potassium t-butoxide in t-butanol (RINGOLD and MALHOTRA, 1962) gave a mixture of ca. 22% of starting material (10) and ca. 78% of the 5-en-3-one (11). The mixture was reduced with sodium borotritide in diglyme solution to give the 5-en-3-ol together with some 20% of the allylic 4-en-3-ol. Acid treatment of this mixture removed the ketal and dehydrated the allylic alcohol to the 3,5-dien-20-one which was easily separated from the 3α-^{3}H-pregnenolone (12a) by thin-layer chromatography.

It was now necessary to determine the distribution of the isotopic hydrogen in the tritiated pregnenolone (12a). Material recovered from base catalyzed equilibration of pregnenolone

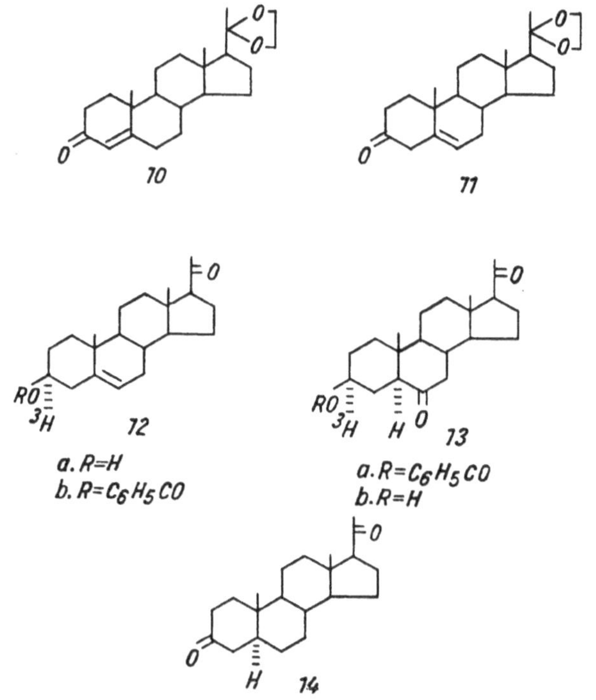

Fig. 3

retained all the tritium showing that none was present at C-17 and C-21. Hydroboration of the benzoate (12b) (BROWN, 1962) and oxidation of the intermediate with Jones' reagent gave 3β-benzoyloxy-5ξ-pregnane-6,20-dione (13) which showed the same specific activity as the starting material (12b). Equilibration of the dione (13a) with base gave (13b) and proceeded without the loss of label. These results prove the absence of tritium from C_6 and C_7. Catalytic hydrogenation of the pregnenolone (HERSHBERG et al., 1951) provided a mixture of isomeric 5α-pregnane-3β,20ξ-diols which was oxidized with Jones' reagent to the dione (14). The resulting dione (14) retained 13% of the label indicating that 87% of the tritium had the 3α configuration. Oxidation at C_3 was shown to be complete by the fact that the tritium content of the dione (14) was unaffected by a second treatment with the Jones' reagent. Equilibration of the dione (14) with base gave a product devoid of tritium. Finally Oppenauer oxidation of pregnenolone and subsequent base equilibration removed all the tritium, confirming the above results. Thus, 87% of the tritium has the 3α-orientation and the remainder is distributed between C_2 and C_4 (Table 5).

Table 5
Distribution of tritium in pregnenolone

Compound	% ^3H Retained	% ^3H at Positions Indicated	
Pregnenolone (12)	100		
Equilibrated pregnenolone	100	C_{17}, C_{21}	0
3β-Benzoyloxy-5α-pregnane-6,20-dione (13a)	100	C_6	0
3β-Hydroxy-5α-pregnane-6,20-dione (13b)	100	C_7	0
5α-Pregnane-3,20-dione (14)	13	C_3	87
Equilibrated 5α-pregnane-3,20-dione	0	C_2, C_4	13
Equilibrated progesterone (5)	0		

The 3α-tritiated and 4-^{14}C-specimens were mixed to give a doubly labeled 3α-^{3}H-4-^{14}C-pregnenolone (^{3}H/^{14}C ratio 13.3) which was administered to a *Digitalis lanata* plant (CASPI and HORNBY 1968,). After three weeks the plant was processed and the cardenolides were recovered. The biosynthesized digitoxigenin had an ^{3}H/^{14}C ratio of 1.16 indicating that it retained ca. 9% of the tritium present in the administered pregnenolone. Oxidation to digi-

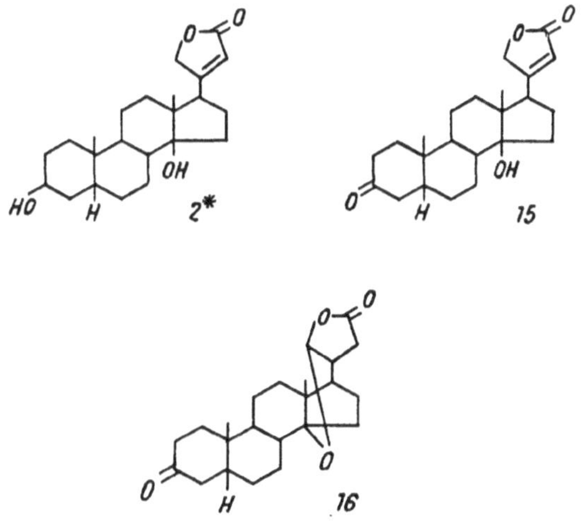

Fig. 4

toxigenone (15) proceeded without loss of tritium, demonstrating the total absence of tritium at C_3 in the cardenolide. Equilibration of the digitoxigenone (15) with base gave isodigitoxigenone (16) devoid of tritium (Table 6). It may be concluded that the biosynthesis of

Table 6

^{3}H/^{14}C ratios[1] of biosynthesized digitoxigenin[2] and its degradation products

	1st Crystallization		2nd Cyrstallization	
	1st Crop	2nd Crop	1st Crop	2nd Crop
Digitoxigenin (3)	1.27		1.14	1.16
Digitoxigenone (15)	1.14	1.13		
Isodigitoxigenone (16)	0[3]			

[1] Ratios based on dpm of ^{3}H and ^{14}C.
[2] Isolated after administration of pregnenolone (^{3}H/^{14}C ratio 13.3) to *Digitalis lanata*.
[3] After purification.

cardenolides must proceed through the intermediacy of a 3-ketone and that the retained tritium (9%) is confined to C_2 and C_4.

Though it seems reasonable that the biosynthetic scheme involves isomerization of the C-5 double bond and formation of a 4-en-3-one (RINGOLD et al., 1965), (OLEINICK and KORITZ, 1966) the hypothesis can not be substantiated on the basis of the present evidence. Definitive experiments on this aspect of the problem are in progress.

Let us now turn to another aspect of the biosynthesis, namely the problem of introduction of the C-14 hydroxyl (TSCHESCHE et al., 1967). One of the characteristic features of cardenolides is the presence of a hydroxyl group at C-14 having the β-configuration. This is of special interest since hydroxylation of steroids at non-vinylogous carbon atoms occurs by a direct displacement of a hydrogen; the incoming hydroxyl assuming the stereochemistry of the removed proton (HAYANO, 1962).

We have indicated above that cholesterol, pregnenolone, progesterone, all of which have a trans junction between rings C and D (14α-H) are incorporated into cardenolides. Hence if the hydroxylation would proceed via the known mechanism instead of the 14β-alcohol a product having a 14α-hydroxyl would be expected. Frequently the contradiction is circumvented by invoking a new, yet unknown, mechanism for this hydroxylation involving a "14β-hydroxylase" (Tschesche et al., 1967). In such instances an ionic reaction similar in principle to a Walden inversion is tacitly inferred. Postulating novel hypotheses can be fruitful in solving problems, provided the hypotheses are amenable to experimental scrutiny. Though it is not unreasonable to assume a Walden type mechanism for the 14β-hydroxylation a critical evaluation of the concept is difficult and at least at present the idea cannot be properly appraised. It is conceivable that eventually the actual mechanism will be proved or disproved by the elimination of *other possibilities* and/or by studies with the pure 14β-hydroxylase.

Because of limitations of the experimental evaluation of the concept of the "14β-hydroxylase", we have considered several alternative possibilities of explaining the reaction (Caspi and Lewis, 1968). The approach we chose was to attempt to rationalize the reaction on the basis of acceptable chemical transformations which must be amenable to testing. Our assumption was that a 14α-H-steroidal precursor of cardenolides could undergo a "normal" hydroxylation and then this hydroxyl could be transposed into a 14β-alcohol. To illustrate the point, several such possibilities will be mentioned.

Let's assume at first that the sequence of events is initiated by the displacement of the 14α-hydrogen in (17) and formation of a 14α-hydroxy intermediate (18). The feasibility of

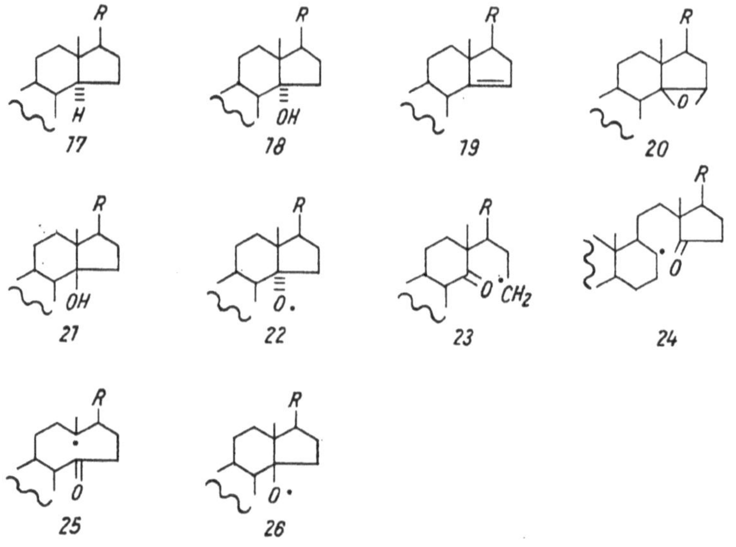

Fig. 5

this reaction is supported by the isolation of 14α-hydroxylated steroids from plant sources (Nakanishi et al., 1966), (Galbraith et al., 1966). The tertiary alcohol (18) on dehydration could give the 14, 15 olefin (19) which upon β-face oxidation would yield a 14β, 15β-epoxide (20). Again the presence of a 14β, 15β-epoxy moiety in bufadienolides lends some indirect support for this view. Finally reductive opening of the epoxide (20) by an attack atC-15 could provide the 14β-alcohol (21). If the dehydration of 2 yields the 8, 14 double bond (not shown in the flow sheet) instead of 14 (15), this would not alter the principle of the concept. The olefin could give an 8β, 14β epoxide which upon reductive attack at C-8 will yield the 14β-alcohol,

The transformation of the 14α-alcohol does not necessarily have to be ionic. In essence a free radical process is equally possible particularly in view of the fact that the biosynthesis of cardenolides occurs in the leaves (TSCHESCHE et al., 1964, 1965, 1967a), (CASPI et al., 1967, 1968, 1968a). Under the influence of sunlight, the 14α-alcohol could give an oxy-radical (22). Such radicals are known to rearrange into keto radicals (HEUSLER et al. ,1963) of the type (23, 24, and 25). Reclosure of the ring may proceed with inversion to yield the 14β-oxy radical (26) and finally the alcohol (21). The operation of this type of mechanism was demonstrated by HEUSLER et al., (loc. cit.) in the free radical transformations of 11β and 4β hydroxy steroids of the 5α-H series. The lead tetraacetat treatment of the 11β alcohol gave the 1β, 11α-oxido-5α-H product while the 4β alcohol provided the 4α,9α-oxido-5β-H compound. For the present argument, it is immaterial which bond is actually broken and which keto radical (23, 24 and 25) is formed. All three intermediates could reclose to (26) and eventually give (21).

The initial hydroxylation does not have to be limited to carbon 14 and may as well take place at C-15, 8, etc. Such hydroxy compounds could be transformed into the 14β-alcohols either via an olefin or a radical mechanism by processes similar to those described.

We considered it worthwhile to explore several such possibilities and we first tested the role, if any, of 14α-hydroxy progesterone (27) in the biosynthesis of cardenolides. The projected

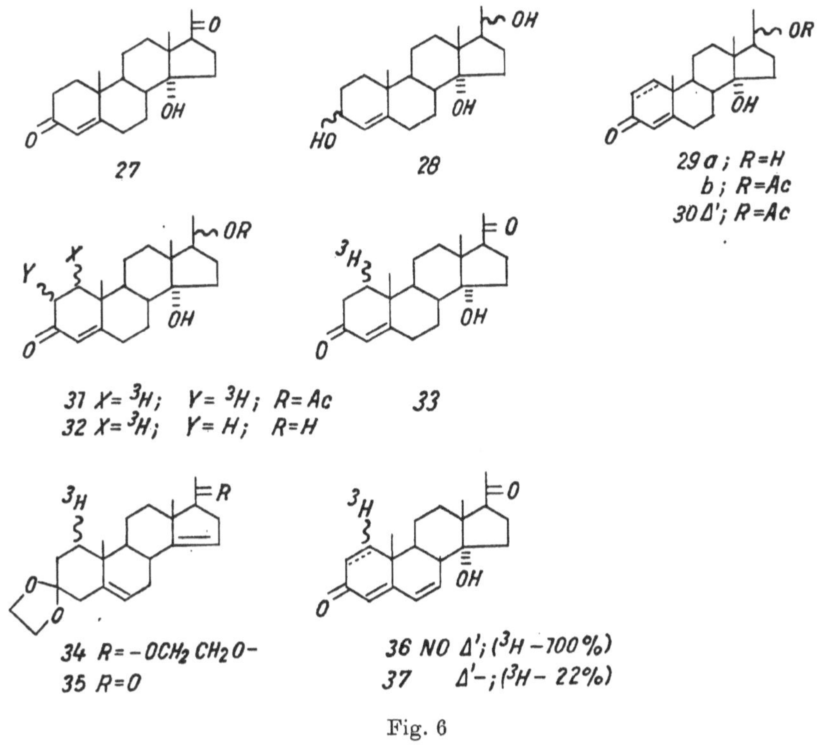

Fig. 6

route was to administer simultaneously 4-^{14}C-progesterone and 1-^{3}H-14α-hydroxy-progesterone to a *Digitalis lanata* plant and to establish whether both isotopes or only ^{14}C is incorporated into digitoxigenin. The experimental design was based on the observation that the progesterone which is incorporated into cardiac aglycones (ca. 2%) will serve as an internal control.

The initial problem we were faced with was the preparation of 1-^{3}H-14α-hydroxyprogesterone. A convenient starting material for the synthesis was 14α-hydroxyprogesterone[1] (27)

[1] We thank Dr. J. A. HOGG, The Upjohn Co., Kalamazoo, Michigan, for a very generous supply of this compound.

which was first reduced with sodium borohydride to (28) and then the C-3 allylic alcohol was reoxidized with 2,3-dichloro-5,6-dicyano-benzoquinone (DDQ) (BURN et al., 1966) to yield (29a). The acetate (29b) was transformed to the dien-one (30) which was then selectively tritiated in the presence of tris-triphenylphosphine rhodium chloride (BIRCH and WALKER, 1966) to (31). The C-2 tritium was removed by equilibration with aqueous methanolic potassium hydroxide and the resulting, 32, 32 was rigorously purified by thin–layer chromatography (tlc). The purified (32) which gave a single radioactive zone on tlc was oxidized (chromium trioxide-pyridine) to (33). The 1-^{3}H-14α-hydroxyprogesterone (33) gave a single radioactive spot on tlc and the specific activity of a sample diluted with cold material remained unchanged in several crystallizations.

Though by this criteria the product (33) appeared homogeneous, it was considered desirable to test further its purity by preparing a derivative. Consequently a diluted sample of (33) (3.86×10^{6} dpm/mM) was converted to the 14-dehydro-diketal (34) by treatment with ethylene glycol and p-toluenesulfonic acid. At this point rather unexpectedly (34) showed a 10% increase in specific activity (4.24×10^{6} dpm/mM). Conversion of the diketal (34) to the monoketal (35) did not further change the specific activity. A comparable increase in specific activity (14%) was observed when (33) (6.87×10^{6} dpm/mM) was treated with chloranil (AGNELLO and LAUBACH, 1959, 1960) to yield the 4,6-dien-3-one (36) (7.83×10^{6} dpm/mM).

In view of these results the 1-^{3}H-14α-hydroxyprogesterone (33) was further purified by tlc but the specific activity of diluted samples remained unchanged. Consequently the observations were viewed as involving retained solvent (of crystallization ?) in (17) and this product was used in the biosynthetic investigations.

To determine the stereochemistry of the isotopic hydrogen at C-1 two sets of experiments were performed. Introduction of a C-1 double bond by dehydrogenation of a 3-ketone with DDQ was shown to proceed via the trans diaxial elimination of the 1α and 2β hydrogens (RINGOLD et al., 1962a, 1963). The 1,4,6-trien-3-ones are known to rearrange in the presence of acids to 1-methyl-3-hydroxy benzenes (CASPI et al., 1963, 1964, 1964a). This rearrangement proceeds by the migration of the C-10 methyl to C-1 and elimination of the C-1 hydrogen (CASPI, loc cit.) (BRODIE, 1962).

Treatment of the 4,6-dien-3,20-dione (36) with DDQ in refluxing dioxan yielded the 1,4,6-trien-3,20-dione (37) which retained 22% of tritium. On the other hand when the trien-dione (37) was exposed to acetic anhydride and p-toluenesulfonic acid a mixture of aromatic products was formed as evidenced by tlc and nmr spectroscopy. The formation in this instance of the mixture of aromatic products can probably be traced to the 14α-alcohol. Presumably in addition to the $\Delta^{8(14)}$ and Δ^{15} aromatic products 14α-acetoxy and other phenols were formed. In any event the aromatic mixture retained ca. 5.5% of the tritium initially present in (33). Based on these results it may be concluded that ca. 78—80% of the tritium in (33) has the 1α-stereochemistry and a minimum of ca. 15% has the 1β-orientation.

A mixture of the 1-^{3}H-14α-hydroxyprogesterone (154 μc) and 4-^{14}C-progesterone (18.9 μc) (^{3}H/^{14}C ratio 8.1) was then administered to a *Digitalis lanata* plant as described above. After two weeks the plant was harvested and processed. The distribution of the tracers in the various fractions is summarized in Table 7. It is evident that the ^{3}H/^{14}C′ ratios of the ligroin and chloroform extracts approach that of the administered mixture of "precursors". On the other hand, proportionally much larger amounts of tritiated products were present in the remaining fractions. The fact that considerably larger amounts of tritiated materials were found in the washings of the leaf could indicate a "better" absorption of the 4-^{14}C progesterone or its transformation products. Processing of the plant led to the overall recovery of ca. 40.2% and 32.5% of the administered tritium and ^{14}C respectively.

As in previous instances the isolation of digitoxigenin was carried out on the chloroform extract. Of interest is the fact that acid hydrolysis of the glycosides (^{3}H/^{14}C 7.29) in this frac-

Table 7

Radioactivity in the fractions obtained by processing[1] of the *Digitalis lanata* plant after the application of 1-^{3}H-14α-hydroxy progesterone (154 μc) and 4-^{14}C-progesterone (18.9 μc) ($^{3}H/^{14}C$ 8.1)

Extract	^{3}H (× 10^{7} dpm)	^{14}C (× 10^{6} dpm)	$^{3}H/^{14}C$ ratio
Leaf washings (Unabsorbed radioact.)	3.49	1.17	30.0
Ligroin	0.78	0.20	8.8
Chloroform	7.17	9.85	7.29
Chloroform Ethanol (2:1)	2.62	1.18	22.2
Chloroform Ethanol (3:2)	2.4	2.02	11.9
Corrected Recovery[2] (%)	40.2	32.5	
Acid hydrolysis of chloroform extract[1]	6.67	9.34	7.14

[1] For the description of the fractions see refs. 6 and 18.
[2] Corrected for unabsorbed radioactivity found in the leaf washings.

tion did not influence the $^{3}H/^{14}C$ ratio of the aglycones which remained unchanged (7.14). On the other hand, upon chromatographic fractionation of the aglycones the $^{3}H/^{14}C$ ratio of the products in the "digitoxigenin" zone fell off rapidly (Table 8). After four chromato-

Table 8

Chromatographic isolation of Digitoxigenin from the acid hydrolyzed chloroform fraction[1] ($^{3}H/^{14}C$ 7.14)

Chromatographic System[2]	Digitoxigenin zone $^{3}H/^{14}C$
TLC-Benzene ethyl-acetate (1:4)	4.44
Ligroin/Toluene/t.BuOH/MeOH/H_2O) (5:5:3:4:6)	1.51
Benzene/MeOH/H_2O (2:1:1)	0.76
Toluene-propylene glycol	0.45

Co-crystallization of material from last chromatography with cold digitoxigenin.

crop — 1	3.47 × 10^{5} dpm/mM-^{14}C	No ^{3}H
crop — 1 — recryst.	3.48 × 10^{5} dpm/mM-^{14}C	No ^{3}H

[1] For description of the fraction see Table 1 and refs. 6 and 18.
[2] The systems were used sequentially. TLC was carried out on silica gel HF_{254} (Merck A.G.), paper chromatography on Whatman No. 1 filter paper. Solvents — in volumes.

graphic separations a small amount of tritiated material was still associated with the area corresponding to digitoxigenin ($^{3}H/^{14}C$ ratio 0.45). The tritiated impurity was finally removed by cocrystallization with cold material whereby only ^{14}C-digitoxigenin was obtained. If the reasonable assumption is made that the ^{14}C-recovered from the last chromatography was indeed incorporated into digitoxigenin then ca. 1.88% of the absorbed 4-^{14}C-progesterone was metabolized to digitoxigenin. The usual transformation of progesterone into digitoxigenin is of the order of 2%. Hence it may be concluded that as far as progesterone is concerned its metabolism was not impaired.

In analyzing the results certain fundamental limitations must be kept in mind. From the outset it was obvious that the two different exogeneously supplied steroids would be absorbed, transported and metabolized at different rates. A mathematical treatment of several biological models of this type was provided and the difficulties and limitations of interpreting such results were pointed out (Burstein et al., 1962, 1967). Hence any conclusions derived from the present investigation should only be viewed "as a first approximation" of the events. With this limitation in mind, it may be stated that 14α-hydroxyprogesterone *is not* a precursor of cardenolides. In reaching this conclusion it is assumed that the metabolism of the 14α-hydroxyprogesterone was not blocked *totally* by the simultaneously administered progesterone. Differently stated, the observed lack of incorporation of isotopic hydrogen into digitoxigenin is inherent in the nature of the tritiated steroid and is not due to the interference of progesterone. On the other hand, it is amply evident that the progesterone was incorporated in nearly normal amounts into digitoxigenin.

Whether other 14α-hydroxy intermediates are involved in the formation of cardenolides can obviously not be answered at this point. It is conceivable for example that the "early" introduction of the 14α-hydroxyl inhibited the metabolic transformations in the direction of digitoxigenin. The role of 14α-hydroxy intermediates, if any at all, in the biosynthesis of cardenolides requires further investigation.

Acknowledgement

This work was very skilfully carried out by Drs. G. M. Hornby, D. O. Lewis, D. M. Piatak, J. A. F. Wickramasinghe and A. Winter to whom the author is greatly indebted for their efforts.

References

[1] Agnello, E. J., and G. D. Laubach, J. Am. Chem. Soc. **79**, 1257 (1959).
[2] Agnello, E. J., and G. D. Laubach, J. Am. Chem. Soc. **82**, 4293 (1960).
[3] Bennett, R. D., and E. Heftmann, Phytochemistry **4**, 475 (1965).
[4] Birch, A. J., and K. A. M. Walker, J. Chem. Soc. C 1894 (1966).
[5] Brodie, H. J., M. Hayano, and M. Gut, J. Am. Chem. Soc. **84**, 3766 (1962).
[6] Brown, H. C., Hydroboration, W. A. Benjamin, Inc., New York, 1962 p. 93.
[7] Burn, D., V. Petrov, G. O. Weston, Tetrahedron Letters **9**, 14, (1966).
[8] Burstein, Sh., and R. I. Dorfman, Acta Endocrin. **40**, 188 (1962).
[9] Burstein, Sh., and B. Bhavnani, Endocrinology **8**, 351 (1967).
[10] Caspi, E., and P. K. Grover, Tetrahedron Letters 591 (1963).
[11] Caspi, E., P. K. Grover and Y. Shimizu, J. Am. Chem. Soc. **86**, 2463 (1964).
[12] Caspi, E., Rev. Portuguesa Quimica **6**, 145 (1964a).
[13] Caspi, E., D. O. Lewis, D. M. Piatak, K. V. Thimann and A. Winter, Experientia **22**, 506 (1966).
[14] Caspi, E., and D. O. Lewis, Science **156**, 519 (1967).
[15] Caspi, E., and G. M. Hornby, Phytochemistry **7**, 423 (1968).
[16] Caspi, E., and D. O. Lewis, Phytochemistry **7**, 683 (1968a).
[17] Cornforth, J. W., R. H. Cornforth, C. Donninger, G. Popjak, Y. Shimizu, S. Ichii, E. Forchielli and E. Caspi, J. Am. Chem. Soc. **87**, 3224 (1965).
[18] Dorfman, R. I., and F. Ungar, Metabolism of Steroid Hormones, Academic Press, New York, (1965).
[19] Euw von J. and T. Reichstein, Helv. Chim. Acta **47**, 711 (1964).
[20] Euw von J. and T. Reichstein, Helv. Chim. Acta **49**, 1468 (1966).
[21] Frantz, I. D., and G. J. Schroepfer, Jr., Ann. Rev. Biochem. **36**, 691 (1967).
[22] Galbraith, M. N., and D. H. S. Horn, Chem. Comm. 905 (1966).
[23] Gregory, H., and E. Leete, Chem. and Ind. 1242 (1960).
[24] Gross, E., and E. Leete, J. Am. Chem. Soc. **87**, 3479 (1965).
[25] Heftmann, E., Lloydia **30**, 209 (1967).
[26] Hershberg, E. B., E. Oliveto, M. Rubin, H. Staeudle and L. Kuhler, J. Am. Chem. Soc. **73**, 1144 (1951).
[27] Heusler, K., J. Kalvoda, G. Anner and A. Wettstein, Helv. Chim. Acta **46**, 352 (1963).
[28] Hayano, M. in Oxygenases; O. Hayashi ed. for Academic Press New York, 1962 p. 182.
[29] Leete, E., H. Gregory and E. Gross, J. Am. Chem. Soc. **87**, 3475 (1965).
[30] Meyer, B. S., D. B. Anderson and R. H. Bohning, Introduction to Plant Physiology, van Nostrand, Princeton, N.J. 1960 p. 324.
[31] Nakanishi, K., M. Koreda, S. Sasaki, M. L. Chang, and H. Y. Hsu, Chem. Comm. 915 (1966).
[32] Oleinick, N. L., and S. B. Koritz, Biochemistry **5**, 3400 (1966).
[33] Ringold, H. J., and S. K. Malhotra, Tetrahedron Letters, **15**, 609 (1962).
[34] Ringold, H. J., and A. Turner, Chem. and Ind. 211 (1962a).

[35] RINGOLD, H. J., M. HAYANO, and V. STEFANOVIC, J. Biol. Chem. **238**, 1960 (1963).
[36] RINGOLD, H. J., and S. K. MALHOTRA, J. Am. Chem. Soc. **87**, 3228 (1965).
[37] TALALAY, P., Phys. Rev. **37**, 362 (1957).
[38] TOMKINS, G. M., Recent Progr. Hormone Res. **12**, 125 (1956).
[39] TOMKINS, G. M., Ann. N.Y. Acad. Sci. **82**, 836 (1959).
[40] TSCHESCHE, R., Fortschr. Chem. Org. Naturstoffe, Springer Verlag, Berlin—Heidelberg **12**, 131. (1955).
[41] TSCHESCHE, R., and G. LILIENWEISS, Z. Naturforsch. **19b** 265 (1964).
[42] TSCHESCHE, R., and B. BRASSAT, Z. Naturforsch. **22b**, 894 (1966).
[43] TSCHESCHE, R., H. HULPKE and H. SCHOLTEN, Z. Naturforsch. **22b**, 677 (1967).
[44] TSCHESCHE, R., and B. BRASSAT, Z. Naturforsch. **22b**, 679 (1967a).

Studies on the Pathways of Brain Cholesterol Biosynthesis after Squalene Cyclization

G. Galli, E. Grossi Paoletti and J. F. Weiss
Institute of Pharmacology, University of Milan, Milan, Italy

Summary

The biosynthetic pathways leading to cholesterol after squalene cyclization have been extensively studied in animal tissues, but a number of steps remain to be elucidated, and it is possible that in different tissues and in various organisms preferential pathways may exist. There are methodological difficulties in attempting to obtain an extensive characterization of sterol intermediates, because they are present in trace amounts and are difficult to separate because of similarity of structure. In the present investigation, sterols from brain unsaponifiable material were separated as acetate derivatives, using columns containing Silica Gel G-Celite-$AgNO_3$. The eluted sterols were qualitatively and quantitatively analyzed by gas liquid chromatography (GLC) and the structures elucidated by GLC-mass spectrometry. The incorporation of labeled mevalonic acid into brain sterols of the developing chick embryo and of the rat was also studied. C-29, C-28 and C-27 sterols were identified in both immature and mature nervous tissue, although the amounts, and radioactivity of the intermediates relative to cholesterol varies, depending on the degree of brain maturation. The results obtained allow us to postulate in detail a biosynthetic sequence for cholesterol in nervous tissue.

Introduction

The last stages of cholesterol biosynthesis in biological tissues after squalene cyclization, as investigated in the classic papers by Bloch (Bloch, 1957 and 1959), have been recently reviewed by Olson (1966). Few data are available for sterol biosynthesis in tissues other than liver and skin, and data reported on the possible biosynthetic pathways are mostly based on very few intermediate compounds isolated or synthesized in radioactive form. In our investigations on the biosynthesis of cholesterol in mature and growing mammalian and avian brain, we have used a new methodological approach. A quantitative procedure for complete sterol analysis of the tissue has been developed: the compounds were separated with very high efficiency with the aid of argentation column chromatography and identified using the new combination of gas chromatography and mass-spectrometry.

The incorporation of labeled mevalonate into cholesterol and its sterol precursors has been measured.

The aim of our work was to clarify the routes for cholesterol biosynthesis in mature and growing brains and to determine whether the nervous tissue shows pathways of cholesterol synthesis comparable to that of other tissues.

Materials

Silica Gel-G (Merck); Celite 535 acid washed; $AgNO_3$ (Carlo Erba, Milan); 2-^{14}C mevalonate (Amersham, U. K.); benzene (B. D. H., Analar); hexane (Merck); acetic anhydride and pyridine (Merck) freshly distilled; petroleum ether (b.p. 60—80 °C, B. D. H., Analar).

The sterols were tested for purity using GLC, UV and IR methods and the acetate derivatives were prepared. The following standards have been investigated: (I) cholest-5-en-3β-ol (cholesterol), Merck AG., Darmstadt; (II) cholest-5,24-dien-3β-ol (desmosterol), Merrell Inc., Cincinnati; (III) cholest-7-en-3β-ol (lathosterol), gift of Dr. J. SJÖVALL, Karolinska Institutet, Stockholm; (IV) cholest-7-en-3β-ol-4-methyl (lophenol) gift of Prof. C. DJERASSI, Stanford University; (V) cholest-7-en-3β-ol-4,4'-dimethyl gift of Dr. J. L. GAYLOR, Cornell University, Ithaca; (VI) cholest-8,24-dien-3β-ol (zymosterol) gift of Prof. L. CANONICA, Institute of Organic Chemistry, Milan; (VII) cholest-8-en-3β-ol (zymostenol) obtained by catalytic hydrogenation of (VI); (VIII) cholest-8-en-3β-ol-4,4'-dimethyl gift of Dr. J. L. GAYLOR; (IX) cholest-5,7-dien-3β-ol (7-dehydrocholesterol), Fluka AG Buchs SG.

Methods

Preparation of sterol mixtures

Brains were obtained from chick embryos (11 and 18 days incubation) and mature Sprague-Dawley rats. The excised brains were pooled and saponified in ethanolic KOH. The unsaponifiable material was extracted in petroleum ether (b.p. 60—80 °C) and the extracts washed and dried over anhydrous Na_2SO_4. The filtered extract, after evaporation under nitrogen, was acetylated with acetic anhydride in pyridine in the dark for 12 hours at room temperature. The sterol acetates were extracted with ethyl ether and successively washed with dilute HCl, 5% $KHCO_3$ and water. The extract was dried over anhydrous Na_2SO_4 and the acetylated sterols redissolved in hexane.

Separation of sterol acetates

The hexane solution of sterol acetates was applied to a chromatographic column containing celite–silica gel — $AgNO_3$ (10—10—4 by weight) as previously described (GALLI and GROSSI PAOLETTI, 1967). The fractions eluted with hexane–benzene of increasing polarity were controlled individually by GLC and the fractions containing the same sterols were combined.

The screening of the fractions and their quantitation was carried out with a Fractovap Model C gas chromatograph (Carlo Erba Co.) with PhSi 1% columns. Quantitation was obtained using androst-5-en-17-one-3-acetate as internal standard (FUMAGALLI et al., 1965).

Gas chromatography — mass spectrometry

Identification of the single sterol acetates was performed with the LKB model 9000 gas chromatograph — mass spectrometer. The ion current was 60 μA, the energy of the electrons 70 eV and the scan speed 4—6 seconds.

We have used glass columns filled with 1% PhSi and with 1% Se 30. Temperature of molecule separator and the ion source were 240 °C and 290 °C.

The identification of sterol from biological tissues is carried out by comparison with authentic sterols when available or by analogy of characteristc fragmentation patterns (GALLI and MARONI, 1967). The comparison of the spectra of sterols obtained from biological material and from standards is illustrated in fig. 1 using as example desmosterol acetate.

Results

Separation of acetylated brain sterols by $AgNO_3$ column chromatography was performed using:

1. adult rats injected intravenously with 250 μC/Kg of 2-^{14}C mevalonate 1 hour before sacrifice;

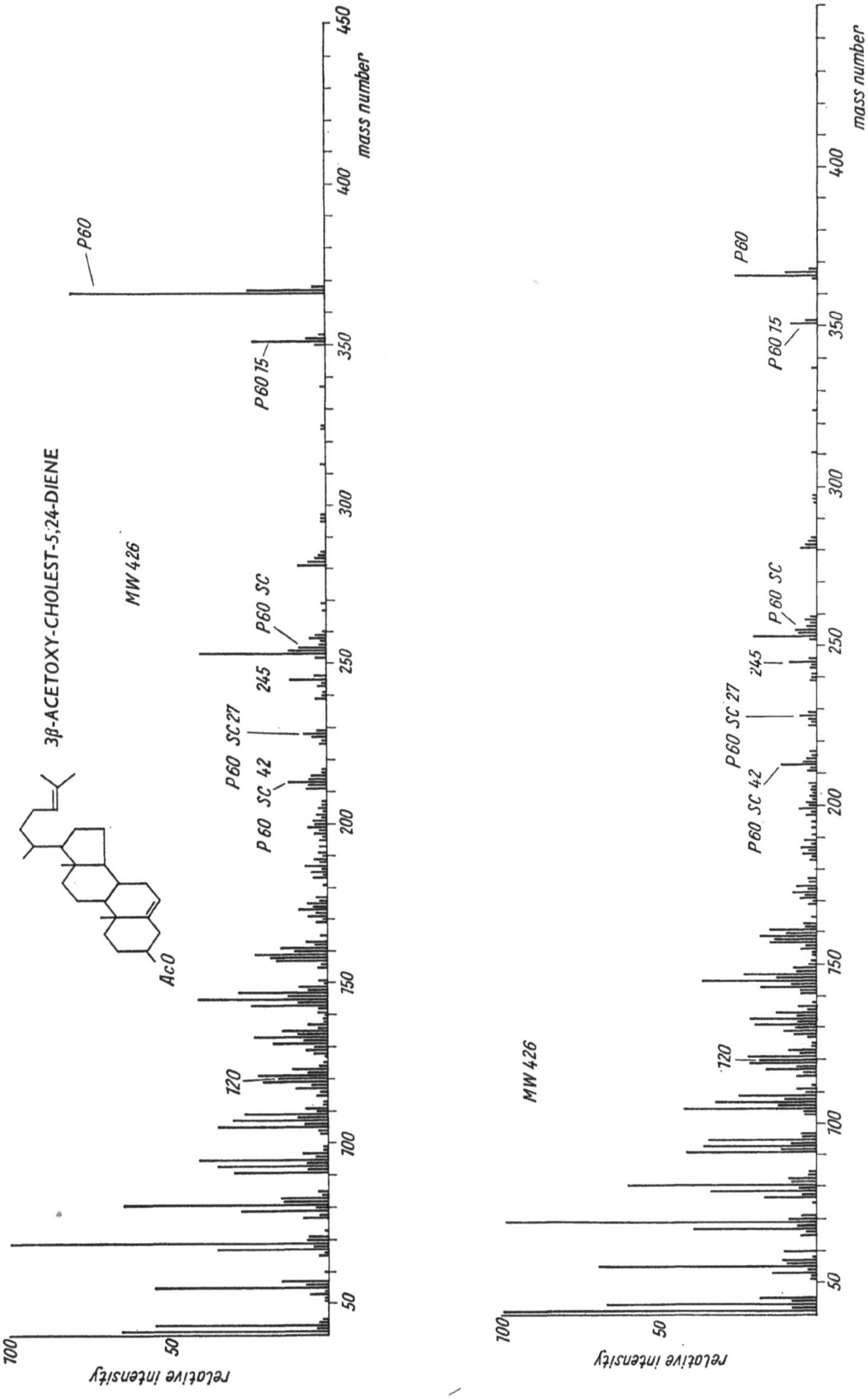

Fig. 1. Mass spectra of acetate derivative of desmosterol; above: standard, below: biological sample; p = molecular ionpeak; sc = side chain

2. chick embryos, 11 and 18 days of incubation, injected 4 hours before harvest with 2 μC 2-^{14}C mevalonate into the yolk sac (51 brains were collected for the 11 day experiment and 26 for the 18 day).

Table 1 shows the percentage of identified 3β-hydroxy sterols found in brain of chick embryo and adult rat. As reported previously, desmosterol occurs in developing nervous tissue and is quantitatively the next important sterol after cholesterol (Fumagalli et al., 1963). But we have also found desmosterol in the adult rat brain (0.8%). It can be seen that all the other minor sterol components detected in the developing brains were also present in the adult brain but generally at lower percentages.

Table 1

Brain sterol composition of chick embryo and adult rat (%)

	11 day embryo	18 day embryo	adult rat
C29 Δ^{8}	0.033	0.027	0.007
C28 Δ^{8}	0.015	0.246	0.011
C27 Δ^{8}	0.016	—	0.115
C29 Δ^{8-24}	0.155	0.322	0.082
C28 Δ^{8-24}	0.142	0.203	0.050
C27 Δ^{8-24}	0.067	0.108	0,047
C27 Δ^{7-24} (?)	0.013	0.125	0.070
C27 Δ^{5-24}	7.574	6.865	0.797
C27 Δ^{5}	89.330	90.590	98.000
C27 Δ^{0}	0.454	0.437	0.122
others	2.208	1.983	0.699

The elution sequence of the acetylated sterols isolated from adult brain is presented in Table 2. The amounts of sterols are reported in μg and radioactivity as DPM/fraction. First cholestanol (C-27 Δ^{0}) was eluted from the column. Then the Δ^{8} series was eluted: C-29, C-27 and C-28. Cholestanol and the Δ^{8} homologues were less radioactive than some of the sterols eluted later. Cholesterol was then separated in large amounts and this fraction contained the largest amount of total radioactivity. The Δ^{8-24} series is eluted in the order C-29, C-28, C-27. The specific activity of the Δ^{8-24} sterols is very high, approximately 100 times greater than the specific activity of cholesterol, as can be easily calculated, because these sterols are represented by very small pools, (only μg quantities). The next sterol was tentatively identified as C-27 Δ^{7-24}.

Desmosterol, the last identified sterol, has a specific activity intermediate between that of cholesterol and the Δ^{8-24} sterol series.

More polar sterols (poly hydroxy) were eluted from the column and their structures are under investigation.

The eluted sequence of the acetylated sterols isolated from developing chick brain was similar, except that relatively larger amounts of the sterol intermediates were found and the incorporation of radioactivity is much more efficient.

The same important intermediates were isolated from both adult and immature nervous tissue — that is, those highly labeled intermediates occurring in significant quantities.

Conclusions

The described methods of $AgNO_3$ column chromatography and the use of the combination of gas chromatography and mass spectrometry has allowed us to separate and identify many possible sterol intermediates of brain cholesterol biosynthesis. The data permit us to postulate a pathway for brain cholesterol synthesis after squalene cyclization. From the radioactivity

Table 2

Brain sterols of adult rat

Elution sequence of the sterol-acetates from $AgNO_3$-Silica Gel-G column

fraction	solvent system hexane: benzene	C27-Δ^0	C27-Δ^8	C29-Δ^8	C28-Δ^8	C27-Δ^5	C29-Δ^{8-24}	C28-Δ^{8-24}	C27-Δ^{8-24}	C27-Δ^{7-24} (?)	C27-Δ^{5-24}	Sterols µg/fraction	D.P.M./fraction
11–12	90:10	35.123		1.234								36.357	15
13	90:10	34.371	26.726	3.008								64.105	24
14	90:10		39.785		6.125							46.010	88
15–16	90:10					2,093.733						2,093.733	690
17–39	85:15					54,571.100				17.810		54,571.100	9.020
40–43	85:15					36.763	1.937			14.820		56.507	35
44–45	80:20					14.666	2.770			7.290		32.256	40
46–48	80:20					15.195	24.165	2.832				49.477	695
49–50	80:20					3.595	7.542	6.720				17.857	315
51–52	80:20					2.233	3.233	7.024	4.143			16.653	275
53–54	75:25						2.342	11.592	11.710			31.461	410
55	75:25						1.578		5.940			38.140	268
56–57	75:25						3.137		5.130			61.257	390
58–59	75:25										128.430	172.250	230
60–68	75:25										215.250	219.250	225
69–72	70:30										69.087	80.580	75
73–80	70:30										41.820	136.210	90
81–95	50:50											172.773	1.134

The figures indicate µg of sterol acetates in each fraction pool.

data it could be proposed that cholesterol synthesis occurs in brain by a preferential route. A scheme is shown in fig. 2.

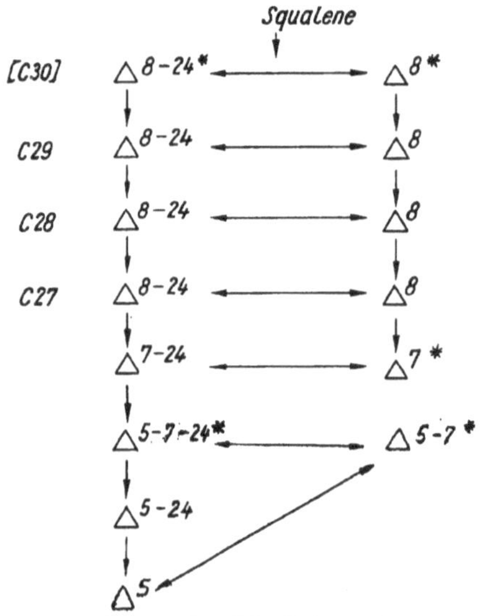

Fig. 2. Possible routes of brain cholesterol synthesis

We have not detected experimentally sterols with 30 carbon atoms but the sequence of demethylation indicates that lanosterol, C-30 Δ^{8-24}, is probably present but in undetectable amounts. The pathway from C-29 to C-27 occurs through sterols retaining the double bonds in the 8, 9 position. Δ^{8-24} intermediates, which occur in larger quantities and are highly radioactive, appear to be the preferential intermediates in the demethylation sequence.

On the other hand, demethylation through the Δ^8 intermediates (C-29, C-28 and C-27) with saturated side chain, probably represents a minor route in the nervous tissues. It is proposed that in brain the isomerization of the double bonds from the Δ^8 to Δ^7 position may occur only at the 27 carbon atom stage and is followed by the formation of the Δ^5 sterols. The saturation of the side chain apparently occurs as the last step of the pathway.

Our results give a picture of the biosynthetic routes of cholesterol in the mature and growing nervous tissues. It is now easier, by using this approach, to detect at molecular levels modifications which may occur in nervous tissue in pathological conditions and to detect the points of action of drugs interfering with brain sterols. The same approach may be also applied to a variety of biological systems.

Acknowledgement

This research has been partially supported by the Association for the Aid of Crippled Children. Giovanni Galli holds a fellowship from the Commission for Scientific Research of Italian Switzerland. J. F. Weiss, is a United States Public Health Service Post-doctoral Fellow (No. 1-F2-CA-31,598-01) from the National Cancer Institute.

References

[1] Bloch, K., Vitam. and Horm., **15**, 119 (1957).
[2] Bloch, K., "Biosynthesis of terpenes and sterols" p. 4 (Edit. G. E. W. Wolstenholme and M. O'Conner) Boston: Little, Brown & Co., (1959).
[3] Fumagalli, R. and R. Paoletti, Life Sci., **2**, 291 (1963).
[4] Fumagalli, R., P. Capella, and W. G. A. VandenHeuvel, Analyt. Biochem., **10**, 377 (1965).
[5] Galli, G., and E. Grossi Paoletti, Lipids, **1**, 72 (1967).
[6] Galli, G., S. Maroni, Steroids, **10**, 189 (1967).
[7] Olson, I. A., Ergebn. Physiol, biol. Chem. exp. Pharmakol. **56**, 173 (1965).

The Function of Sterols in Mycoplasma

P. F. Smith
University of South Dakota
Vermillion, South Dakota, USA

Summary

Organisms of the genus Mycoplasma either require planar, hydroxylated polyterpenes for growth or synthesize carotenoid pigments. The former are deficient in enzymes of the biosynthetic pathway to polyterpenols. The polyterpenol requirement can be met with planar, 3-hydroxy steroids. Sterols are incorporated into the cell membrane by all Mycoplasma. Sterols spare the synthesis of carotenoid pigments and overcome inhibition of growth caused by blockage of the biosynthetic pathway to carotenoids prior to the formation of phytoene. Thus, sterols and carotenoids appear to carry out analogous functions. The only alterations of the structure of the sterol are esterification and glycosidation of the 3-hydroxy group. Both the esterified fatty acids and the glucose turn over during growth and metabolism. One group of Mycoplasma is incapable of hexose fermentation and does not contain the steryl glucoside; rather, these organisms metabolize short-chained fatty acids. The fatty acids of the esterified sterol are principally butyric and acetic. Another group of Mycoplasma degrade glucose to acetate. These organisms contain the steryl glucoside and steryl esters containing principally acetic acid. Mycoplasma containing sterol are sensitive to digitonin induced lysis while those containing carotenoids are not. All Mycoplasma are easily lysed and their cell membrane solubilized by amphipathic compounds. These and other data suggest that sterols and carotenols are involved in (1) membrane structure and (2) as carriers for substrates in membrane transport phenomena or as precursors to some other cellular component. Models of membrane structure and of permeability function have been devised to account for the available evidence concerning the lipids of Mycoplasma.

The genus, Mycoplasma, is comprised of small (ca. 200 mμ dia.) plastic organisms with a spheroidal or filamentous morphology. They are completely devoid of a morphologically detectable cell wall and contain none of the usual chemical components of bacterial cell walls. The external envelope encompassing the cell is a typical unit membrane composed almost entirely of lipoprotein. Within this genus two nutritional types are found, those species which require a sterol for growth and those species which are nutritionally sufficient for lipids. Among the sterol requiring species are organisms which derive their energy by fermentative metabolism of hexoses and organisms which are incapable of hexose fermentation. These latter species obtain their energy by β-oxidation of short chain fatty acids followed by utilization of the formed acetate via the tricarboxylic acid cycle. All sterol nonrequiring mycoplasmas appear to ferment glucose to acetate.

The nature of the lipids in these three types of mycoplasmas is shown in Table 1. Essentially all these lipids are found associated with the cell membrane. The principal neutral lipid of sterol requiring species is the sterol in which the organisms are grown (Rothblat and Smith, 1961; Smith, 1962). These organisms are incapable of polyterpene synthesis and fail to alter the sterol provided in the culture medium except for esterification and in the case of fermentative species glycosidation. Sterol nonrequiring species grown in the absence of sterol contain neurosporene, an all trans 3,3′ dihydroxy carotene presumably containing 8 conjugated double bonds, and esters and the glucoside of this carotenol (SMITH, 1963a). Sterol, steryl esters and steryl glucoside are found together with the carotenoids if grown in the presence of a sterol.

Table 1
Nature of lipids found in Mycoplasma

	Sterol-requiring, fermentative	Sterol-requiring, non-fermentative	Sterol-nonrequiring, fermentative
Neutral lipids	Cholesterol (cholestanol, ergosterol) Steryl esters (primarily acetate)	Cholesterol (cholestanol, ergosterol) Steryl esters (primarily butyrate and acetate)	Neurosporene Dihydroxycarotene Carotenyl esters (primarily acetate)
	Steryl-β-D-glucoside		Carotenyl-[α,β]-D-glucoside
	(Glycosyl diglycerides)		Glucosyl diglycerides
Phospholipids	Glycerophospholipids	Glycerophospholipids	Glycerophospholipids Phosphatidyl glucose

The sterol requirement of mycoplasmas has been studied extensively (SMITH, 1964; SMITH and LYNN, 1958). Only planar, 3-hydroxy sterols adequately support growth (Table 2). Alteration or removal of the 3-hydroxyl group destroys the growth promoting ability of a sterol. Esterification with fatty acids of chain length up to 10 carbon units results in decreasing growth response. Cis-fused rings and the absence of aliphatic side chain likewise render sterols incapable of supporting growth. Unsaturation in the ring or side chain appears to have little effect on the adequacy of the sterol for the organisms although the fully saturated compound, cholestanol, has proven to give the best growth response. Sterols are incorporated into the cell membrane by a physical process best described by an adsorption isotherm (SMITH and ROTHBLAT, 1960). This uptake occurs both in resting and growing cells. In order for incorporation to occur some amphipathic compound is required to increase the aqueous solubilization or micellarization of the sterol together with the protein moiety of certain lipoproteins (e.g., α_1-lipoprotein of serum) which regulate the rate of incorporation (SMITH and BOUGHTON, 1960). Neither amphipathic compound nor protein are incorporated or

Table 2
List of steroids which support or fail to support growth of Mycoplasmas

Support growth	do not support growth
Δ^5-Cholesten-3β-ol	Δ^5-Cholesten-3α-ol
Cholestan-3β-ol	Cholestan-3α-ol
β-Sitosterol	Coprostan-3α-ol
Ergosterol	Coprostan-3β-ol
Stigmasterol	Cholesteryl esters (myristate through linoleate)
Cholesteryl esters (propionate through laurate)	Cholestane
	Bicholesteryl ether
	Cholestan-3-one
	Δ^5-Cholesten-3-one
	Estrone
	Estradiol
	Testosterone
	Bile acids
	Progesterone

degraded by the mycoplasmas. The only structural requirement for incorporation is the aliphatic side chain of the sterol (Table 3). Thus, sterols which do not exhibit planarity are incorporated as easily as planar molecules even though they fail to support growth. Once bound in the organism the sterol exhibits no turnover.

Table 3

List of steroids which are incorporated into the cell membranes of mycoplasmas

Incorporated	Not incorporated
Δ^5-Cholesten-3β-ol	Progesterone
Δ^5-Cholesten-3α-ol	Testosterone
Cholestan-3β-ol	
Cholestan-3α-ol	
Coprostan 3α-ol	
Coprostan-3β-ol	
Δ^5-Cholesten-3-one	
Cholestan-3-one	

Growth of sterol nonrequiring mycoplasmas in the presence of cholesterol results in a reduction of the biosynthesis of carotenoid pigments (Table 4) (SMITH, 1963a). Compounds

Table 4.

Sparing action of cholesterol on carotenoid biosynthesis by *Mycoplasma laidlawii*

Serum fraction (Cholesterol source) %	Pigment OD_{438}/mg dry wt cells	^{14}C acetate incorporation into unsaponifiable lipids (cpm/mg lipid)
none	0.014	169
0.05	0.012	184
0.10	0.011	166
0.50	0.010	119
1.00	0.009	63

which inhibit enzymic steps in the biosynthetic pathway of polyterpenes prior to the formation of phytoene, (e.g., farnesenic acid, β-diethylaminoethyldiphenyl propyl acetate and phenethyl biguanide) inhibit the growth of these mycoplasmas. Cholesterol reverses this inhibitory action (SMITH and HENRIKSON, 1966). Organisms grown under cultural conditions in which carotenoid synthesis is inhibited but in which growth inhibition is reversed by the presence of cholesterol contain cholesterol rather than carotenoid pigments (Table 5). The results imply that carotenoids and sterols carry out analogous functions in the mycoplasmas. They further suggested that the sterol requirement of certain mycoplasmas might be a reflection of their incapacity to synthesize polyterpenes.

Table 5

Unsaponifiable lipids of *Mycoplasma laidlawii* grown in the presence of phenethyl biguanide and cholesterol

Culture medium plus	Dry wt of culture (μg/ml)	Cholesterol (μg/mg dry wt)	Pigment (OD_{438}/mg dry wt)
No addition	27.2 ± 0.36	0.003	0.013
Cholesterol (2.8×10^{-5} M)	37.8 ± 0.06	53.3 ± 1.3	0.010
Cholesterol (2.8×10^{-5} M) + phenethyl biguanide (10^{-3} M)	30.4 ± 1.14	70.9 ± 1.7	0.002

Three species of *Mycoplasma* representative of the three physiological types described previously were examined for specific enzymes involved in the synthesis of dimethylallyl pyrophosphate from acetate. Table 6 summarizes the results showing that (1) all three strains

Table 6

Enzymes in mycoplasmas involved in the biosynthesis of polyterpenes

Enzyme	Reaction	Organism		
		M. laidlawii B	*M. hominis* strain 07	*Mycoplasma sp.* strain J
Acetate kinase	Acetate → Acetyl-P	+	0	+
Phosphateacetyltransferase	Acetyl-P → Acetyl CoA	+	0	+
Acetyl CoA-synthetase	Acetate → Acetyl CoA	0	+	+
Propionate CoA-transferase	Acetyl CoA → Butyryl CoA	0	+	+
3-Ketoacid CoA-transferase	Succinyl CoA → AcAcCoA	+		+
Acetoacetyl CoA-thiolase	2 Acetyl CoA → AcAcCoA	+	+	+
Hydroxymethylglutaryl CoA-synthase	AcCoA + AcAcCoA → → HMGCoA	+	0	+
Hydroxymethylglutaryl CoA-reductase	HMGCoA → MVA	+	0	+
Mevalonate kinase	MVA → MVA-P	+	0	0
Phosphomevalonate kinase	MVA-P → MVAPP	+	0	0
Pyrophosphomevalonate decarboxylase	MVAPP → IPP	+	0	0
Isopentenylpyrophosphate isomerase	IPP → DMAPP	+	0	+

Data compiled from HENRIKSON and SMITH, 1966a; SMITH and HENRIKSON, 1965b; and VAN DEMARK and SMITH, 1964.

Table 7

Ability of polyterpenoid precursors to substitute for the sterol requirement of certain mycoplasmas

	M. hominis strain 07	*Mycoplasma sp.* strain J
Acetate	0	0
Mevalonic acid	0	0
5-Phosphomevalonic acid	0	0
5-Pyrophosphomevalonic acid	0	0
Isopentenyl pyrophosphate	0	+
Geranyl pyrophosphate	0	0
Farnesyl pyrophosphate	0	0
Squalene	0	0
Phytoene	±	+
Phytofluene	0	+
ζ-carotene	0	+
Neurosporene	±	+
Lycopene	0	0
β-carotene	0	0

+ – growth; ± – minimal growth; 0 – no growht.

are able to synthesize acetoacetyl CoA, (2) strain 07 lacks all the remaining enzymes, (3) strain J lacks the three enzymes which carry out the conversion of mevalonic acid to isopentenyl pyrophosphate and (4) *M. laidlawii* is sufficient for all the enzymes (HENRIKSON and SMITH, 1966a; SMITH and HENRIKSON, 1965a). Nutritional experiments employing precursors of polyterpenes confirm these enzymic results (Table 7) (HENRIKSON and SMITH, 1966b). The inability of geranyl and farnesyl pyrophosphates to replace the sterol requirement is attributed to impermeability of the cells to long chain phosphorylated compounds. Growth of strain J in the presence of isopentenyl pyrophosphate in lieu of sterol resulted in the synthesis of polyterpenes which yield lycopersane on complete hydrogenation and of hydroxylated derivatives of these polyterpenes. These results further confirm an interrelation between sterols and carotenols and imply that the sterol requirement of certain species is fortuitous reflecting enzymic deficiencies for the biosynthesis of carotenoid pigments.

The sterol requirement of all mycoplasmas examined could be met by all trans 3,3′-dihydroxy carotenes (Table 8). Mono hydroxylated compounds were ineffective. Growth of strain J in the presence of the carotenol from *M. laidlawii* in place of sterol permitted the isolation of the unchanged carotenol from the organism.

Table 8
List of polyterpenes which support or fail to support growth of sterol requiring mycoplasmas

Support growth	do not support growth
M. laidlawii carotenol	Geraniol
Lutein	Farnesol
Sarcinaxanthin	Nerolidol
	Phytol
	Geranyl garaniol
	Chloroxanthin
	Rubixanthin
	Solanesol
	Dolichol

Establishment of the analogy of sterols and carotenols in mycoplasmas led to studies on the possible function(s) of these compounds. Sterol comprises about half of the total lipids of sterol requiring species which contain 12 to 20% of their dry weight as lipid (RAZIN et al., 1963; SMITH and ROTHBLAT, 1962). In sterol nonrequiring species carotenoid pigments and glucosyl diglycerides account for half of the total lipid, the latter components predominating. This large proportion of neutral lipid and its location in the cell membrane suggest a structural role. Compatible with this possibility is the lack of turnover of sterol in the organisms and the absence of turnover of glucose in the glycosyl diglycerides during pulse labeling of growing or metabolizing organisms. Mycoplasmas grown in the presence of cholesterol are extremely sensitive to lysis by digitonin. The absence of sterol in the organisms renders them refractory to digitonin lysis (RAZIN and ARGAMAN, 1963; SMITH and ROTHBLAT, 1960). Identical results are obtained with streptolysin 0 (BERNHEIMER and DAVIDSON, 1965) and the large polyene antibiotics, such as filipin (LAMPEN et al., 1963). Thus specific removal or interference with the sterol in the organisms results in breakdown of the cell membrane.

Sterols are not the sole lipoidal components involved in maintenance of membrane structure. Detergents, phospholipase A, and phosphomonoesterases also cause cell lysis and solubilization of membranes (KOOSTRA and SMITH, 1968). These effects imply a role also for phospholipids. Glycerophospholipids are present in membranes of all mycoplasmas and include (bis)-phosphatidyl glycerol, phosphatidyl glycerophosphate and phosphatidyl glycerol (SMITH and KOOSTRA, 1967; SMITH and HENRIKSON, 1965b; PLACKETT, 1967). These lipids are partially hydrolyzed by the aforementioned enzymes. The effect of detergents, particularly the anionic compounds, on purified mycoplasmal membranes is solubilization or disaggregation to form homogeneous lipoprotein subunits with an s value of 2.5 and an estimated molecular weight of 30,000 to 40,000 (RAZIN et al., 1965; KOOSTRA and SMITH, 1968). Removal of detergent by

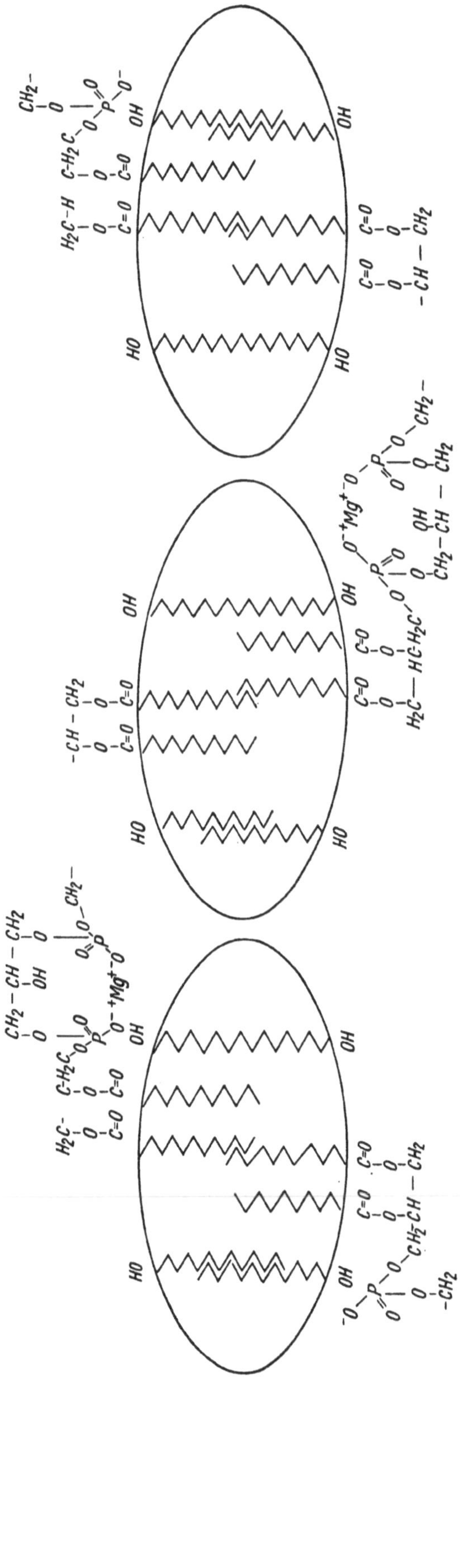

Fig. 1. Structural model for the cytoplasmic membrane of mycoplasmas

dialysis followed by exposure to di- or polyvalent cations results in reaggregation of these subunits. Monovalent cations have no such effect, nor do chelating agents cause dissolution of the membranes. The effect of the latter on intact cells is the production of an increase in permeability, i.e., leaky cells (Rodwell, 1965). Experimental results then indicate a role for sterols, glycosyl diglycerides and glycerophospholipids, in the maintenance of membrane structures.

Several problems remain: 1. the structural fit of these lipids in the membrane which is compatible with the experimental data; 2. the establishment of an analogous function for sterols and carotenoid pigments; 3. an explanation of the requirement for planarity in the sterol and carotenol molecules; 4. a reason for the necessity of terminal hydroxyl groups on the sterol and carotenol. The unit membrane comprising the cell envelope of mycoplasmas is a triple layered structure 75 to 100 Å in width (Van Iterson and Ruys, 1960). The external layers are electron dense and hence must contain charged groups. Sandwiched between these layers is an electron transparent zone about 35 Å in width. The electron transparency presumably signifies uncharged moleular careas. The usual pictorial view of such a membranous structure is lamellar in form. Our data conflict with such a representation, but are compatible with a corpuscular structure. The cell membrane of mycoplasma is pictured (figure 1) as interconnected subunits bound together by phospholipids, in particular (bis)-phosphatidyl glycerol. Each subunit represents a protein, coiled so that the apolar segments are contained within the globule and the polar segments are located at the surface. The fatty acid chains of phospholipids and glycolipids are bound in the apolar region of the protein by hydrophobic bonds. (Bis)-phosphatidyl glycerol serves as a bridge between two subunits, the terminal fatty acid residues being bound in separate subunits. Divalent cation is bound in the phosphate ester bridge to the negatively charged oxygen atoms causing condensation of the subunits. Bridging of subunits by other glycerophospholipids and glycolipids by means of both polar and apolar bonding can be visualized. Sterols and carotenols lie lengthwise across the width of the subunits being bound by hydrophobic bonds. The terminal hydroxyl groups lie at the surface of the protein subunits and are free for enzymic use. The length of the apolar segment of all trans 3,3'-dihydroxy carotenes approximates 30 Å, compatible with the electron transparent zone of the mycoplasmal membrane. A dimer of planar 3-hydroxy sterol formed by overlapping aliphatic side chains produces a molecular structure similar in conformation and dimension to that of the carotenol. Non-planarity of sterol molecules would not interfere with their binding in the subunit but would result in burying the hydroxyl groups in the subunit making them inaccessible to enzyme action. Removal of sterol would disrupt the hydrophobic interactions of lipid and protein resulting in cell lysis. Enzymic removal of fatty acid chains or phosphate groups would break the bridges between subunits. The large number of uncharged molecular segments in the central portion of the membrane would cause the appearance of electron transparency. The dimensional fit and the orientation of sterol and carotenol are almost identical. The structural requirements of sterol capable of supporting growth thus are explained.

The necessity for free hydroxyl groups rather than any charged radical at the 3 position on sterol and carotenol requires further examination. Additional data suggest that sterols and carotenols possess another function in addition to the structural role, namely, membrane transport (SMITH, 1963b). As previously stated the nature of the lipids of a given species of *Mycoplasma* bears a relationship to its metabolic character. Species which ferment hexoses to acetate contain steryl or carotenyl glucosides and acetate esters of sterols and carotenols. Species incapable of hexose fermentation contain no glycosides but contain butyrate and acetate esters of sterol. These nonfermentative species derive their energy by oxidation of fatty acids. All mycoplasmas examined contain a membrane associated esterase capable of hydrolytic or thiolytic cleavage of steryl and carotenyl esters of short chain fatty acids (SMITH, 1959). Only hexose utilizing mycoplasmas contain a membrane associated glucosidase (HENRIKSON and SMITH, 1964). The enzymes associated with ester synthesis also are membrane associated. No information is available yet on the site of glucoside synthesis but presumably this also is located in the cell membrane.

Turnover studies employing pulse labeling technics with ^{14}C-glucose have been performed with some fermentative species (SMITH, 1967). As previously stated the total content of sterol and carotenoids does not exhibit fluctuation during growth and metabolism. These lipids appear irreversibly bound in the membrane. On the contrary the acyl and glucosidic derivatives do exhibit fluctuations. The glucoside content is depressed somewhat in actively growing cells. Pulse labeling with ^{14}C-glucose results in rapid labeling of the carotenyl glucoside followed by rapid disappearance of label upon removal of exogenous radioactivity. The carotenyl ester becomes labeled slightly after the glucoside and likewise quickly loses its radioactivity. The appearance of radioactive acetate and carbon dioxide in supernatant fluid follows the pattern of the pulse. Kinetic studies and measurement of the fate of ^{14}C rules out the possibility that the aforementioned lipids are serving as biosynthetic precursors. The rate of evolution of radioactive carbon dioxide is equivalent to the rate of disappearance of radioactivity from the total lipids. Compounds which inhibit glucosidase activity permit the limited accumulation of glucoside but prevent glucose degradation. Pulse labeling is abolished by inhibitors of glucose utilization, such as iodoacetic acid.

A permeability model has been devised using the structural model previously described. The membrane subunits are not so closely packed that easy diffusion of water molecules is impeded. Expansion of these pores can be achieved by removal of cation bound to the negatively charged oxygen atoms. Specific removal of cation would be dependent upon the substance to be transported. In some instances specific enzymes would use this cation as a cofactor. Portions of the subunits would have certain enzymic activity. The subunits may represent a mosaic of enzymic activities, so that only certain areas of the membrane would be available to permit the entry or exit of a given substance. Membrane transport of glucose into the cell and acetate out of the cell in the particular example in question could proceed as follows (Fig. 2): Glucose approaching the cell from the outside would be covalently bonded

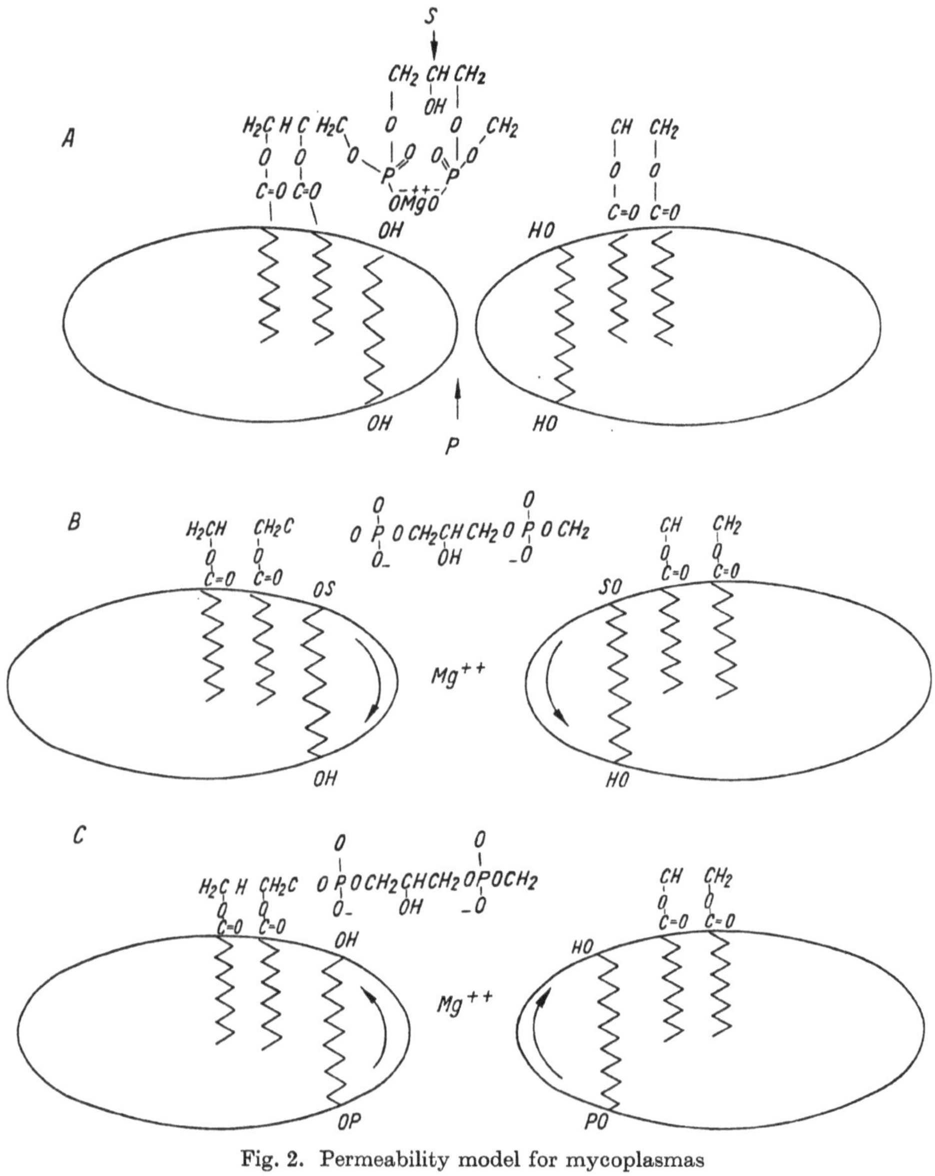

Fig. 2. Permeability model for mycoplasmas

to sterol or carotenol forming the glucoside. The increase in polarity of the molecule could trigger a rotational movement toward the interior of the cell exposing the glucose to the cytoplasm. The glucosidase would cleave the glucose radical from the sterol or carotenol. After the degradation of glucose by the cytoplasmic enzymes, the end product, acetate, would become covalently linked to the free hydroxyl of the sterol or carotenol. The ensuing decrease in polarity would trigger a rotational movement to the outside of the cell at which location the esterase would remove the acetate expelling it to the supernatant medium. Theoretically the entry of glucose and exit of acetate would occur during one rotational movement of the dihydroxy carotenol or dimer of sterol, since in each case one free hydroxyl group is facing the interior and the other the exterior of the cell. Fatty acid transport into the cells of nonfermentative species could be effected through the esterification of sterol. Carbon dioxide and water could diffuse out of the cells through contracted pores. Although this model may exhibit too much simplicity, it explains the data in mycoplasmas available at the present time and is compatible with the structural model.

The structural model also allows for other mechanisms of permeability. Certain pores formed by adjoining subunits may be lined with specific enzymes or nonenzymic proteins which regulate entry and/or exit from the cell; others may possess certain charges and be concerned with transport of other substances. But a discussion of these mechanisms is beyond the scope of this symposium. These models and mechanisms are considered to be working hypotheses only and undoubtedly will require modification or rejection as our knowledge of membranes and lipids increases.

Acknowledgments

This work was supported in part by a contract (No. 4898) from the Office of Naval Research and research grant AI 04410 from the National Institute of Allergy and Infectious Diseases, U.S. Public Health Service.

References

[1] Bernheimer, A. W., and M. Davidson, Science **148**, 1229 (1965).
[2] Henrikson, C. V., and P. F. Smith, J. Gen. Microbiol. **37**, 73 (1964).
[3] Henrikson, C. V., and P. F. Smith, J. Bacteriol. **92**, 701 (1966a).
[4] Henrikson, C. V., and P. F. Smith, J. Gen. Microbiol. **45**, 73 (1966b).
[5] Koostra, W. L., and P. F. Smith, Biochim. Biophys. Acta (in prep.) (1968).
[6] Lampen, J. O., J. W. Gill, P. M. Arnow, and I. Magana-Plaza, J. Bacteriol. **86**, 945 (1963).
[7] Plackett, P., Ann. N.Y. Acad. Sci. (in press) (1967).
[8] Razin, S., and M. Argaman, J. Gen. Microbiol. **30**, 155 (1963).
[9] Razin, S., M. Argaman, and J. Avigan, J. Gen. Microbiol. **33**, 477 (1963).
[10] Razin, S., H. J. Morowitz, and T. M. Terry, Proc. Nat. Acad. Sci. **54**, 219 (1965).
[11] Rodwell, A. W., J. Gen. Microbiol. **40**, 227 (1965).
[12] Rothblat, G. H., and P. F. Smith, J. Bacteriol. **82**, 479 (1961).
[13] Smith, P. F., J. Bacteriol. **77**, 682 (1959).
[14] Smith, P. F., J. Bacteriol. **84**, 534 (1962).
[15] Smith, P. F., J. Gen. Microbiol. **32**, 307 (1963a).
[16] Smith, P. F. In "Recent Progress in Microbiology" (N. E. Gibbons, ed.) Vol. VIII, pp. 518–525, University of Toronto Press, Toronto (1963b).
[17] Smith, P. F., J. Lipid Res. **5**, 121 (1964).
[18] Smith, P. F., Ann. N.Y. Acad. Sci. (in press) (1967).
[19] Smith, P. F., and J. E. Boughton, J. Bacteriol. **80**, 851 (1960).
[20] Smith, P. F., and C. V. Henrikson, J. Bacteriol. **89**, 146 (1965a).
[21] Smith, P. F., and C. V. Henrikson, J. Lipid Res. **6**, 106 (1965b).
[22] Smith, P. F., and C. V. Henrikson, J. Bacteriol. **91**, 1854 (1966).
[23] Smith, P. F., and W. L. Koostra, J. Bacteriol. **93**, 1853 (1967).
[24] Smith, P. F., and R. J. Lynn, J. Bacteriol. **76**, 264 (1958).
[25] Smith, P. F., and G. H. Rothblat, J. Bacteriol. **80**, 842 (1960).
[26] Smith, P. F., and G. H. Rothblat, J. Bacteriol. **83**, 500 (1962).
[27] Van Demark, P. J., and P. F. Smith, J. Bacteriol. **88**, 1602 (1964).
[28] Van Iterson, W., and A. C. Ruys, J. Ultrastructure Res. **3**, 282 (1960).

Formation of Sterol Esters in Mycobacteria

K. Schubert, G. Kaufmann and C. Hörhold

Institut für Mikrobiologie und experimentelle Therapie, Jena,
der Deutschen Akademie der Wissenschaften zu Berlin, DDR

Summary

Mycobacterium smegmatis and other species of *Mycobacterium* are capable of taking up and esterifying cholesterol from culture medium. Sterols related to cholesterol are esterified too. The esters formed are equal to other lipids with respect to their composition of fatty acids in *Mycobacterium smegmatis*. With growing cells two more esters were formed by *M. smegmatis*. Each of these esters proved to be a succinate of cholesterol with a hydroxylated Δ^4-cholestene-3-one and of cholesterol with a hydroxylated $\Delta^{1,4}$-cholestadiene-3-one respectively. The hydroxy group was shown to be located in the side chain at position 26. These hydroxylated fermentation products of cholesterol are formed via Δ^4-cholestene-3-one.

Our investigations were based on quite a simple finding: we observed an accumulation of cholesterol in cells of *Mycobacterium smegmatis*, when we added cholesterol to cells of this particular bacterium suspended in a medium of mineral salts. Simultaneously, the oxidative degradation of cholesterol to carbon dioxide took place. In the culture medium we could find — besides cholestenone — only small quantities of some intermediates arising in the degradation of the steroid skeleton. Meanwhile, other research teams could bring under control the rapid microbiological oxidation of cholesterol by addition of chelating agents or other inhibitors [1—3].

We were interested to know what happened with the cholesterol taken up by the cells. We found that cholesterol existed in a free state only at the beginning. After a fermentation period of some hours, esters of cholesterol were detectable, and two to four days later, nearly all the cholesterol had been esterified [4].

The quantity of esters formed was about 0.4% of the dry weight of bacteria, corresponding to about 2% of the lipid extract. The fatty acids from lipids are shown in Fig. 1.

The pattern of fatty acids in the range of 12 to 20 carbon atoms is similar to that of other bacteria. Besides the even-numbered fatty acids, also odd-numbered and methyl branched ones are to be found. Methyl-octadecanoic acid, which certainly is tuberculostearic acid, is even the major constituent. The dark right columns indicate the percentage of each acid in the fraction of sterol esters, the hatched middle columns indicate the percentage of these acids in the other lipids of the same culture, and the dotted left columns indicate for comparison the pattern of acids obtained from a culture without cholesterol. It can be seen that the esters of cholesterol with regard to their fatty acids have nearly the same pattern as the other lipids.

The sterol isolated after saponification from the mixture of esters contained cholesterol and another compound, a product of hydrogenation, which proved to be 5α-cholestane-3β-ol. Cholestanol was found in esters and in the fraction of free sterols isolated from cells. However, the culture filtrate contained only unaltered cholesterol.

After we had found also esters of cholestanol in *Mycobacterium smegmatis* it was interesting to know, whether further steroids are esterified by this mycobacterium or not.

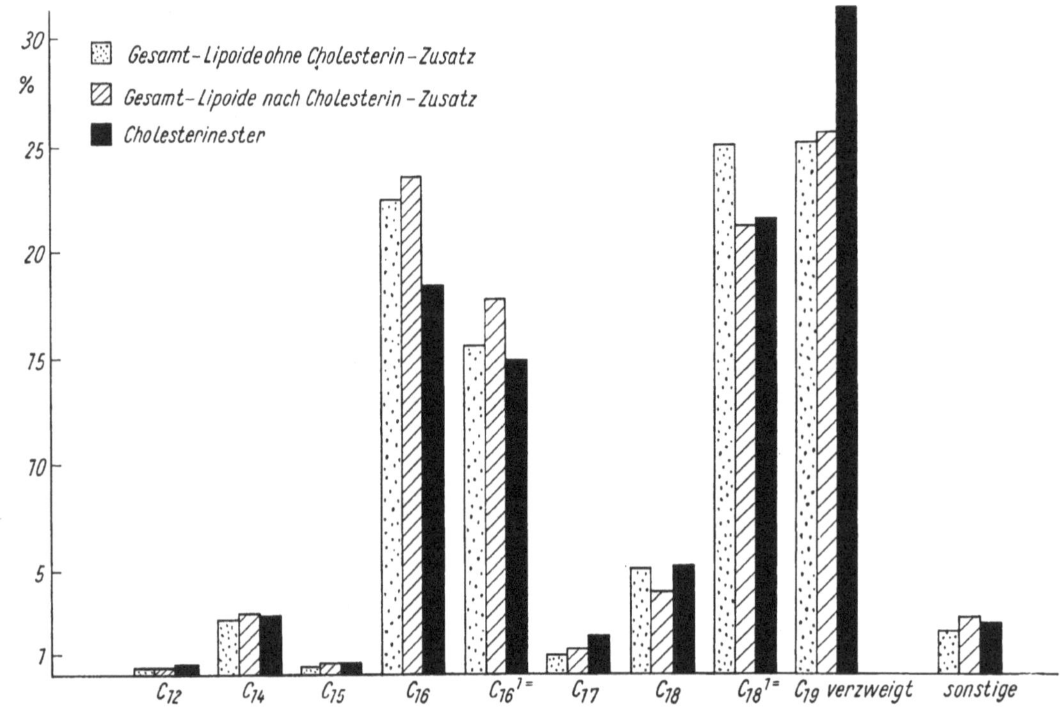

Fig. 1

The steroids investigated are shown in Fig. 2.

The left column contains those compounds which are esterified in similar yields as cholesterol or even better. The middle column contains the steroids esterified in lower yields, and in the case of compounds on the right of the figure, practically no esterification was observed. *Mycobacterium smegmatis* esterified sterols, which only differed from cholesterol by alkyl groups or double bounds. When sterols with changed oxygen functions were used, esterification was lower, and with absence of the long side chain no esterification became visible at all.

Furthermore, we wished to know, whether the esterification of sterols is due to this particular bacterium or whether other bacteria can do the same. We tested thirty different bacteria. Some of them degradated cholesterol, but esterification was observed only in cases of mycobacteria. Nevertheless, these findings do not exclude the possibility that some bacteria may give other results under modified conditions of culture.

Fermentations reported so far in this paper were made with resting cells. With growing cells of *Mycobacterium smegmatis*, we could detect two more products, at first with labelled cholesterol. Fig. 3 shows a thin-layer chromatogram of such a bacterial extract, measured by a radio scanner. The highest peak near the solvent front comes from esters of fatty acids, the lower from free cholesterol. Peaks 1 and 2 between these two are newly formed substances, which we investigated further.

The main component — like other esters — could be saponified by alkali, but two products of saponification were obtained. Their R_f-values on the thin-layer chromatogram are equal to those of cholesterol and a more polar derivative (Fig. 4).

Therefore, the whole ester isolated later as a uniform, crystalline substance should consist of two steroids. A third component should hold the two steroids together. The mass spectrum gave a mol weight higher than 800 and thus established this assumption. But neither sulfate nor phosphate were detectable. Finally, succinic acid was isolated as the expected binding substance between cholesterol and the other steroid.

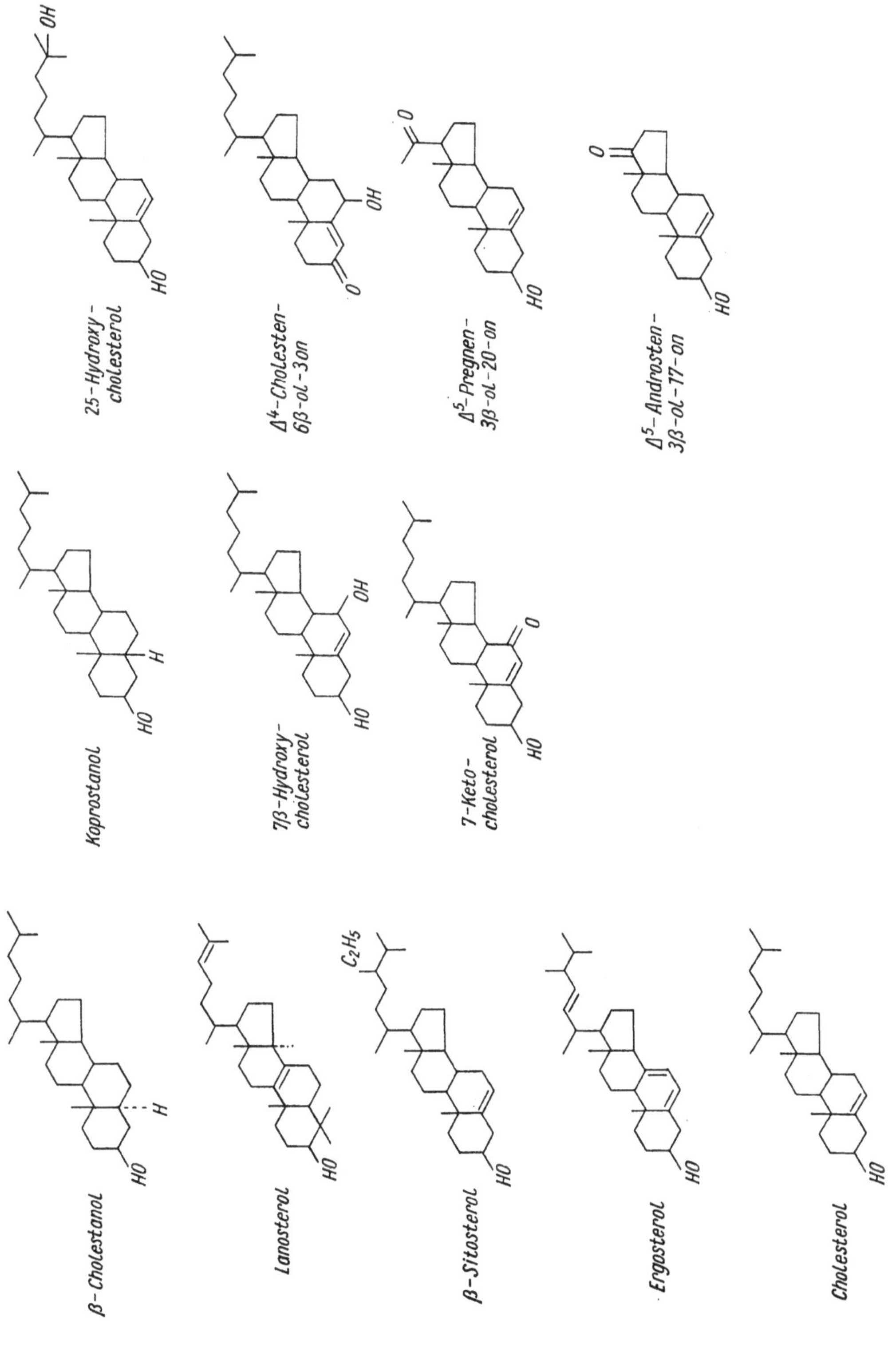
25-Hydroxy-cholesterol
Δ4-Cholesten-6β-ol-3on
Δ5-Pregnen-3β-ol-20-on
Δ5-Androsten-3β-ol-17-on
Koprostanol
7β-Hydroxy-cholesterol
7-Keto-cholesterol
β-Cholestanol
Lanosterol
β-Sitosterol
Ergosterol
Cholesterol

Fig. 2

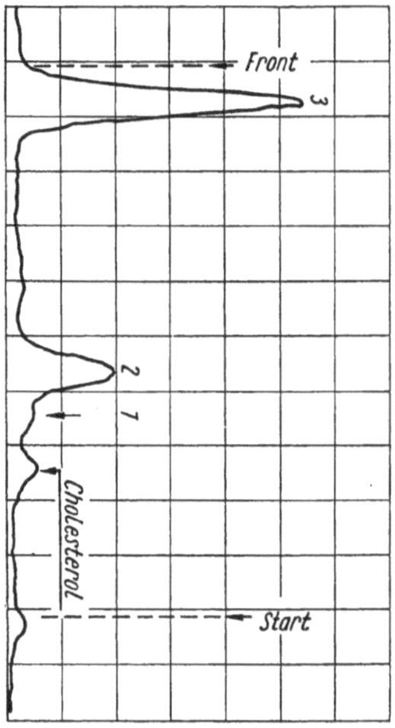

Fig. 3

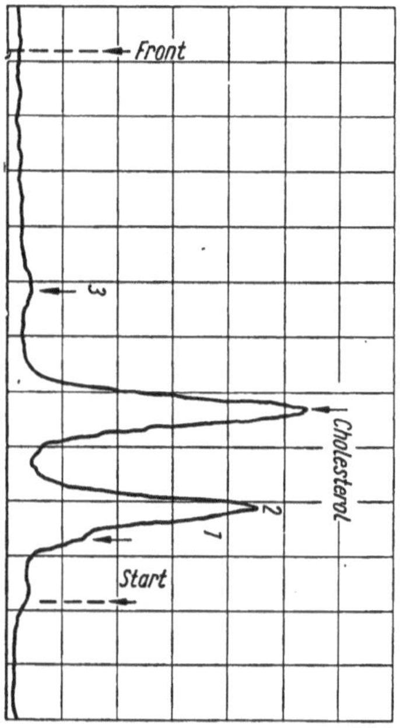

Fig. 4

This other steroid proved to be a cholesten-ol-one. The hydroxy group of cholesterol had been oxidized to the Δ^4-3-oxo group, and another hydroxy group had been introduced in the molecule.

Saponification of the second new compound which was formed from cholesterol by *Mycobacterium smegmatis* in lower yields, likewise yielded cholesterol, succinic acid and a new steroid, in this case a hydroxylated $\Delta^{1,4}$-cholestadiene-3-one.

Thus, the two new esters are mixed succinates, each consisting of cholesterol and one of its further oxidized metabolites.

Then the position of the newly introduced hydroxy group in these two metabolites had to be find out. We could exclude the positions on the ring skeleton by various methods Finally, NMR and mass spectra showed that the hydroxy group was in position 26.

At this point there is to be seen a connection to the microbiological cleavage of the cholesterol side chain, which was investigated mainly by Whitmarsh [1] and by Sih and coworkers [5, 6] in the last few years. They found a breakdown of the side chain in positions 24/25, 22/23 and 17/20, differing from the degradation in mammalian organism, which includes a hydroxylation in position 20 and 22 followed by breakdown to isocaproic acid and pregnenolone. Whitmarsh isolated Δ^4-bisnorcholen-3-one-22-oic acid and the corresponding $\Delta^{1,4}$-compound as intermediates in the formation of androstenedione and androstadienedione from cholesterol, using 8-hydroxy-quinoline as an inhibitor. Sih and coworkers identified the cleavage products of the side chain, viz. propionic acid from carbon atoms 25 to 27, acetic acid from carbon atoms 23 and 24 and another propionic acid from carbon atoms 20 to 22. As steroidal counterparts the corresponding acids were formed.

Our two side chain hydroxylated metabolites isolated from the succinates, were degradated in their free form by *M. smegmatis*, but we were unable to identify products from the small amounts available. The dehydrogenation at ring A suggests a comparison with the degradation products isolated by Whitmarsh (but also with the metabolites in mammalian organism). The introduction of the hydroxy group could be interpreted as a first step before the splitting off of the side chain.

Now the question was, whether the esters formed in the cell, are themselves intermediates in the microbiological degradation or not.

In Fig. 5 is demonstrated the process of conversion of tritium labelled cholesterol by growing cultures of *Mycobacterium smegmatis*. The percentage of free cholesterol in the cell-fraction decreased rapidly, and simultaneously esters of cholesterol with fatty acids were formed. A little later the mixed succinates appeared. It was common to all esters that their percentage decreased only very slowly.

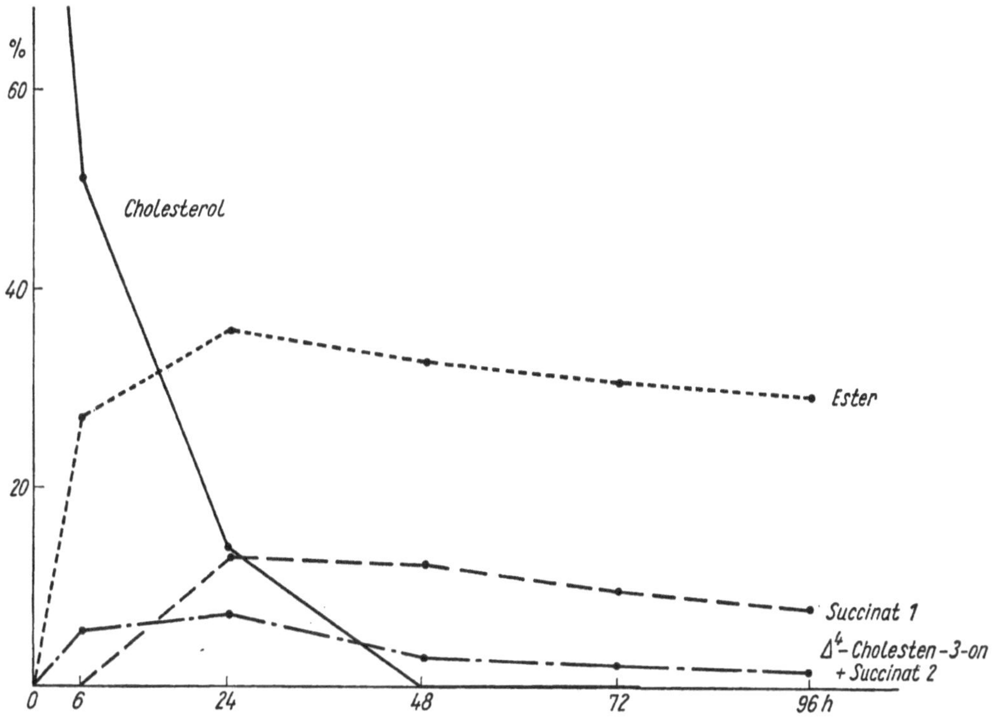

Fig. 5

In a second experiment ^{14}C-labelled cholesterol was added to a culture which had formed tritium labelled esters already.

In its left part figure 6 demonstrates the findings mentioned above, the decrease of ^{3}H-cholesterol, the formation of its esters with fatty acids and the formation of the main succinate. After an incubation period of 48 hours, the cell material was separated and suspended in fresh culture medium. Unfortunately, centrifugation resulted in a marked loss of material in this case. To the growing culture cholesterol was added again, which was ^{14}C-labelled this time.

If one of the esters was an intermediate in the degradation of cholesterol, its percentage of ^{14}C would increase and its percentage of tritium would decrease simultaneously. Let us look at the clearest period between 54 and 72 hours, during which the free ^{14}C-cholesterol decreases from 15 to 1 per cent. We notice an equal, slight lowering both of ^{3}H- and of ^{14}C-cholesterol esters of fatty acids. In the succinate the percentage of ^{3}H and ^{14}C was not changed either. From this follows that the degradation of cholesterol led neither via its esters nor via the succinate of cholestenolone and cholesterol.

Only Δ^4-cholestene-3-one was found to be an intermediate in the formation of the cholestenolone or of its cholesteryl-succinate respectively. The succinate-fraction from *M. smegmatis*, grown in a culture medium containing tritium labelled Δ^4-cholestene-3-one and non-radio-

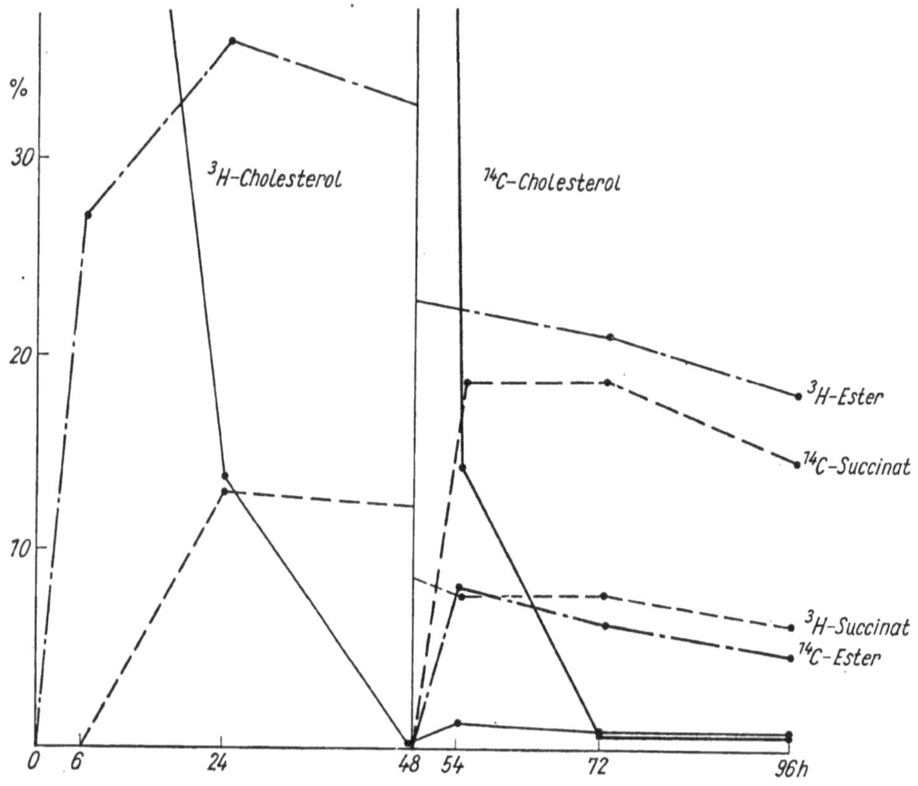

Fig. 6

active cholesterol, was radioactive and, upon saponification, yielded labelled cholestenolone beside cholesterol with only weak radioactivity.

So the esterifications reported here seem to be additional reactions, which, beside the usual degradation of the steroid skeleton, take place with these particular bacteria. It is interesting to note that in this esterified state two otherwise not isolated products of microbiological conversion of cholesterol remain in cells. In most other cases the addition of chelating agents or other inhibitors or the use of chemically modified steroids was necessary in order to isolate steroid fermentation products of the cholestane series.

The function of the reported sterol esters in the cells of mycobacteria has yet to be cleared up. These bacteria which form high amounts of lipids may use sterols, too, if they happen to be available. But it is also possible that the sterol esters formed have a definite function in the membrane.

References

[1] J. M. WHITMARSH, Biochem. J. **90**, 23P (1964).
[2] Neth. Appl. 6502883, Dec. 3, 1965.
[3] W. F. VAN DER WAARD, J. DOODEWAARD, J. d. FLINES, and S. VAN DER WEELE, these papers.
[4] K. SCHUBERT, and G. KAUFMANN, Biochim. Biophys. Acta **106**, 592 (1965).
[5] C. J. SIH, K. C. WANG, and H. H. TAI, J. Amer. Chem. Soc. **89**, 1956 (1967).
[6] C. H. SIH, H. H. TAI, and Y. Y. TSONG, J. Amer. Chem. Soc. **89**, 1957 (1967).

Addendum

In the original lecture the side chain hydroxy group of the cholestenolone was reported to be in position 23, based on one mass spectrum. But further investigations — the NMR spectrum, melting points and mass spectra of some derivatives — proved the position 26, according to a contribution made by Dr. E. CASPI in the discussion.

Sterols in Bacteria

K. Schubert, G. Rose and C. Hörhold

Institut für Mikrobiologie und experimentelle Therapie, Jena,
der Deutschen Akademie der Wissenschaften zu Berlin, DDR

Summary

The contradictory results in the literature concerning the evidence of sterols in bacteria led to the widespread opinion that bacteria are not capable of synthesizing sterols. Recently we succeeded in isolating and identifying sterols in *Escherichia coli*, *Azotobacter chroococcum*, and *Streptomyces olivaceus*. The results show that bacteria may not differ from more highly developed organisms with regard to such a characteristic feature as is indicated by the occurrence of sterols.

It is a widespread opinion that bacteria, contrary to fungi, are not capable of synthesizing sterols. References that bacteria probably do contain sterols are to be found already in publications of the 1930's. According to the stage of technical development of that time the findings were based on colour reactions and on the precipitation reaction with digitonin. The results were contradictory. Besides this, the possibility of mixing sterols from the usual culture media could not be excluded. The publications concerning the isolation of a sterol mixture from *Azotobacter chroococcum* by Sifferd and Anderson, 1936 [1] and from *Escherichia coli* by Dauchy and coworkers, 1956, 1958 [2, 3] have to be pointed out. But in both cases the further identification of sterols was impossible. We have set ourselves the task to follow the problems of sterols in bacteria with modern techniques.

Ten strains of bacteria have been investigated so far: *E. coli*, containing seven sterols, *Azotobacter chroococcum* containing six sterols and *Streptomyces olivaceus*, containing only one sterol. Sterols could not be detected in: *Aerobacter cloacae*, *Bacillus subtilis*, *Lactobacillus bulgaricus*, *Corynebacterium pseudodiphtheriticum*, in two strains of *Mycobacterium smegmatis* and in the Bacterium-Calmette-Guerin-Strain used in Jena [4, 5, 6, 7].

In order to avoid mixing of sterols from the usual culture media, subcultures of each strain were made on fully synthetic culture agar. The main culture was made submers in carefully cleaned glass vessels. *E. coli* was cultured in a medium with lactose and later with glucose as source of carbon. Ikekawa in Tokio confirmed the results with these media.

Azotobacter chroococcum was cultured in a fully synthetic medium with mannite and glucose as sources of carbon. The culture medium of *Streptomyces olivaceus* was a solution with the organic components glycerol, citric acid and aspartic amide.

In order to control our results, we made the following experiments. We first tested the components of the culture medium as control for sterols, but no sterols were detectable. Then we tested the small mass of bacteria used as vaccine. Sterols were undetectable in such small quantities.

Naturally, the most exact and elegant evidence is the biosynthesis of sterols from labeled intermediates, but low percentage of sterols in bacteria requires the application of intermediates with an especially high specific radioactivity. Besides this, we must find conditions which are more favourable for the synthesis of sterols than for the microbial degradation of the inserted intermediates.

Procedure: The wet separated cell material was repeatedly extracted with ethanol and then with chloroform exhaustively. The vacuum dried residues were saponified. The dried ether extract was a neutral oil. By application of gradient elution it was purified by chromatography on alumina. The unified Liebermann-Burchard positive fractions containing sterols were crystallized. The sterol mixtures which do not separate easily by column chromatography or those which make such separation altogether impossible were examined by means of gaschromatography both in the form of free compounds and via the trimethylsilylethers.

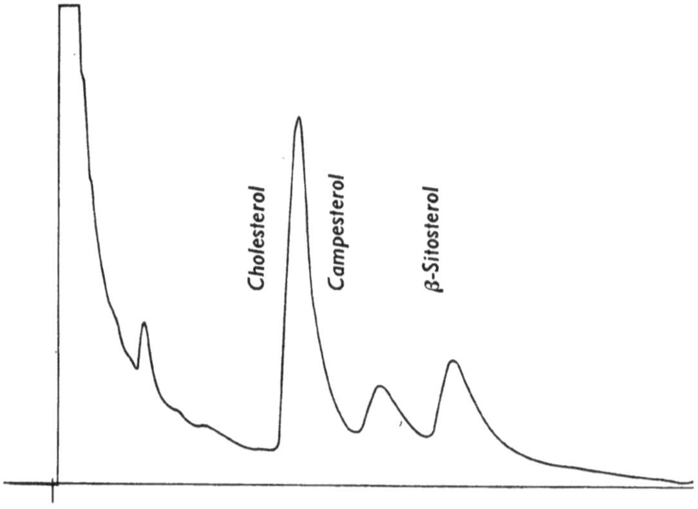

Fig. 1. Gaschromatical separation of sterols in *Escherichia coli*

Figure 1 shows the gaschromatographical separation of three main components of the sterols in *Escherichia coli*: Cholesterol, Campesterol and β-Sitosterol.

For the determination of the mass numbers and for the semi-quantitative inquiry of the percentage of the compounds of the sterol mixture from *Escherichia coli* and *Azotobacter chroococcum* the method of mass molecular spectrometry with ionization by addition of

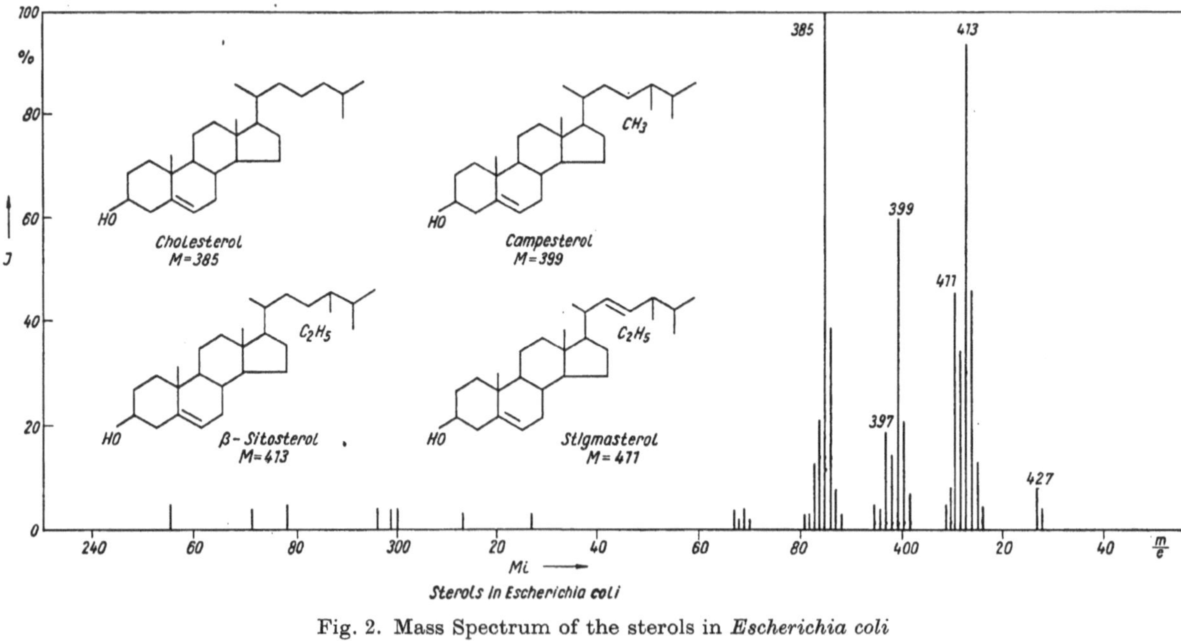

Fig. 2. Mass Spectrum of the sterols in *Escherichia coli*

negative ions was used. As a result of H-separation the mass number is one unit below the molecular weight.

The mol weights established by mass spectrometry of the sterol mixture of *Escherichia coli* (Fig. 2) could be assigned to the following sterols: cholesterol, campesterol, β-sitosterol and stigmasterol. The retention times established by gas-chromatography agree with these findings. Another prominent peak at mass number 397 points to a dehydrocampesterol. A sterol with a characteristic UV-Absorption corresponding to a $\Delta^{5,7}$-diene, which could be deduced from ergosterol, dehydro-campesterol or stigmasterol was separated chromatographically from the mixture on a column of alumina. In the meantime the results have been confirmed by IKEKAWA in Tokio [8].

Mixture of several components could also be isolated from *Azotobacter chroococcum*. An essential difference between the two isolated mixtures however is the failure of the occurrence of cholesterol in *Azotobacter chroococcum* whereas cholesterol is found to be the main compound in the case of *Escherichia coli*.

The sterols with the mass numbers 397 and 399 were found to be Δ^7-ergostene-3β-ol and $\Delta^{7,22}$-ergostadiene-3β-ol which was confirmed by gaschromatography. The structures of these compounds were also confirmed by the mass spectra obtained by direct combination of gas chromatography and mass spectrometry. Although the structures of the four compounds, with the mass numbers 426, 413 and 395 have yet to be proved with comparative substances, the following relations can be derived from the differences in mass numbers.

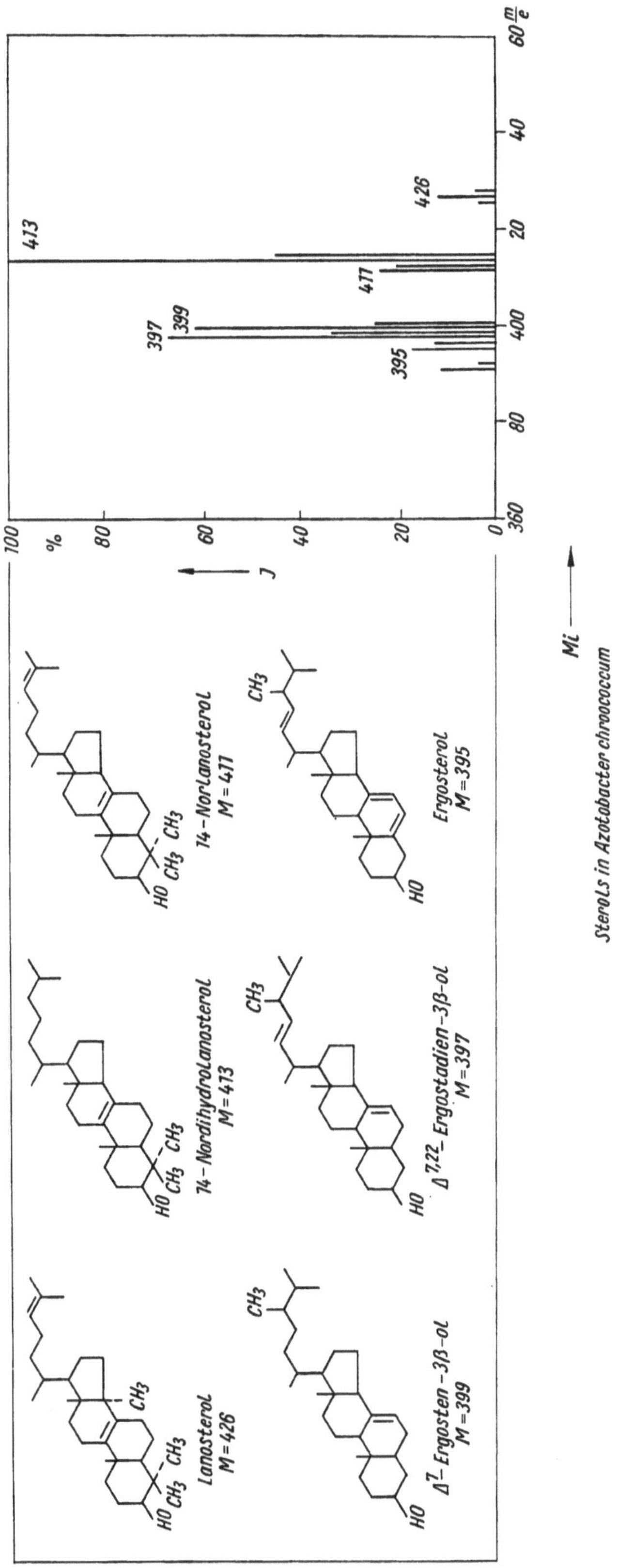

Fig. 3. Mass Spectrum of the sterols in *Azotobacter chroococcum*

The sterol with mass number 395 is distinguished from $\Delta^{7,22}$-ergostadiene-3β-ol by an additional double bond. This must be ergosterol, whose typical UV absorption was measured in the mixture as a $\Delta^{5,7}$-diene. The fraction of this compound determined by UV-spectrophotometry was of the same order. From the Δ^{7}-ergosten-3β-ol the following compounds can be considered: The 14-nordihydrolanosterol upon addition of a methyl group, the 14-norlanosterol upon addition of a further double bond and the lanosterol upon addition of another methyl group. These six sterols show biogenetic relations.

From *Streptomyces olivaceus*, which is a bacteria species of the order Actinomycetales, we isolated cholesterol as the sole sterol. The uniformity of the isolated sterol was established by chromatographic techniques as well as by mass spectrometry. Mol weight, melting point and the infrared spectrum are identical with those of cholesterol.

The quantities of sterols in bacteria are small. In *Escherichia coli* we found the smallest sterol content, viz. 0.0004% related to the dry weight of cells. There were 0.01% of dry bacteria cells in *Azotobacter chroococcum* and 0.0035% in *Streptomyces olivaceus*. In comparison to that the sterol content of the fungus *Polystictus versicolor* was 0.125%. The limit of evidence was 0.0001% sterols in 100 gm of dry bacteria. It is possible that our negative findings should mean that the eventual sterol content is below the present level of sensitivity.

The isolation and identification of sterols in *E. coli*, *Azotobacter chroococcum* and *Streptomyces olivaceus* show that bacteria may not differ from more highly developed organisms with regard to such a basic feature as the occurrence of sterols. In addition to that, the occurrence of cholesterol as the sole sterol in *Streptomyces olivaceus* refuted the opinion that of the numerous sterols existing in the lower organisms only cholesterol has remained in the selection of vertebrates and humans in the course of biogenesis. Studies in the cellfractions of fungi and higher organisms reveal that sterols are constituents of the cytoplasmic membranes. It is wellknown that the uniformity of biochemical principles are to be followed from the cell of a bacterium to the cells of mammals. We have no evidence that the sterols in bacteria are bound to the cytoplasmic membrane, but it is, at least obvious that the sterols in bacteria should be seen in connection with the cytoplasmic membrane. In this connection we wish to refer some recent reports on intracytoplasmic membranes or mesosomes or chondrioids in bacteria, for example in *E. coli* [9].

Finally we should remember the findings that bacteria do not only synthesize sterols but also degrade them. *Streptomyces olivaceus*, which is capable of synthesizing cholesterol, degrades cholesterol to CO_2 via Δ^{4}- and $\Delta^{1,4}$-androstadiene-3,17-dion.

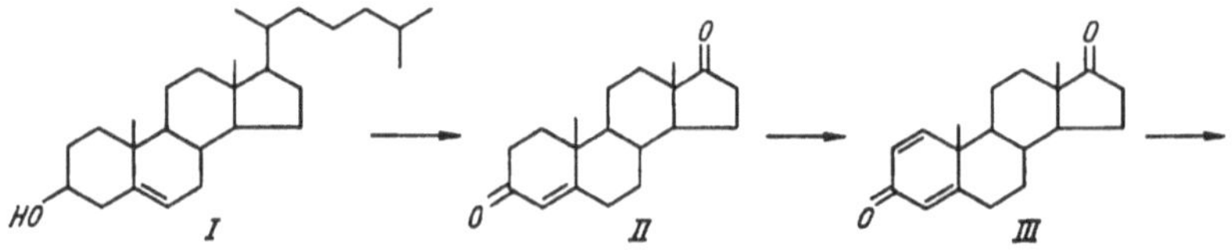

Fig. 4
Degradation of Cholesterol by *Streptomyces olivaceus*

Possibly steroids may play any functional role in the bacterial metabolism.

References

[1] R. H. Sifferd, and R. J. Anderson, Hoppe-Seyler's Z. physiol. Chem. **239**, 270 (1936).
[2] S. Dauchy, F. Kayser, and J. Villoutreix, C. R. Séances Soc. Biol. Filiales **150**, 1974 (1956).
[3] S. Dauchy, and F. Kayser, Bull. Soc. Chim. Biol. **40**, 1533 (1958).
[4] K. Schubert, G. Rose, R. Tümmler, and N. Ikekawa, Hoppe-Seyler's Z. physiol. Chem. **339**, 293 (1964).
[5] K. Schubert, G. Rose, and C. Hörhold, Biochim. Biophys. Acta (Amsterdam) **137**, 168 (1967).
[6] K. Schubert, G. Rose, H. Wachtel, C. Hörhold, and N. Ikekawa, Europ. J. Biochemistry **5**, 246, (1968)
[7] K. Schubert, Z. f. Chemie **7**, 289 (1967).
[8] N. Ikekawa, Privatmitteilung.
[9] C. Schnaitman, and J. W. Greenawalt, J. Bact. **92**, 780 (1966).

Studies on the Degradation of Ring A of Steroids by Microbial Enzymes

A. W. Coulter and P. Talalay

Department of Pharmacology and Experimental Therapeutics,
The Johns Hopkins University School of Medicine, Baltimore,
Maryland, USA

It has been known for over 50 years that certain microorganisms can sustain their growth on media containing steroids as the only source of organic carbon. Söhngen (1913) and subsequently Haag (1926, 1927) showed that culture media in which cholesterol was the only organic compound, were capable of supporting the growth of some *Mycobacteria*, and Tak (1942) demonstrated that the sterol was actually consumed. Subsequently, a large number of bacteria and fungi have been shown to utilize sterols, bile acids and steroid hormones (Schatz et al., 1949; Talalay, 1957). The isolation of *Pseudomonas testosteroni* (Talalay et al., 1952) by enrichment culture from soil, led to the demonstration that both growing and resting cells of this microorganism could effect the complete oxidative degradation of testosterone and related steroids to carbon dioxide and water by means of a system of induced enzymes.

A number of induced enzymes were subsequently obtained in a highly purified or crystalline state from *P. testosteroni* and their properties and mechanism of action described (Talalay, 1957, 1965). Among these enzymes were: 1. *Hydroxysteroid dehydrogenases*, which promote the pyridine nucleotide-dependent, reversible, and stereospecific interconversion of hydroxyl and carbonyl functions on the steroid skeleton; 2. *Δ-Dehydrogenases*, which are concerned with the introduction of double bonds into the Δ^1- and Δ^4-positions of steroid ring A, probably through a mechanism involving flavoproteins; and 3. *Δ^5-3-Ketosteroid isomerase* which promotes the conversion of $\Delta^{5(6)}$- and $\Delta^{5(10)}$-3-ketosteroids to their corresponding conjugated 3-ketosteroids.

The first information on the precise site of oxidative attack on the steroid skeleton itself was provided by the experiments of Dodson and Muir (1958, 1961) who established the conversion of Δ^4-androstene-3,17-dione (I, Scheme 1) to 3-hydroxy-9,10-*seco*-$\Delta^{1,3,5(10)}$-androstatriene-9,17-dione (IV) by a species of *Pseudomonas*. These workers advanced evidence that this transformation resulted from the participation of two enzymatic reactions: 9α-hydroxylation and Δ^1-dehydrogenation, and postulated the formation of the hypothetical β-hydroxydienone III which would be expected to undergo facile conversion to the *seco*-phenol IV. The studies of Schubert, Böhme and Hörhold (1960, 1961a) established an analogous series of reactions in *Mycobacterium smegmatis*. Further investigations by Schubert and colleagues (1961b) led to the isolation of the first low molecular weight degradation product of a steroid which was identified as a perhydroindanpropionic acid, related to V. It seemed logical to assume that the perhydroindane moiety of this compound was derived from rings C and D of the steroid whereas the propionic acid residue originated from C-5, C-6 and C-7 of the steroid. Sih and Wang (1963) isolated the perhydroindanpropionic acid derivative V as a steroid degradation product by *Nocardia restrictus* and established that the absolute configuration of this product is identical to that existing originally at C-8, C-13 and C-14 of the steroid. These studies have been reviewed by Schubert (1967) and by Talalay (1965).

Considerable attention has been directed recently to the metabolic fate of the remainder of ring A and the C-19 angular methyl group. A series of elegant investigations from the laboratory of SIH (SIH et al., 1965; SIH et al., 1966; GIBSON et al., 1966) established that the aromatic ring A of the *seco*-phenol IV undergoes hydroxylation at C-4, and subsequently is cleaved between C-4 and C-5 by a metapyrocatechase type reaction. These workers (SIH et al., 1965; GIBSON et al., 1966) also isolated 2-oxo-4-ethylbutyrolactone (VIII) as a fragmentation product of steroid ring A. This lactone itself was not further metabolized by the microbial enzyme preparations and was assumed to be a side-product formed by the nonenzymatic lactonization of 2-oxo-4-hydroxyhexanoic acid. GIBSON et al. (1966) further demonstrated that synthetic 2-oxo-4-hydroxyhexanoic acid (prepared from its lactone) underwent facile and almost quantitative aldol cleavage to propionaldehyde and pyruvate (Scheme 1). The origin of the hydroxyl group of the C_6 fragment was not established by these studies, and it was suggested that the hydroxyl group was introduced before the C_6 fragment was cleaved from the remainder of the ring system (SIH et al., 1965; GIBSON et al., 1966).

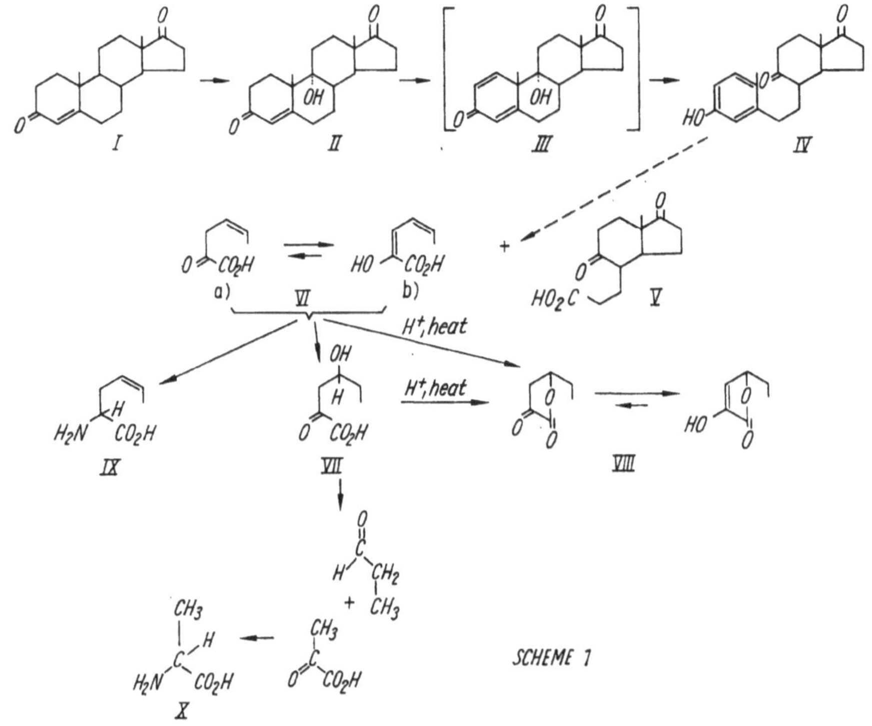

SCHEME 1

SHAW, BORKENHAGEN and TALALAY (1965) reported that ultracentrifuged extracts of steroid-induced cells of *Pseudomonas testosteroni* when supplemented with NAD and a NADPH generating system were capable of oxidizing Δ^4-androstene-3,17-dione-4-^{14}C to $^{14}CO_2$ in high yield. If the medium was acidified before the reaction was complete, the radioactive metabolites could be fractionated into those that were extracted into ethyl acetate and those which were retained in the aqueous phase. By the addition of EDTA to the reaction system, the formation of $^{14}CO_2$ was markedly suppressed, and most of the steroid metabolites accumulated in the aqueous phase. Two α-amino acids were isolated from this source: L-2-amino-*cis*-4-hexenoic acid-1-^{14}C (IX) and DL-alanine-1-^{14}C (X).

Since the enzyme extracts contain considerable quantities of endogenous alanine, no special difficulty was encountered in isolating sufficient quantities of this amino acid. Identification was accomplished by comparison of the mobility of the isolated product with that of authentic alanine in paper chromatographic systems, on high voltage electrophoresis, and on the column of the amino acid analyzer. Moreover, the solid state infrared spectrum of the isolated compound was identical with that of DL-alanine, and optical rotatory dispersion measurements revealed no rotation. Since the enzyme preparations contain extremely active alanine racemase activity, it was concluded that the primary product was probably an optically active alanine which was subsequently racemized (SHAW et al., 1965).

No significant quantities of L-2-amino-*cis*-4-hexenoic acid are present in the enzyme preparations derived from *P. testosteroni*, but small quantities of this compound were isolated from the aqueous radioactive fraction already described. The identification of this C_6 amino acid was based on the following criteria (SHAW et al., 1965): 1. Comparison of migration of the isolated compound on chromatography and electrophoresis with authentic synthetic DL-2-amino-*cis*-4-hexenoic acid (SKINNER et al., 1961); 2. On catalytic hydrogenation the isolated acid was smoothly converted to norleucine; 3. Identity of the NMR spectrum with that of the synthetic racemic product; and 4. Optical rotatory dispersion measurements indicated that the compound had the L-configuration.

Nearly all of the radioactivity derived from Δ^4-androstene-3,17-dione-4-^{14}C was located in the carboxyl groups of the isolated DL-alanine and the L-2-amino-*cis*-4-hexenoic acid. On the basis of these findings and other evidence, SHAW et al. (1965) suggested that the alanine was derived from C-2, C-3 and C-4 of the steroid, whereas the aminohexenoic acid originated from these carbon atoms as well as from C-1, C-10 and C-19 of the steroid. Since the α-amino acids are not further degraded, it was suggested that they were formed by amination of their corresponding α-oxo acids: pyruvate and 2-oxo-*cis*-4-hexenoic acid, which represented the direct intermediates in the enzymatic degradation of the steroid (Scheme 1).

In an effort to obtain a more detailed picture of the intermediate steps in the degradation of steroid ring A, we undertook the synthesis of 2-oxo-*cis*-4-hexenoic acid-1-^{14}C. This compound was prepared from DL-2-amino-*cis*-4-hexenoic acid-1-^{14}C by way of the N-trifluoroacetyl derivative which was cyclized with N,N-dicyclohexylcarbodiimide to the oxazolone. Hydrolysis of the oxazolone under mild conditions provided the desired 2-oxo-*cis*-4-hexenoic acid (VI) which exists at neutral pH predominantly in the dienolic form (VIb) with an absorption maximum at 274 mμ which shifts to 307 mμ in base (COULTER and TALALAY, 1967, 1968).

When the synthetic 2-oxo-*cis*-4-hexenoic acid was added to the enzyme system derived from *P. testosteroni* that converts Δ^4-androstene-3,17-dione-4-^{14}C to $^{14}CO_2$, a very rapid liberation of $^{14}CO_2$ was observed. This reaction was dependent on enzyme concentration, was almost totally inhibited by 5 mM EDTA, and was somewhat stimulated by NAD but not by the presence of a NADPH generating system. These findings qualified the α-oxo acid as a possible intermediate in the enzymatic degradation of the steroid. When 2-oxo-*cis*-4-hexenoic acid-1-^{14}C was added to cell-free extracts of *P. testosteroni* in the absence of any further additions, its absorption at 274 mμ disappeared rapidly in consequence of a stereospecific enzymatic hydration of the double bond. Although the resultant 2-oxo-4-hydroxyhexanoic acid (VII) was not isolated as such, the reaction mixture on heating in acid gave rise to 2-oxo-4-ethylbutyrolactone (VIII). The 2-oxo-4-ethylbutyrolactone was characterized by the identity of its infrared and ultraviolet absorption spectra with those of a synthetic sample. The quinoxaline derivatives of isolated and synthetic samples were also indistinguishable by a number of criteria (COULTER and TALALAY, 1968). Since, however, 2-oxo-4-ethylbutyrolactone may be formed nonenzymatically by heating of 2-oxo-*cis*-4-hexenoate in acid solution, it became necessary to establish that the product which was isolated from the enzymatic reaction mixture had indeed arisen by an enzyme-catalyzed process. The strongest evidence for this contention was the fact that whereas the 2-oxo-4-ethylbutyrolactone obtained by heating 2-oxo-*cis*-4-hexenoate in acid was optically inactive, that obtained from the enzymatic hydration product (2-oxo-4-hydroxyhexanoic acid) had a strong Cotton effect with a positive maximum at 233 mμ. These findings support the view that 2-oxo-*cis*-

4-hexenoic acid undergoes a stereospecific enzymatic hydration to form optically active 2-oxo-4-hydroxyhexanoic acid, of as yet unestablished configuration (COULTER and TALALAY, 1968). Since Gibson and colleagues (1966) had already demonstrated in cell-free extracts of *Nocardia restrictus* that racemic 2-oxo-4-hydroxyhexanoic acid was cleaved to pyruvate and propionaldehyde (both isolated as their 2,4-dinitrophenylhydrazones) it seems probable that the degradative pathways in *Nocardia restrictus* and in *P. testosteroni* are analogous.

Further support for the participation of 2-oxo-*cis*-4-hexenoic acid as an intermediate in the degradation of Δ^4-androstene-3,17-dione in cell free extracts of *P. testosteroni* was obtained from experiments in which the 2-oxo-*cis*-4-hexenoic acid-1-^{14}C was incubated with extracts of *Pseudomonas* in the presence of EDTA. Under these conditions, formation of $^{14}CO_2$ was suppressed, and examination of the aqueous phase revealed the presence of radioactive alanine and 2-amino-*cis*-4-hexenoic acid, both of which had been obtained earlier (SHAW et al., 1965) as transformation products from the steroid.

In sum, these observations provide strong evidence for the participation of 2-oxo-*cis*-4-hexenoic acid as a key intermediate in the oxidative cleavage of steroid ring A, and suggest that this compound undergoes a stereospecific hydration in position C-4 (corresponding to C-1 of the original steroid). The 2-oxo-4-hydroxyhexanoic acid (VII) then undergoes aldol cleavage to give pyruvate and propionaldehyde, as already shown by GIBSON et al. (1966) for *Nocardia restrictus*. It is therefore reasonable to assume that alanine and 2-amino-*cis*-4-hexenoic acid which accumulate in the presence of EDTA arise by amination of the corresponding α-oxo acids.

Although the precise mechanism for the cleavage of the C_6 fragment from the indanpropionic acid (V) remains unclear, it may be assumed that during this process the *cis* configuration at the C-4(5) double bond of the 2-oxo-*cis*-4-hexenoic acid is established. The further fate of the indanpropionic acid moiety derived from rings B, C and D is the subject of a separate communication in this Symposium from Professor Kurt Schubert's laboratory (SCHUBERT et al., 1968).

Acknowledgments

These studies were supported by Research Grant AM 07422 and Training Grant GM 1183 from the National Institutes of Health of the United States Public Health Service.

References

[1] COULTER, A. W., and P. TALALAY, Biochem. Biophys. Res. Comm., **29**, 413 (1967).
[2] COULTER, A. W., and P. TALALAY, J. Biol. Chem., **243**, 3238 (1968).
[3] DODSON, R. M., and R. D. MUIR, J. Amer. Chem. Soc., **80**, 5004, 6148 (1958).
[4] DODSON, R. M., and R. D. MUIR, J. Amer. Chem. Soc., **83**, 4627, 4631 (1961).
[5] GIBSON, D. T., K. C. WANG, C. J. SIH, and H. W. WHITLOCK, Jr., J. Biol. Chem., **241**, 551 (1966).
[6] HAAG, F., Archiv f. Hygiene, **97**, 28 (1926).
[7] HAAG, F., Zentralbl. f. Bakt. Parasitenk., II, **71**, 1 (1927).
[8] SCHATZ, A., K. SAVARD, and I. J. PINTER, J. Bacteriology, **58**, 117 (1949).
[9] SCHUBERT, K., Z. f. Chemie, **7**, 289 (1967).
[10] SCHUBERT, K., K.-H. BÖHME, and C. HÖRHOLD, Z. f. Naturforsch., **15b**, 584 (1960).
[11] SCHUBERT, K., K.-H. BÖHME, and C. HÖRHOLD, Z. f. Naturforsch., **16b**, 595 (1961a).
[12] SCHUBERT, K., K.-H. BÖHME, and C. HÖRHOLD, Hoppe-Seiler's Z. physiol. Chem., **325**, 260 (1961b).
[13] SCHUBERT, K., K.-H. BÖHME, and C. HÖRHOLD, Symposium über biochemische Aspekte der Steroidforschung (1968), 81.
[14] SHAW, D. A., L. F. BORKENHAGEN, and P. TALALAY, Proc. nat. Acad. Sci., USA., **54**, 837 (1965).
[15] SIH, C. J., and K. C. WANG, J. Amer. Chem. Soc., **85**, 2135 (1963).

[16] SIH, C. J., K. C. WANG, D. T. GIBSON, and H. W. WHITLOCK, Jr., J. Amer. Chem. Soc., **87**, 1386 (1965).
[17] SIH, C. J., S. S. LEE, Y. Y. TSONG, and K. C. WANG, J. Biol. Chem., **241**, 540 (1966).
[18] SKINNER, C. G., J. EDELSON, and W. SHIVE, J. Amer. Chem. Soc., **83**, 2281 (1961).
[19] SÖHNGEN, N., Zentralbl. f. Bakt. Parasitenk., II, **37**, 595 (1913).
[20] TAK, J. D., Antonie van Leeuwenhoek, **8**, 32 (1942).
[21] TALALAY, P., Physiol. Rev., **37**, 365 (1957).
[22] TALALAY, P., Ann. Rev. Biochem., **34**, 347 (1965).
[23] TALALAY, P., M. M. DOBSON and D. F. TAPLEY, Nature, London, **170**, 620 (1952).

Microbiological Degradation of Bile Acids

XX[1]. Some Nitrogen-Containing Compounds as Cholic Acid Metabolites by Microorganisms

S. Hayakawa, S. Hashimoto and T. Fujiwara, Shionogi Research Laboratory, Shionogi & Co. Ltd., Fukushima-ku, Osaka, Japan and T. Onaka Research Foundation, Itsuu Laboratory, Setagaya-ku, Tokyo, Japan

Summary

Cholic acid (I) was converted by *Streptomyces rubescens* into 4α-(2-carboxyethyl)-5-oxo-7aβ,γ(R)-dimethyl-3aα-hexahydroindan-1β-butyric acid (Va), 3-oxo-6aβ,γ(R)-dimethyl-2,3,4,6,6aβ,7,8,9,9aα,9bβ-decahydro-1H-cyclopenta[f]quinoline-7β-butyric acid (VIa), 3-oxo-6aβ,γ(R)-dimethyl-2,3,4,6,6aβ,7,8,9,9aα,9bβ-decahydro-1H-cyclopenta[f]quinoline-7β-butyramide (VIc), 3-oxo-6aβ,α(S)-dimethyl-2,3,4,6,6aβ,7,8,9,9aα,9bβ-decahydro-1H-cyclopenta[f]quinoline-7β-acetic acid (VIIa), 6aβ-methyl-2,3,4,6,6aβ,7,8,9,9aα,9bβ-decahydro-1H-cyclopenta[f]quinoline-3,7-dione (VIII), and 1β-(1(R)-methyl-3-carbamoylpropyl)-5-oxo-7aβ-methyl-3aα-hexahydroindan-4α-propionic acid (IX). 4α-(2-Carboxyethyl)-5-oxo-7aβ,γ(R)-dimethyl-3aα-hexahydroindan-1β-butyric acid (Va) was found to be converted into N-[4α-(2-carboxyethyl)-5-oxo-7aβ,γ(R)-dimethyl-3aα-hexahydroindan-1β-butyryl]-L-alanine (X) by *Corynebacterium equi*. All structures of the nitrogen-containing compounds VIa, VIc, VIIa, VIII, IX, and X were conclusively established by partial synthesis of each compound. On the basis of the structures of these compounds and previous reports, some hypothetical pathways for the cholic acid degradation by microorganisms were proposed as shown in Chart 1.

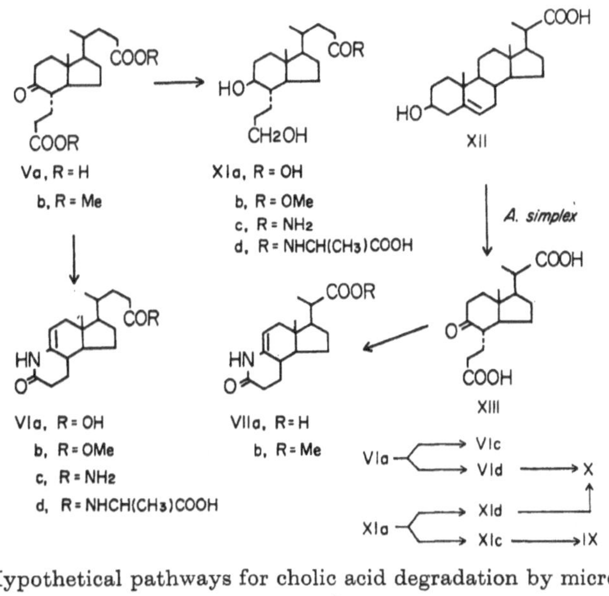

Chart 1. Hypothetical pathways for cholic acid degradation by microorganisms

Since the first report (Hayakawa, 1954) on the microbiological degradation of cholic acid (I), additional investigations have allowed the proposal of some possible pathways for the cholic acid (I) degradation (Hayakawa et al., 1957, 1958c, and 1967). During the past few years, a considerable number of reports on the microbiological degradation of steroids into

[1] For paper XIX in this series, see Hayakawa et al., 1958b.

the perhydroindan derivatives and the related problems has been published, and several intermediates in this complex degradative pathway have been indetified (Schubert et al., 1961, 1964, 1965, and 1966; Sih et al., 1962, 1963a, 1963b, 1965a, 1965b and 1966; Wang et al., 1963; Lee et al., 1964 and 1967; Tsong et al., 1964; Chang et al., 1964; Shaw et al., 1965; Gibson et al., 1966; Coombe et al., 1966). We have also found that cholic acid (I) is converted into the perhydroindan derivative, 4α-(2-carboxyethyl)-5-oxo-7aβ,γ(R)-dimethyl-3aα-hexahydroindan-1β-butyric acid (Va), by *Arthrobacter simplex* (Hayakawa et al., 1967). Our continued interest in defining the intermediates and reaction sequence, involved in the complete oxidation of the cholic acid molecule, has prompted us to reinvestigate the metabolism of cholic acid (I) by *Streptomyces rubescens* which had been found to be a strong cholic acid utilizer by us (Saburi et al., 1955 and Hayakawa et al., 1958a, 1958b, and 1958c). Furthermore, our interest in this problem has prompted us to search the further degradative products of the acid Va obtained in a previous paper (Hayakawa et al., 1967).

This paper reports the isolation and characterization of 4α-(2-carboxyethyl)-5-oxo-7aβ,γ(R)-dimethyl-3aα-hexahydroindan-1β-butyric acid (Va) and some nitrogen-containing compounds, 3-oxo-6aβ,γ(R)-dimethyl-2,3,4,6,6aβ,7,8,9,9aα,9bβ-decahydro-1H-cyclopenta[f]quinoline-7β-butyric acid (VIa), 3-oxo-6aβ,γ(R)-dimethyl-2,3,4,6,6aβ,7,8,9,9aα,9bβ-decahydro-1H-cyclopenta[f]quinoline-7β-butyramide (VIc), 3-oxo-6aβ,α(S)-dimethyl-2,3,4,6,6aβ,7,8,9,9aα,9bβ-decahydro-1H-cyclopenta[f]quinoline-7β-acetic acid (VIIa), 6aβ-methyl-2,3,4,6,6aβ,7,8,9,9aα, 9bβ-decahydro-7H-cyclopenta[f]quinoline-3,7-dione (VIII), and 1β-(1(R)-methyl-3-carbamoyl-propyl)-5-oxo-7aβ-methyl-3aα-hexahydroindan-4α-propionic acid (IX) from the culture filtrate of *S. rubescens*, cultured in a medium containing cholic acid (I) as the sole carbon source. This paper also reports that *Corynebacterium equi*, cultured in a medium containing 4α-(2-carboxyethyl)-5-oxo-7aβ,γ(R)-dimethyl-3aα-hexahydroindan-1β-butyric acid (Va) as the sole source of carbon, produces a nitrogen-containing compound, N-[4α-(2-carboxyethyl)-5-oxo-7aβ,γ(R)-dimethyl-3aα-hexahydroindan-1β-butyryl]-L-alanine (X). *C. equi* also yields besides the amino acid conjugate (X) several products. The isolation and characterization of these metabolites are in progress and will be reported in the future.

When cholic acid (I) was incubated with *S. rubescens*, several products containing a cholane nucleus were produced as reported previously (Hayakawa et al. 1958a and 1958b). In continuing the incubation, it was found that additional products were newly produced in parallel with a disappearance of these products from the incubation mixture as shown in Fig. 1. An acidic fraction obtained from the incubation mixture was purified by chromatography on silicic acid and gave four products Va, VIa, VIIa, and IX. It was quite difficult to isolate the product VIIa free from an unknown product which showed a strong peak at 232 nm in the ultraviolet absorption spectrum and the characteristic infrared absorption bands at 1677 and 1620 cm^{-1} in the 1700—1600 cm^{-1} regions. But this unknown product was not obtained in pure form in this experiment. The first product Va, m.p. 164.5—167°, was characterized as 4α-(2-carboxyethyl)-5-oxo-7aβ,γ(R)-dimethyl-3aα-hexahydroindan-1β-butyric acid on the basis that its infrared spectrum was indentical to an authentic sample and mixed melting point determination gave no depression. Both the second product VIa, m.p. 233° (dec., sintered at 224°), and the third product VIIa m.p. 241—247.5° (dec.), showed a similar ultraviolet absorption at around 231 nm to that of the product VIII, which had been isolated from a neutral fraction in this experiment and then identified as 6aβ-methyl-2,3,4,6,6aβ,7,8,9-9aα,9bβ-decahydro-1H-cyclopenta[f]quinoline-3,7-dione by partial synthesis. The elementary analyses of both products were consistent with $C_{18}H_{27}O_3N$ and $C_{16}H_{23}O_3N$, respectively. On the basis of these physical constants the structures of both products were assigned as 3-oxo-6aβ,γ(R)-dimethyl-2,3,4,6,6aβ,7,8,9,9aα,9bβ-decahydro-1H-cyclopenta[f]quinoline-7β-butyric acid (VIa) and 3-oxo-6aβ,α(S)-dimethyl-2,3,4,6,6aβ,7,8,9,9aα,9bβ-decahydro-1H-cyclopenta-[f]quinoline-7β-acetic acid (VIIa), respectively. These assignments were confirmed by partial

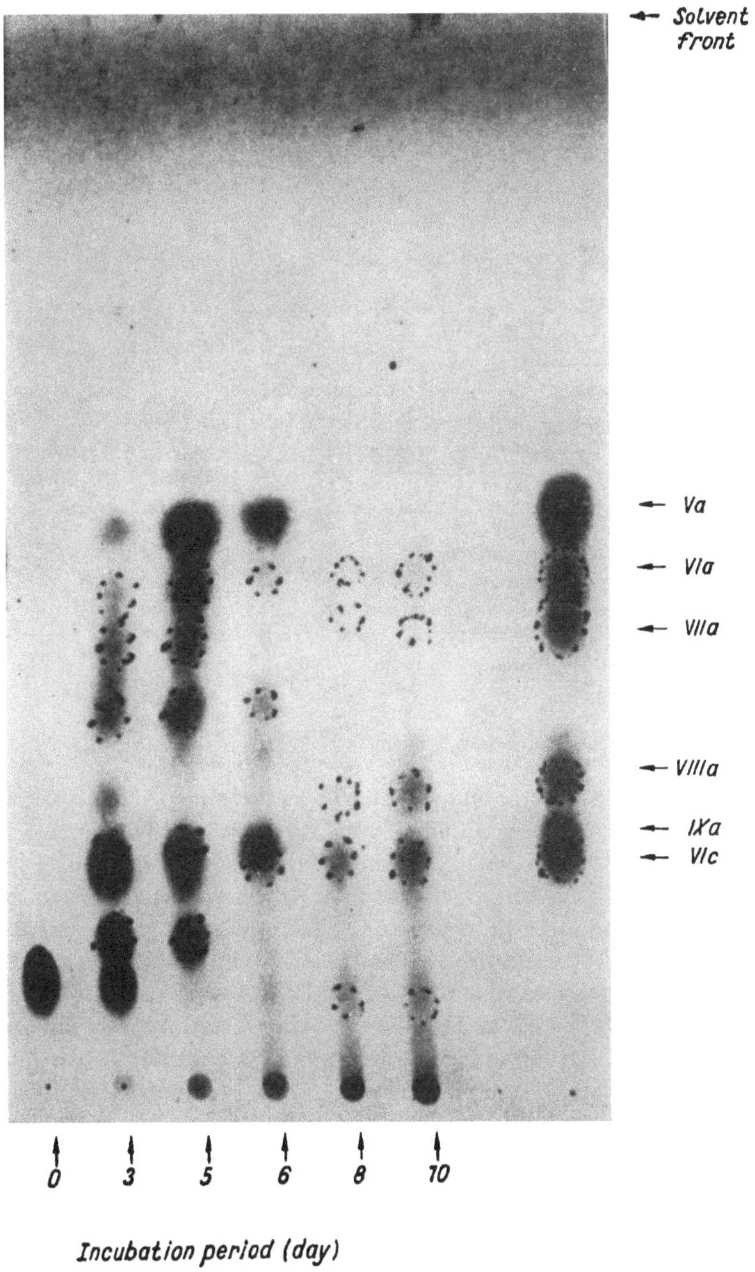

Fig. 1. Thin-layer chromatography of the cholic acid metabolites by *S. rubescens*. An aliquot of the incubation mixture was aseptically separated from the shaking flask and lyophilized in the usual manner. The residue was suspended in a minimum volume of water, acidified with dilute hydrochloric acid and extracted with ethyl acetate. The ethyl acetate extract was applied to silica gel GF_{254} plate and developed with isooctane–ethyl acetate–acetic acid (5:15:1 v/v). The spots on the plate were detected by using ultraviolet lamp (253.7 μm) and concentrated sulfuric acid. The spots detected by ultraviolet absorption were encircled with a dotted line. Arabian figures refer to incubation period in day and roman figures refer to the authentic samples (see Chart 1)

synthesis as follows: The synthetic method of the corresponding enamine lactam by heating a δ-keto-carboxylic acid with ammonia-saturated ethanol, reported by DOORENBOS et al. (1961a), was applied to the acid Va to yield the compound VIa, m. p. 238° (dec., sintered at 225°), as shown in Chart 2. The melting behaviour of this compound on a Kofler block

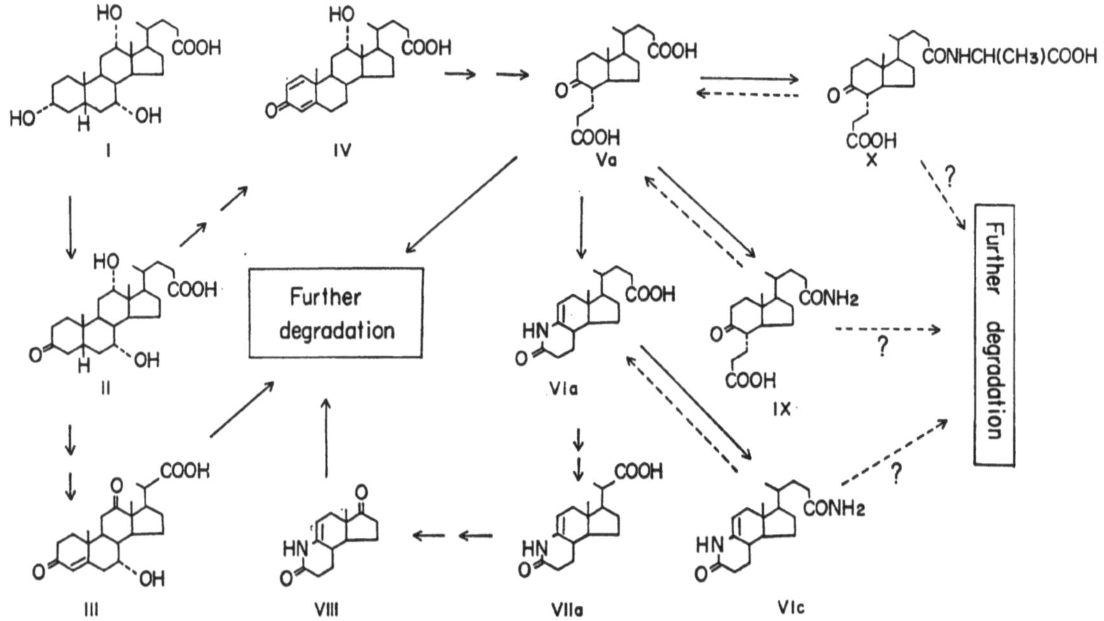

Chart 2

was quite similar to that of the compound VIII and was not sharp. But its methyl ester gave a fairly sharp melting point. The compound VIa was identical with a sample of VIa obtained from the incubation mixture. *A. simplex*, cultured in a medium containing 3β-hydroxy-23-dinorchol-5-enic acid (XII) as the sole carbon source, produced 1β-(1(S)-carboxyethyl)-5-oxo-7aβ-methyl-3aα-hexahydroindan-4α-propionic acid (XIII) in yield of about 70%[1]. As shown in Chart 2, the compound XIII was subjected to the above enamine lactam synthesis method to produce VIIa, identical with a sample of VIIa obtained from the incubation mixture. The fourth product IX, m.p. 183–185°, was initially recognized as a monoamide of the acid Va, the structure of IX or 4α-(2-carbamoylethyl)-5-oxo-7aβ,γ(R)-dimethyl-3aα-hexahydroindan-1β-butyric acid, on the basis of the following data: The elementary analysis was in good agreement with $C_{18}H_{29}O_4N$; its ultraviolet spectrum showed no characteristic peak in the 210—340 nm regions; its infrared spectrum in Nujol showed bands at 3438, 3318, and 3238 (NH), 2660, 2520, and 2360 (COOH), 1711 (COOH and 6-membered ring ketone), and 1650 and 1591 cm^{-1} ($CONH_2$); hydrolysis with alkali gave the acid Va. To clarify which one of the two carboyxl groups in the acid Va molecule is converted into the amide, partial synthesis of the compound IX was first performed according to the following sequence. Reduction of Vb with an excess of sodium borohydride in 90% ethanol at room temperature gave a mixture of methyl 4α-(3-hydroxypropyl)-5β-hydroxy-7aβ,γ(R)-dimethyl-3aα-hexahydroindan-1β-butyrate (XIb) and its 5α-epimer[2]. The acid XIa, after hydrolysis of XIb, was converted into the corresponding amide by the synthetic method of conjugated bile acid with carbonic-carboxylic acid anhydride reported by NORMAN (1955). The acid XIa was first converted into tri-n-butyl ammonium salt in dioxane and then treated with ethyl chlorocarbonate to yield a carbonic-

[1] HAYAKAWA et al., Unpublished data.

[2] A detail discussion of these results will be published in another part of this series.

carboxylic anhydride. The mixed anhydride was decomposed with aqueous ammonia to produce the amide XIc. The amide XIc was oxidized with chromic anhydride in acetic acid to produce IX, identical with a sample of IX obtained from the incubation mixture.

A neutral fraction from the incubation mixture was crystallized to yield the crude product VIc and the mother liquor was subjected to preparative thin-layer chromatography to yield the product VIII, m.p. 255° (dec., sintered at 238°). The melting point was not sharp, but no impurity was detected on the thin-layer chromatography. The elementary analysis afforded figures in good agreement with $C_{13}H_{17}O_2N$. The ultraviolet spectrum showed a strong peak at 230.5 nm (ε, 13,800), characteristic of enamine lactams (DOORENBOS et al., 1961a and 1961b). The infrared spectrum in KBr disk exhibited bands at 3188 and 3084 (NH), 1733 (5-membered ring ketone), and 1674 and 1623 cm^{-1} (enamine lactam). The nuclear magnetic resonance n.m.r.) spectrum exhibited bands at 0.95 (singlet, 3H, one tertiary CH_3), 4.95 (unresolved multiplet, 1H, one vinylic proton), and 8.54 ppm (broad singlet, 1H, CONH-proton). On the basis of these facts the structure of the product VIII was assigned as 6aβ-methyl-2,3,4,6,6aβ,-7,8,9,9aα,9bβ-decahydro-1H-cyclopenta[f]quinoline-3,7-dione. This assignment was confirmed by partial synthesis as shown in Chart 3: 17β-Hydroxy-estra-4,9-dien-3-one (XIVa)

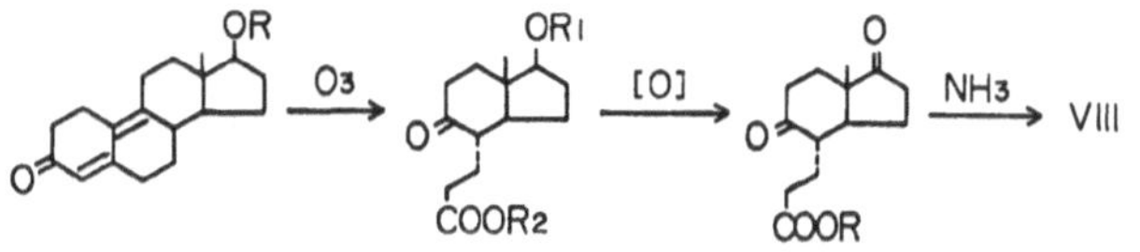

XIVa, R = H
b, R = Ac

XVa, R1 = Ac, R2 = H
b, R1 = R2 = H
c, R1 = H, R2 = Me

XVIa, R = Me
b, R = H

Chart 3

and its acetate XIVb were prepared according to the known method (PERELMAN et al., 1960 and 1963). Ozonolysis of XIVb afforded 1β-acetyloxy-5-oxo-7aβ-methyl-3aα-hexahydroindan-4α-propionic acid (XVa). The compound XVa was treated with aqueous potassium hydroxide to give 1β-hydroxy-5-oxo-7aβ-methyl-3aα-hexahydroindan-4α-propionic acid (XVb), m.p. 151—152° and $[\alpha]_D^{20} + 10.4 \pm 0.7°$ (C, 0.454 in ethanol). These physical constants were quite consistent with those reported in the literature (VELLUZ et al., 1963). After the acid XVb was converted into the methylester XVc with diazomethane, the ester was oxidized with chromic acid-pyridine complex to yield methyl 1,5-dioxo-7aβ-methyl-3aα-hexahydroindan-4α-propionate (XVIa). The ester XVIa was hydrolysed with methanolic potassium carbonate to yield 1,5-dioxo-7aβ-methyl-3aα-hexahydroindan-4α-propionic acid (XVIb), m.p. 107—107.5°, $[\alpha]_D^{20} + 100.7 \pm 1.2°$ (C, 1.004 in chloroform), and $\nu_{max}^{CHCl_3}$ 3080 and 2600 (COOH), 1740 (1-ketone), and 1710 cm^{-1} (COOH and 5-ketone). These constants were almost consistent with those reported in the literature (SIH et al. 1963). The synthetic method of the enamine lactam reported by DOORENBOS et al. (1961a) was applied to this acid XVIb to produce VIII, identical with a sample of VIII obtained from the incubation mixture.

The crude product VIc, obtained by direct crystallization of the neutral fraction as described before, was purified by chromatography on silicic acid to yield the second neutral product VIc, m.p. 218.5—221° (dec.). The elementary analysis was consistent with $C_{18}H_{28}O_2N_2$. The ultraviolett spectrum showed a strong preak at 232 nm (ε, 13,200), characteristic of enamine lactams (DOORENBOS et al., 1961a and 1961b), and its infrared spectrum in KBr disk exhibited bands at 3384 (broad), 3185, and 3050 (NH) and 1679 and 1614 (shoulder) cm^{-1} (amide and enamine lactam). The n.m.r. spectrum exhibited bands at 0,73 (singlet, 3H, one tertiary CH_3), 0.95 (doublet, J = 4.4 cps, 3H, one secondary CH_3), 4.82 (unresolved multiplet, 1H, one vinylic proton), 5.77 (unresolved multiplet, 2H, $CONH_2$-protons), and 8.27 ppm (broad singlet, 1H, CONH-proton). On the basis of these physical constants the structure of the product VIc was assigned as 3-oxo-6aβ,γ(R)-dimethyl-2,3,4,6,6aβ,7,8,9,9aα,9bβ-decahydro-1H-cyclopenta[f]quinoline-7β-butyramide. This assignment was confirmed by partial synthesis as shown in Chart 2. The amidation method which had led from the acid XIa to the corresponding amide XIc was applied to the compound VIa to yield VIc, which was found to be identical with a sample of VIc obtained from the incubation mixture.

A. simplex, capable of transforming cholic acid (I) into 4α-(2-carboxyethyl)-5-oxo-7aβ,γ(R)-dimethyl-3aα-hexahydroindan-1β-butyric acid (Va) (HAYAKAWA et al., 1967), was not able to utilize the compound Va as the sole source of carbon at a significant rate. However, it was found that *C. equi*, precultured in a medium containing only the acid XIa (0.1%) and yeast extract (Difco, 0,05%) as the carbon source, was capable of utilizing the compound Va as the sole source of carbon and of accumulating the intermediates. Without the preculture, *C. equi* seemed to be unable to utilize the compound Va as the sole source of carbon. Furthermore, when *C. equi* was cultured in a medium containing yeast extract (Difco, 0,05%) besides the compound Va (0.1—0.3%) as the carbon source, the compound Va was rapidly metabolized without the apparent accumulation of degradative intermediates which gave an insight as to the mechanism of the compound Va degradation. The exposure of the compound Va to *C. equi*, precultured as described above, produced a complex mixture of metabolic products and the yield of each component in the mixture was extremely low. However, one of them was isolated as crystals, m.p. 161—163.5° after silicic acid chromatography. The elementary analysis of this compound was in good agreement with $C_{21}H_{33}O_6N$ (M. W. 395.48) and non-aqueous titration gave a neutralization equivalent of 380. The infrared spectrum in KBr disk showed the presence of a carboxyl group at 2500 and 1738 cm^{-1}, a six-membered ring ketone at 1698 cm^{-1}, and an amide group at 3358 and 3178 cm^{-1} (NH) and 1620 and 1553 cm^{-1} (amide I and II bands). The amide bond in this compound was easily cleaved by heating with azeotropic hydrochloric acid to yield the original acid Va and a substance which gave a purple color with ninhydrin reagent for amino acids. The latter substance was assigned as alanine by using an amino acid autoanalyzer. This assignment was confirmed by converting this substance into the crystalline N-2,4-dinitrophenyl derivative by Sanger's method (1945), identical with that of L-alanine reported in the literature (RAO et al., 1954). On the basis of these data the structure of the product X was assigned as N-[4α-(2-carboxyethyl)-5-oxo-7aβ, γ(R)-dimethyl-3aα-hexahydroindan-1β-butyryl]-L-alanine or its isomer, 4α-{2-[N-(1(S)-carboxyethyl) carbamoyl]ethyl}-5-oxo-7aβ,γ(R)-dimethyl-3aα-hexahydroindan-1β-butyric acid. To clarify which isomer is identical with a sample of X obtained from the incubation mixture, partial synthesis of the compound X was first performed according to the following sequences: The acid XIa was first converted into tri-n-butyl ammonium salt in dioxane and then treated with ethyl chlorocarbonate to yield a carbónic-carboxylic anhydride as described before. The mixed anhydride was condensed with sodium salt of L-alanine to yield N-[4α-(3-hydroxypropyl)-5β-hydroxy-7aβ,γ(R)-dimethyl-3aα-hexahydroindan-1β-butyryl]-L-alanine (XId). As shown in Chart 2, the compound VId was also prepared from the enamine lactam carboxylic acid VIa in a similar manner. Although the yield was quite low in both cases, the chromic acid oxidation of XId and the hydrolysis with hydrochloric acid of VId gave the same compound X. Fortunately, this compound X was found to be identical with a sample of X obtained from the incubation mixture.

From the foregoing experimental results, all structures of six nitrogen-containing compounds VIa, VIc, VIIa, VIII, IX, and X have been conclusively established. This comfirms that the original steric configuration of the cholic acid (I) molecule remains undisturbed in the degradative sequence. And also this suggests that a possible pathway for the cholic acid degradation by *S. rubescens* may be I → Va → VIa → VIIa → VIII → further degradation as judged by their chemical structures, although there are some possibilities that the reaction of an enamine lactam formation is probably a side metabolic reaction rather than a direct step in the degradative sequence. This can be assumed from the fact that the compounds VIIa and VIII are newly produced in parallel with a disappearance of the compounds Va and IX, as seen in Fig. 1. We have not any experimental evidence on the fate of the compounds VIc, IX, and X in the cholic acid metabolism by microorganisms, but they are probably the side metabolites and they might be first converted back into the acids Va or VIa before further degradation as shown in Chart 1. The isolation and identification of the compound X demonstrate the presence of a reaction in microorganisms which corresponds to a well known detoxication mechanism in animals; the conjugation of a foreign compound with amino acids. This evidence might be a quite interesting finding in view of comparative biochemistry, because it has been believed so far that the conjugation of a foreign compound with amino acids occurs only in animals (WILLIAMS et al., 1963 and WOLLEN et al., 1959).

Exposure of cholic acid (I) to *Streptomyces gelaticus* has produced the acids II and III and other metabolites (HAYAKAWA, 1954 and HAYAKAWA et al., 1957). This result indicates that the β-oxidation of the butyric acid side chain in cholic acid (I), *via* a manner analogous to the conventional fatty acid oxidation mechanism, occurs before a conversion of the cholane nucleus into the perhydroindan nucleus. Alternatively, the β-oxidation (VIa $\rightarrow$ VIIa $\rightarrow$ VIII) of the butyric acid side chain in cholic acid (I) by *S. rubescens* in this experiment perhaps occurs after a formation of the perhydroindan nucleus. Furthermore, it has been already reported that exposure of cholic acid (I) to *A. simplex* produces the acids II, IV, and Va and other metabolites (HAYAKAWA et al., 1964 and 1967). These observations suggest that the initial degradative sequence of cholic acid (I) by microorganisms occurs probably *via* the acid II and then the acid II may be further degraded *via* the acids III or Va, although some microorganisms such as *Alcaligenes faecalis* and *Escherichia coli* initially oxidize the 7α-hydroxyl group in the cholic acid (I) molecule to a 7-ketone group as discussed in a previous paper (HAYAKAWA et al., 1957). Although inconclusive, some possible pathways for the cholic acid (I) degradation by microorganisms may be proposed as shown in Chart 1. SHAW et al., (1965) have reported that cell-free extracts of *Pseudomonas testosteroni* are capable of catalyzing the formation of L-2-amino-*cis*-4-hexenoic acid-1-C^{14} which originates from the ring A of androst-4-ene-3,17-dione-4-C^{14}. Our proposed pathways also show the microbiological formation of new types of nitrogen-containing compounds from cholic acid (I). Much of the representation of our pathways is still largely hypothetical. However, further work along this line is being pursued.

Experimental

General. — Cultivation of microorganisms was worked up according to the procedure described previously (HAYAKAWA et al., 1958a and 1964). All melting points were taken on a Kofler block and were uncorrected. All optical rotations were measured in a 1-dm tube with a Perkin–Elmer model 141 polarimeter. The infrared spectra were recorded with a Nihonbunko model DS201-B spectrometer or/and a Nihonbunko model IR-S spectrometer. The ultraviolet absorption spectra were recorded with a Hitachi model ESP-2 spectrometer. The nuclear magnetic resonance spectra were taken with a Varian A-60 spectrometer in deuteriochloroform containing tetramethylsilane as an internal reference at room temperature. Thin-layer chromatography was carried out essentially as reported by ENEROTH (1963) with the use of silica gel G or GF_{254} (according to Stahl, Merck A. G., Germany) and the solvent systems were as follows: Solvent system 1; isoactane—ethyl acetate—acetic acid (5:15:1 v/v); solvent system 2; cyclohexane—ethyl acetate—acetic acid (5:5:1 v/v). Ultraviolet lamp and concentrated sulfuric acid were used for the detection of spots. Reversed phase partition chromatography was performed by the procedure reported by SJÖVALL (1964).

Isolation and Characterization of Cholic Acid Metabolites by S. rubescens. — According to a similar method reported previously (HAYAKAWA et al., 1958a), 6 g. of sodium cholate were incubated with *S. rubescens* for 9 days. The culture broth was filtered to remove mycelium and the culture filtrate was concentrated to about 200 ml. without warming the solution to more than 45°. The concentrated culture filtrate was acidified with dilute hydrochloric acid to a pH of about 2 and thoroughly extracted with ethyl acetate. The acidic aqueous layer was not subjected to further treatment. The ethyl acetate layer was extracted with 5% sodium bicarbonate solution. The organic layer was washed with saturated sodium chloride solution, dried over sodium sulfate and evaporated to dryness to yield 790 mg. of semi-crystalline residue (neutral fraction). The basic sodium bicarbonate layer was acidified with dilute hydrochloric acid to a pH of about 3 and extracted with ethyl acetate. This extract was washed with saturated sodium chloride solution, dried over sodium sulfate and taken down to dryness to give 540 mg. of residue (acidic fraction).

4α-(2-Carboxyethyl)-5-oxo-7aβ,γ(R)-dimethyl-3aα-hexahydroindan-1β-butyric Acid (Va), 3-Oxo-6aβ,γ(R)-dimethyl-2,3,4,6,6aβ,7,8,9aα,9bβ-decahydro-1H-cyclopenta[f]quinoline-7β-butyric Acid (VIa), 3-Oxo-6aβ,α(S)-dimethyl-2,3,4,6,6aβ,7,8,9aα,9bβ-decahydro-1H-cyclopenta[f]quinoline-7β-acetic Acid (VIIa), and 1β-(1(R)-Methyl-3-carbamoylpropyl)-5-oxo-7aβ-methyl-3aα-hexahydroindan-4α-propionic Acid (IX). — 540 mg. of the above acidic fraction were chromatographed on 20 g. of silicic acid (Mallinckrodt) and eluted with dichloromethane with progressively increasing amount of acetone. Each fraction was tested by thin-layer chromatography using the solvent system 1. The fractions which possessed the same ultraviolet absorption, mobility, and color were pooled. The eluate with 10% acetone in dichloromethane was crystallized from acetone to yield 73 mg. of crude acid Va. The crude material was esterified with diazomethane and chromatographed on 2.5 g. of alumina (Woelm, neutral, grade III). The eluates with 10% benzene in petroleum ether and benzene gave 64 mg. of an oily mass. The infrared spectrum was identical with that of methyl 4α-(2-methoxycarbonylethyl)-5-oxo-7aβ,γ(R)-dimethyl-3aα-hexahydroindan-1β-butyrate (Vb). The oily mass was refluxed with 2 ml. of 2.5% methanolic potassium carbonate solution for 1.5 hours and worked up in the usual manner to yield 50 mg. of pale yellowish solid. Two times recrystallization from acetone gave 13 mg. of the product Va, m.p. 164.5—167°; $[\alpha]_D^{24}$ + 22.6 ± 0,7° (c, 0.852 in chloroform); ν_{max}^{Nujol} 3200—2600 (COOH) and 1703 cm^{-1} (COOH and 5-ketone). This compound was identical in all respects (m.p., mixed m.p., and infrared spectrum) with an authentic sample.

Anal. Calcd. for $C_{18}H_{28}O_5$: C, 66.64; H, 8.70. Found: C, 66.55; H, 8.68.

The eluates with 20 and 40% acetone in dichloromethane gave 98 mg. of a mixture of the products VIa and VIIa and some unknown metabolites. Attempts to isolate the products VIa and VIIa from this mixture were not successful.

The more polar fractions containing a substance, which gave a purple coloration with sulfuric acid and gave no ultraviolet absorption on thinlayer chromatography, were eluated with 40% acetone in dichloromethane and acetone. Recrystallization from acetone gave 29 mg. of the product IX, m.p. 183—185°; $[\alpha]_D^{25}$ + 24.8 ± 2.0° (c, 0.323 in ethanol); ν_{max}^{Nujol} 3438, 3318, and 3238 (NH), 2660, 2520, and 2360 (COOH), 1711 (COOH and 5-ketone), and 1650 and 1591 cm^{-1} ($CONH_2$).

Anal. Calcd. for $C_{10}H_{30}O_4N$: C, 66.84; H, 9.04; N, 4.33. Found: C, 66.78; H, 9.02; N, 4.64.

The methyl ester was prepared with diazomethane in the usual manner. The infrared spectrum in chloroform showed the reasonable bands at 3532 and 3438 (NH), 1732 (COOMe), 1704 (5-ketone), and 1680 and 1594 cm^{-1} ($CONH_2$), although an attempt to crystallize the ester was not successful.

Since attempts to isolate the products VIa and VIIa from the acidic fraction were not successful as described before, 13.2 g. of sodium cholate were incubated with *S. rubescens* for 11 days in a similar manner and gave 1.210 g. of an acidic fraction and 1.666 g. of a neutral fraction. Recrystallization of the neutral fraction from ethyl acetate gave 1.018 g. of the crude product VIc. The acidic fraction (1.210 g.) was chromatographed on 50 g. of silicic acid (Mallinckrodt). Each fraction with 20 and 40% acetone in dichloromethane was tested by thin-layer chromatography using the solvent system 1. The fractions which possessed the same ultraviolet absorption, mobility, and color were pooled to produce three new fractions. The first fraction (195 mg.) mainly consisted of the acid Va which had been isolated in this experiment as described before. Recrystallization from acetone gave 134 mg. of crude Va, m.p. 159—160°. The second fraction (191 mg.) mainly consisted of a product which gave the spot just below that of the product Va on the thin-layer chromatography. Recrystallization from acetone gave 112 mg. of crude VIa. Repeated recrystallization from methanol gave 9 mg. of VIa, m.p. 233° (dec., sintered at 224°); $[\alpha]_D^{24} + 149.0 \pm 37.4°$ (c, 0.051 in ethanol); λ_{max}^{alc} 231 nm (ε, 13,300); ν_{max}^{KBr} 3185 and 3045 (NH), 2600 and 1706 (COOH), and 1676 and 1668 (sh) cm^{-1} (enamine lactam).

Anal. Calcd. for $C_{18}H_{27}O_3N$: C, 70.79; H, 8.91; N, 4.59. Found: C, 70.74; H, 8.62; N, 4.44.

From the mother liquor additional 35 mg. of crude VIa were recovered.

The third fraction (225 mg.) was a complex mixture and fractionated by reversed phase partition chromatography. The moving phase (825 ml. of methanol and 65 ml. of water) and the stationary phase (225 ml. of chloroform and 25 ml. of n-heptane) were equilibrated for 12 hours at 23°. 21 ml. of the less polar phase were supported on 31.5 g. of Hostalen. 4 ml. of each fraction were collected at the flow rate of 0.13 ml. per minute. An aliquot of each fraction collected from the column was titrated with 1/50 N-NaOH to give a titration curve having three peaks. The titrated solution was first concentrated to a minimum volume with nitrogen stream and then subjected to thin-layer chromatography using the solvent system 1. The first peak (fraction No. 6—13) centered at 32 ml. eluate was consisted of a complex mixture and not further treated. The third peak (fraction No. 44—70) centered at 232 ml. eluate was a quite low and broad peak and not subjected to further treatment. Fractions No. 28—34 in the second peak (fraction No. 25—36) centered at 128 ml. eluate were pooled and evaporated to dryness with nitrogen stream at room temperature. The residue was treated with hot acetone to remove a material which was almost insoluble in acetone. The material was recrystallized from methanol and gave 1 mg. of an unknown carboxylic acid, m.p. 285° (dec., sintered at 274°); λ_{max}^{alc} 232 nm ($E_{1\,cm}^{1\%}$, 890); ν_{max}^{KBr} 3190, 3140 (sh), 3036, 2500, 1677, and 1620 cm^{-1}. The mother liquor after removal of this material was evaporated to dryness and recrystallized from acetone to yield 39 mg. of crude VIIa. Repeated recrystallization from acetone and then ethyl acetate gave 3 mg. of pure VIIa, m.p. 241—247.5° (dec.); $[\alpha]_D^{24} + 151.0 \pm 12.8°$ (c, 0.149 in ethanol); λ_{max}^{alc} 231 nm (ε, 12,900); ν_{max}^{KBr} 3190 and 3028 (NH), 2580 (COOH), and 1725, 1677, and 1628 cm^{-1} (COOH and enamine lactam).

Anal. Calcd. for $C_{16}H_{23}O_3N$: C, 69.28; H, 8.36; N, 5.05. Found: C, 69.40; H, 8.40; N, 4.85.

3-Oxo-6aβ,γ(R)-dimethyl-2,3,4,6,6aβ,.7.8.9.9aα,9bβ-decahydro-1H-cyclopenta[f]quinoline-7β-butyramide (VIc) and 6aβ-Methyl-2,3,4,6,6aβ,7,8,9,9aα,9bβ-decahydro-1H-cyclopenta[f]quinoline-3,7-dione (VIII). — Recrystallization of the neutral fraction (790 mg.) from ethyl acetate yielded 518 mg. of crude VIc. Since attempts at direct recrystallization for further purification failed, the crystals were chromatographed on 20 g. of silicic acid (Mallinckrodt) and eluted with dichloromethane with progressively increasing amount of acetone. Elution with 40% acetone in dichloromethane was crystallized from methanol to give 228 mg. of the product VIc, m.p. 218.5—221° (dec.); $[\alpha]_D^{22} + 156.0 \pm 3,6°$ (c, 0.545 in chloroform); λ_{max}^{alc} 232 nm (ε, 13,200); ν_{max}^{KBr} 3384 (br), 3185, and 3050 (NH), and 1679 and 1614 (sh) cm^{-1} ($CONH_2$ and enamine lactam); n.m.r. 0.73 (singlet, 3H, 6a-Me), 0.95 (doublet, J = 4.4 cps, 3H, γ(R)-Me in the side chain), 4.82 (unresolved multiplet, 1H, 5-H), 5.77 (unresolved multiplet, 2H, $CONH_2$-protons), and 8.27 ppm (broad singlet, 1H, 4-H).

Anal. Calcd. for $C_{18}H_{28}O_2N_2$: C, 71.01; H, 9.27; N, 9.20. Found: C, 70.82; H, 9.29; H, 9.09.

The mother liquor was evaporated to dryness to yield 272 mg. of residue. The residue was dissolved in a minimum volume of chloroform and the solution was applied to nine plates (20 × 20 cm, 0.75 mm thickness) of silica gel GF_{254}. Separation of at least three distinct bands, as detected by ultraviolet absorption, was achieved by developing with the solvent system 1. The band just above that of the product VIc was scraped from the plates and extracted with a mixture of chloroform and methanol (9:1 v/v) to yield 78 mg. of semi-crystalline solid. Two times recrystallization from methanol gave 25 mg. of the product VIII, m.p. 255° (dec., sintered at 238°); $[\alpha]_D^{23} + 340.5 \pm 10.0°$ (c, 0.385 in chloroform); λ_{max}^{alc} 230.5 nm (ε, 13,800); ν_{max}^{KBr} 3188 and 3084 (NH), 1733 (7-ketone), and 1674 and 1623 cm^{-1} (enamine lactam); n.m.r. 0,95 (singlet, 3H, 6a-Me), 4.95 (unresolved multiplet, 1H, 5-H) and 8.54 ppm (broad singlet, 1H, 4-H).

Anal. Calcd. for $C_{13}H_{17}O_2N$: C, 71.20; H, 7.82; N, 6.39. Found: C, 71,18; H, 7,77; N. 6.67.

3-Oxo-6aβ,γ(R)-dimethyl-2,3,4,6,6aβ,7,8,9,9aα,9bβ-decahydro-1H-cyclopenta[f]quinoline-7β-butyric Acid (VIa). — 3.5 g. of 4α-(2-carboxyethyl)-5-oxo-7aβ,γ(R)-dimethyl-3aα-hexahydroindan-1β-butyric acid (Va), obtained by exposure of cholic acid to *A. simplex*, was dissolved in 70 ml. of absolute ethanol, which had been saturated previously with dry ammonia gas at 0°. The solution was heated in a sealed reaction vessel at 86° for 14 hours. The white crystalline solid which separated on cooling was filtered and the filtrate was concentrated down to dryness. The residue was combined with the crystalline solid, suspended in dilute hydrochloric acid and extracted well with chloroform. The chloroform extracts were washed with water, dried over sodium sulfate and evaporated to dryness. The residue was crystallized from methanol and gave 2.239 g. of VIa as prisms, m.p. 238° (dec., sintered at 225°); $[\alpha]_D^{20} + 149.8 \pm 1.8°$ (c, 1.023 in ethanol); λ_{max}^{alc} 231 nm (ε, 13,100); ν_{max}^{KBr} 3185 and 3050 (NH), 2600 and 1706 (COOH), and 1676 and 1668 (sh) cm^{-1} (enamine lactam); n.m.r. 0.72 (singlet, 3H, 6a-Me), 0.93 (doublet, J = 4.0 cps, 3H, γ(R)-Me in the side chain), 4.83 (unresolved multiplet, 1H, 5-H), and 8.48 ppm (broad singlet, 1H, 4-H). Identity with the product VIa, obtained from the incubation mixture, was determined by m.p., $[\alpha]_D$, λ_{max}^{alc}, and infrared comparisons.

Anal. Calcd. for $C_{18}H_{27}O_3N$: C, 70.79; H, 8.91; N, 4,59. Found: C 70.72; H, 8,95; N, 4.57.

From the mother liquor additional 321 mg. of the compound VIa, m.p. 236.5° (dec., sintered at 225°), were recovered.

Methyl Ester (VIb). — 89 mg. of the acid VIa were suspended in 5 ml. of dichloromethane, and excess ethereal diazomethane was added to this suspension. After the solid dissolved completely into the solution, the mixture was allowed to stand at room temperature for 2 hours and concentrated to dryness. The residue was recrystallized from methanol to yield 57 mg. of crude VIb. Further recrystallization from methanol gave plates of pure VIb, m.p. 187.5—191.5°; $[\alpha]_D^{22} + 146.6 \pm 3,8°$ (c, 0.444 in ethanol); λ_{max}^{alc} 230.5 nm (ε, 13,600); ν_{max}^{KBr} 3186 and 3084 (NH), 1743 (COOMe), and 1688, 1670, and 1623 cm^{-1} (enamine lactam); $\nu_{max}^{CHCl_3}$ 3390, 3200 (NH), 1728 (COOMe), and 1660 cm^{-1} (enamine lactam); n.m.r. 0.72 (singlet, 3H, 6a-Me), 0.93 (doublet, J = 4.6 cps, 3H, γ(R)-Me in the side chain), 3.66 (singlet, 3H, COOMe), 4.80 (unresolved multiplet, 1H, 5-H), and 8.07 ppm (broad singlet, 1H, 4-H).

Anal. Calcd. for $C_{19}H_{29}O_3N$: C, 71.44; H, 9.15; N, 4.39. Found: C, 71.40; H, 9.20; N, 4.26.

3-Oxo-6aβ,γ(R)-dimethyl-2,3,4,6,6aβ,7,8,9,9aα,9bβ-decahydro-1H-cyclopenta[f]quinoline-7β-butyramide (VIc). — 200 mg. of the enamine lactam carboxylic acid VIa was dissolved in 4 ml. of anhydrous dioxane containing 0.15 ml. of tri-n-butylamine by heating. The solution was cooled to 10° and 0.78 ml. of anhydrous dioxane containing 0.06 ml. of ethyl chlorocarbonate was added. After 15 minutes at this temperature 0.5 ml. of concentrated aqueous ammonia (28%) was added at once and the mixture was rapidly mixed. After 30 minutes the reaction mixture was evaporated to dryness. The residue was suspended in a small volume of water and extracted with chlorform. The chloroform extracts were washed with dilute hydrochloric acid, 5% sodium bicarbonate solution and then water, dried over sodium sulfate and evaporated to give a syrup. Recrystallization from methanol gave 127 mg. of VIc as plates, m.p. 215.5—218.5° (dec); $[\alpha]_D^{23} + 156.8 \pm 7.4°$ (c, 0.266 in chloroform); λ_{max}^{alc} 232 nm (ε, 13,500) ν_{max}^{KBr} 3350(br), 3180, and 3045 (NH), and 1679 and 1614(sh)cm^{-1} ($CONH_2$ and enamine lactam); n.m.r. 0,73 (singlet, 3H, 6a-Me), 0.95 (doublet, J = 4.4 cps, 3H, γ(R)-Me in the side chain), 4.80 (unresolved multiplet, 1H, 5-H), 5.60 (unresolved multiplet, 2H, $CONH_2$-protons), and 7.99 ppm (broad singlet, 1H, 4-H). Identity with the product VIc, obtained from the incubation mixture, was determined by m.p., $[\alpha]_D$, λ_{max}^{alc}, n.m.r., and infrared comparisons.

Anal. Calcd. for $C_{18}H_{28}O_2N_2$: C, 71.01; H, 9.27; N, 9.20. Found: C, 70.77; H, 9,29 N, 9.09.

3-Oxo-6aβ,α(S)-dimethyl-2,3,4,6,6aβ,7,8,9aα,9bβ-decahydro-1H-cyclopenta[f]quinoline-7β-acetic Acid (VIIa). — 506 mg. of 1β-(1(S)-carboxyethyl)-5-oxo-7aβ-methyl-3aα-hexahydroindan-4α-propionic acid (XIII), m.p. 158.5—159.5 and $[\alpha]_D^{23} + 0.9 \pm 0,2°$ (c, 1,176 in ethanol), obtained by exposure of 3β-hydroxy-23-dinorchol-5-enic acid to *A. simplex*, was dissolved in 80 ml. of absolute ethanol, which had been saturated previously with dry ammonia gas at 0°. The solution was heated in a sealed reaction vessel at 90° for 15 hours. The reaction mixture was concentrated to dryness. The residue was suspended in dilute hydrochloric acid and thoroughly extracted with a mixture of chloroform and methanol (19:1 v/v). The extracts were washed with water, dried over sodium sulfate and evaporated to dryness to give 432 mg. of solid. Recrystallization from methanol and then chloroform-petroleum ether gave 300 mg. of VIIa as prisms, m.p. 243—247° (dec.). Further recrystallization from acetone gave the analytical sample, m.p. 246--250 (dec.) $[\alpha]_D^{24} + 127.6 \pm 2.3°$ (c, 0.734 in; ethanol); λ_{max}^{alc} 231 nm (ε,12,100); ν_{max}^{KBr} 3180 and 3028 (NH), 2580 (COOH) and 1725, 1677, and 1628 cm^{-1} (COOH and enamine lactam). Identity with the product VIIa, obtained from the incubation mixture, was determined by m.p., $[\alpha]_D$, λ_{max}^{alc}, and infrared comparisons.

Anal. Calcd. for $C_{16}H_{23}O_3N$: C, 69.28; H, 8,36; N, 5.05. Found: C, 69.20; H, 8.40; N, 5,08.

Methyl ester (VIIb). — 62 mg. of the acid VIIa were treated with ethereal diazomethane and the reaction mixture was worked up in the usual manner. Recrystallization from acetone gave 40 mg. of VIIb, m.p. 245—246° (dec.). Further recrystallization from acetone gave the analytical sample, m.p. 250.5—251° (dec.); $[\alpha]_D^{25} + 138.2 \pm 2.1°$ (c, 0.617 in ethanol); λ_{max}^{alc} 232 nm (ε, 12,500); ν_{max}^{KBr} 3184 and 3050 (NH), 1743 (COOMe), and 1678 and 1667(sh)cm^{-1} (enamine lactam); $\nu_{max}^{CHCl_3}$ 3390 and 3200 (NH), 1729 (COOMe), and 1671 and 1660(sh)cm^{-1} (enamine lactam); n.m.r. 0.74 (singlet, 3H, 6a-Me), 1.19 (doublet, J = 6.9 cps, 3H, 1(S)-Me in the side chain), 3.66 (singlet, 3H, COOMe), 4.80 (unresolved multiplet, 1H, 5-H), and 7.88 ppm (broad singlet, 1H, 4-H).

Anal. Calcd. for $C_{17}H_{25}O_3N$: C, 70.07; H, 8.65; N, 4.81. Found: C, 70.27; H, 8.58; N, 4.77.

1β-(1(R)-Methyl-3-carbamoylpropyl)-5-oxo-7aβ-methyl-3aα-hexahydroindan-4α-propionic Acid (IX). — 400 mg. of 4α-(3-hydroxypropyl)-5β-hydroxy-7aβ,γ(R)-dimethyl-3aα-hexahydroindan-1β-butyric acid (XIa), m.p. 133—134° and $[\alpha]_D^{20} + 37.7 \pm 2,0°$ (c, 1.053 in ethanol), obtained from the sodium borohydride reduction of Vb in 90% ethanol, was dissolved in 2 ml. of anhydrous dioxane containing 0.34 ml. of tri-n-butylamine. The solution was cooled to 10° and 0.135 ml. of ethyl chlorocarbonate was added. After 30 minutes at this temperature 0.5 ml. of concentrated aqueous ammonia (28%) was added at once and the mixture was rapidly mixed. After 90 minutes the reaction mixture was treated by the procedure described in the case of the preparation of VIc. The resulting crude amide was recrystallized from acetone to yield 207 mg. of pure 4α-(3-hydroxypropyl)-5β-hydroxy-7aβ,γ(R)-dimethyl-3aα-hexahydroindan-1β-butyramide (XIc), m.p. 124.5—125.5°; $[\alpha]_D^{25} + 37.3 \pm 0.9°$ (c, 0.820 in ethanol); ν_{max}^{Nujol} 3324 and 3200 (OH and NH) and 1670, 1627(sh), and 1606 cm^{-1} ($CONH_2$).

Anal. Calcd. for $C_{18}H_{33}O_3N$: C, 69.41; H, 10.68; N, 4.50. Found: C, 69.60; H, 10.92; N, 4.23.

From the mother liquor additional 115 mg. of XIc, m.p. 121—125°, were recovered.

A solution of 140 mg. of chromic anhydride in 2 ml. of 80% acetic acid was added dropwise over a period of 10 minutes to a stirred solution of 200 mg. of XIc in 3 ml. of 90% acetic acid at room temperature. After the solution was stirred for 3 hours at room temperature, about 10 ml. of water and a small amount of sodium bisulfite were added to destroy the excess chromic acid and the mixture was evaporated to dryness. The residue was suspended in a small volume of water and extracted with ethyl acetate. The organic layer was washed with saturated sodium chloride solution, dried over sodium sulfate and evaporated to dryness to yield 165 mg. of crude compound IX. The crude material was chromatographed on 7 g. of silicic acid (Mallinckrodt) and eluted with dichloromethane with progressively increasing amount of acetone. Elution (40 mg.) with 40% acetone in dichloromethane was recrystallized from acetone to give the analytically pure IX, m.p. 179—182°; $[\alpha]_D^{24}$ + 24.4 ± 0,8° (c, 0.840 in ethanol); ν_{max}^{Nujol} 3450, 3336, and 3230 (NH), 2654, 2510, and 2350 (COOH), 1711 (COOH and 5-ketone), and 1650 and 1591 cm^{-1} ($CONH_2$). Identity with the product IX, obtained from the incubation mixture, was determined by mixed m.p., $[\alpha]_D$, and infrared comparisons.

Anal. Calcd. for $C_{18}H_{29}O_4N$: C, 66.84; H, 9.04; N, 4.33. Found: C, 66.54; H, 9.24; N, 4.06.

1β-Hydroxy-5-oxo-7aβ-methyl-3aα-hexahydroindan-4α-propionic Acid (*XVb*). — A solution of 943 mg. of 17β-acetyloxy-estra-4,9-dien-3-one (XIVb), prepared by the procedure reported by PERELMAN et al., (1960 and 1963), in a mixture of 85 ml. of ethyl acetate and 8.5 ml. of acetic acid was ozonized with a stream of 3% ozonized oxygen at 0° for 2 hours. After 25 ml. of water were added and the mixture was stirred at room temperature for 1 hour, 5 ml. of 30% hydrogen peroxide solution were added and the mixture was allowed to stand for 16 hours in the refrigerator. The mixture was then diluted with water and extracted well with ether. The ethereal solution was extracted several times with 5% sodium bicarbonate solution. The combined alkaline extracts were acidified with dilute hydrochloric acid and extracted with ether. The extracts were washed with saturated sodium chloride solution, dried over sodium sulfate and evaporated to dryness to give 789 mg. of solid. Recrystallization from ether–petroleum ether, followed by acetone, yielded 399 mg. of 1β-acetyloxy-5-oxo-7aβ-methyl-3aα-hexahydroidan-4α-propionic acid (XVa), m.p. 136—138°. Further recrystallization from acetone–petroleum ether led to the analytical sample, m.p. 140—142°; $[\alpha]_D^{20}$ + 6.8 ± 0.3° (c, 1.010 in ethanol); ν_{max}^{Nujol} 3200—2600 (COOH), 1739 (acetate), and 1713 and 1696 cm^{-1} (COOH and 5-ketone).

Anal. Calcd. for $C_{15}H_{22}O_5$: C, 63.81; H, 7.85. Found: C, 63.83; H, 7.93.

A solution of 184 mg. of XVa in 1.7 ml. of N-KOH was heated at 70° for a few minutes. After cooling to room temperature, the mixture was allowed to stand for 2 hours at this temperature. The white crystals which precipitated on acidifying with dilute hydrochloric acid were collected by filtration, washed with water to remove hydrochloric acid and dried to give 111 mg. of crude compound XVb, m.p. 149.5°. Recrystallization from acetone–petroleum ether gave 95 mg. of pure 1β-hydroxy-5-oxo-7aβ-methyl-3aα-hexahydroindan-4α-propionic acid (XVb), m.p. 151—152°; $[\alpha]_D^{20}$ + 10.4 ± 0.7° (c, 0.454 in ethanol); ν_{max}^{Nujol} 3448 (OH), 3128(br), 2600 and 2446 (COOH), and 1723 and 1700 cm^{-1} (COOH and 5-ketone).

Anal. Calcd. for $C_{13}H_{20}O_4$: C, 64.98; H, 8.39. Found: C, 64.81; H, 8.29.

The similar experiments were repeated to produce about 2 g. of XVb.

VELLUZ et al., (1963) reported m.p. 153° and $[\alpha]_D$ + 12° (in ethanol) for this compound.

1,5-Dioxo-7aβ- methyl-3aα-hexahydroindan-4α-propionic Acid (*XVIb*). — The acid XVb was converted to its methyl ester XVc with ethereal diazomethane in the usual manner. A solution of 1.900 g. of XVc in 10 ml. of pyridine was added dropwise to a stirred slurry of the chromic acid–pyridine complex, prepared from 2.280 g. of chromic acid and 22.8 ml. of pyridine by Poos method (1963), at room temperature. After stirring for 1 hour at this temperature the mixture was allowed to stand overnight at room temperature. The reaction mixture was diluted with 140 ml. of water and thoroughly extracted with ether. The ethereal solution was washed with dilute hydrochloric acid and water, dried over sodium sulfate and evaporated to dryness to yield 1.527 g. of an oily mass. The oil was chromatographed on 45 g. of alumina (Woelm, neutral, grade III) and elutions with 10 and 50% benzene in petroleum ether gave 1.056 g. of methyl 1,5-dioxo-7$a\beta$-methyl-3$a\alpha$-hexahydroindan-4α-propionate (XVIa), $\nu_{max}^{CHCl_3}$ 1740 (COOMe and 1-ketone) and 1712 cm^{-1}(5-ketone). Since attempts at crystallization failed, the residue was hydrolyzed by refluxing with 25 ml. of 5% potassium carbonate in 60% aqueous methanol for 1 hour. After neutralization with dilute hydrochloric acid, the mixture was evaporated to a syrup and dissolved in a small volume of water. The white crystalline solid which precipitated on acidifying with dilute hydrochloric acid was filtered, washed with water to remove hydrochloric acid and dried to yield 627 mg. of XVIb, m.p. 107—108°. Repeated recrystallization from aceton-petroleum ether gave the analytical sample, m.p. 107—107.5°, $[\alpha]_D^{20}$ + 100.7 ± 1.2° (c, 1.004 in chloroform); ν_{max}^{KBr} 3197, 2580, and 2420 (COOH), 1733 (1-ketone), and 1708 cm^{-1} (COOH and 5-ketone); $\nu_{max}^{CHCl_3}$ 3080—2600 (COOH), 1740 (1-ketone), and 1710 cm^{-1} (COOH and 5-ketone).

Anal. Calcd. for $C_{13}H_{18}O_4$: C, 65.53; H, 7.61. Found: C, 65.40; H, 7.87.

The filtrate was combined with the washings and thoroughly extracted with a mixture of ether and dichloromethane (4:1 v/v). The extracts were washed with water, dried over sodium sulfate and evaporated to dryness to yield 305 mg. of pale yellow residue. Recrystallization from acetone—petroleum ether gave additional 264 mg. of XVIb, m. p. 106—107°.

Sih et al. (1963) reported m.p. 110—111.5°, $[\alpha]_D^{25}$ + 121° (in chloroform), and $\nu_{max}^{CHCl_3}$ 3.85, 5.75, and 5.85 μ for this compound.

6aβ-Methyl-2,3,4,6,6aβ,7,8,9,9aα,9bβ-decahydro-1H-cyclopenta[f] quinoline-3,7-dione (VIII). 200 mg. of XVIb was dissolved in 40 ml. of absolute ethanol, which had been saturated previously with dry ammonia gas at 0°. The solution was heated in a sealed reaction vessel at 80° for 15 hours. The reaction mixture was evaporated to dryness and extracted with chloroform. The pinkish chloroform solution was washed with water, dried over sodium sulfate and evaporated to dryness to yield 152 mg. of semicrystalline solid. Since attempts at purification by direct recrystallization failed, the solid residue was chromatographed on 6.3 g. of silica gel (Woelm). The elutions with 10 and 15% acetone in dichloromethane gave 122 mg. of crystalline solid. Recrystallization from methanol gave 88 mg. of VIII, m. p. 252° (dec., sintered at 235°); $[\alpha]_D^{22}$ + 320.4 ± 7.1° (c, 0.515 in chloroform); λ_{max}^{alc} 230.5 nm (ε, 13,200); ν_{max}^{KBr} 3190 and 3086 (NH), 1733 (7-ketone), and 1674 and 1623 cm^{-1} (enamine lactam); n.m.r. 0.95 (singlet, 3H, 6a-Me), 4.94 (multiplet, 1H, 5-H) and 8.54 ppm (broad singlet, 1H, 4-H). This compound was completely identical in all respects (m.p., $[\alpha]_D$, λ_{max}^{alc}, ν_{max}^{KBr}, and n.m.r.) with the product VIII obtained from the incubation mixture.

Anal. Calcd. for $C_{13}H_{17}O_2N$: C, 71.20; H, 7.82; N, 6.39. Found: C, 71.20; H, 7.84; N, 6.31.

Isolation and Characterization of N-[4α-(2-Carboxyethyl)-5-oxo-7aβ,γ(R)-dimethyl-3aα-hexahydroindan-1β-butyryl]-L-alanine (X). — To a basal mineral solution containing $(NH_4)_2SO_4$ (2 g.), KH_2PO_4 (1 g.), $MgSO_4$ 7 H_2O (0.5 g.), KCl (0.5 g.), and $FeSO_4 \cdot 7\, H_2O$ (0.01 g.) in 1 liter of distilled water were added 0.5 g. of yeast extract (Difco) and 1 g. of sodium salt of 4α-(3-hydroxypropyl)-5β-hydroxy-7$a\beta$,γ(R)-dimethyl-3$a\alpha$-hexahydroindan-1β-butyric acid (XIa) and the pH was adjusted to 7.4. After autoclaving at 115° for 20 minutes, four cotton-plugged Sakaguchi flasks, each containing 100 ml. of this medium, were inoculated with *Corynebacterium equi* (IID[1]-Ko-85) from a fresh nutrient slant. The inoculated flasks were incubated on a reciprocal shaker at 28° for 2 days. 2 ml. of this culture broth were inoculated to a Sakaguchi flask containing autoclaved 100 ml. of the following medium: 3 g. of disodium salt of of 4α-(2-carboxyethyl)-5-oxo-6$a\beta$,γ(R)-dimethyl-3$a\alpha$-hexahydroindan-1β-butyric acid (Va)

[1] IID denotes Institute for Infectious Disease, Tokyo University, Japan.

were added to 1 liter of the above basal mineral solution and the pH was adjusted to 7.4. The inoculated flasks were incubated on the reciprocal shaker at 28°. To follow the course of the degradation, an aliquot of the incubation mixture was aseptically separated from the flask and subjected to thin-layer chromatography using the solvent system 2, as described in the case of the exposure of cholic acid to *S. rubescens.* The incubation was stopped when a most part of the acid Va disappeared from the incubation mixture. In this experiment, 60 g. of disodium salt of the acid Va were used and the incubation was allowed to proceed for 2 days. The contents of 200 flasks were pooled and continuously centrifuged at speed of 12,000 r.p.m. at 5° to remove the cells. About 20 liters of the supernatant were acidified with dilute hydrochloric acid to a pH of about 4 and extracted three times with 6 liters portions of chloroform. The chloroform extracts were washed with water, dried over sodium sulfate and evaporated to dryness to yield 6.63 g. of crystalline solid. By recrystallization from acetone, 4.53 g. of crude acid Va were recovered. The mother liquor was chromatographed on 85 g. of silicic acid (Mallinckrodt) and ether eluate gave additional 1.60 g. of the original acid Va. The aqueous washings were combined with the aqueous layer and the pH was adjusted to about 7 with 5% sodium bicarbonate solution. The solution was concentrated to about 1 liter under reduced pressure at less than 50°, acidified with dilute hydrochloric acid to a pH of about 2 and extracted three times with 500 ml. portions of ethyl acetate. The ethyl acetate layer was evaporated to dryness under reduced pressure at room temperature to produce 4.99 g. of an oily mass. To a solution of the residue dissolved in a minimum volume of acetone were added 10 g. of silicic acid (Mallinckrodt) and the solvent was completely removed by evaporation under reduced pressure. The residue was transfered to the top of a column prepared with 200 g. of silicic acid (Mallinckrodt) in ether and the fractions (500 ml.) were collected according to the following pattern:

Pool No.	Fraction No.	Eluting solvent	Eluate (mg.)
A	1 – 2	Ether	1,797
B	3	Ether	385
C	4	0.5% AcOEt in ether	115
D	5 – 7	0.5% AcOEt in ether	482
E	8 – 25	0.5 – 50% AcOEt in ether	1,199
F	26 – 27	AcOEt	47
G	28	Acetone	655
H	29	Methanol	63

Pool D was dissolved in about 5 ml. of water with the use of a minimum volume of 5% sodium bicarbonate solution. The solution was over-layered with ether and acidified with dilute hydrochloric acid to cause incipient cloudiness. The white prisms which crystallized out in the aqueous layer on standing overnight in the ice-box were filtered, washed with water to remove hydrochloric acid and dried to yield 38 mg. of N-[4α-(2-carboxyethyl)-5-oxo-7aβ,γ(R)-dimethyl-3aα-hexahydroindan-1β-butyryl]-L-alanine (X), m.p. 161.5—163.5°; $[\alpha]_D^{24}$ + 11.0 ± 0.5° (c, 1.006 in ethanol); ν_{max}^{KBr} 3358 and 3178 (NH), 2500 and 1738 (COOH), 1698 (5-ketone), and 1620 and 1553 cm^{-1} (CONH); neutralization equivalent 380.

Anal. Calcd. for $C_{21}H_{33}O_6N$: C, 63.77; H, 8.41; N, 3.51. Found: C, 63.79; H, 8.22; N, 3.77.

The mother liquor was combined with the washings, neutralized with 5% sodium bicarbonate solution to a pH of about 7 and evaporated to dryness. The residue was dissolved in a minimum volume of water, treated with ether and dilute hydrochloric acid as described above and gave additional 145 mg. of X, m.p. 157—160°. The dimethyl ester of X was prepared by adding ethereal diazomethane into a solution of X in methanol, but the ester was not crystalline. The infrared spectrum in chloroform gave the following bands: 3434 (NH), 1730 (COOMe), 1704 (5-ketone), and 1669 and 1495 cm^{-1} (CONH).

Hydrolysis of N-[4α-(2-Carboxyethyl)-5-oxo-7aβ,γ(R)-dimethyl-3aα-hexahydroindan-1β-butyryl]-L-alanine (X) to 4α-(2-Carboxyethyl)-5-oxo-7aβ,γ(R)-dimethyl-3aα-hexahydroindan-1β-butyric Acid (Va) and L-Alanine. — 59 mg. of X were dissolved in 5 ml. of azeotropic hydrochloric acid by gentle heating and the solution was heated in a sealed glass tube at 100° for 16 hours. The brownish crystals which separated on cooling were filtered, washed with water and dried to yield 40 mg. of crude Va. The colored crystals were decolorized with charcoal in methanol and recrystallized from acetone to yield 18 mg. of Va, m.p. 164.5—167°; $[\alpha]_D^{25}$ $+ 26.5 \pm 1.6°$ (c, 0.415 in chloroform); ν_{max}^{Nujol} 3200—2600 (COOH) and 1703 cm^{-1} (COOH and 5-ketone). Identity with an authentic sample was determined by mixed m.p., $[\alpha]_D$, and infrared comparisons.

Anal. Calcd. for $C_{18}H_{28}O_5$: C, 66.64; H, 8.70. Found: C, 66.72; H, 8.75.

An aliquot of the aqueous mother liquor after removal of crude Va was subjected to the amino acid autoanalyzer (Hitachi, Type KLA-2), operated with 0.2 M sodium citrate buffer (pH, 3.25) at 50° with 50 cm. column and yielded one peak at the maximum of 107 effluent ml. The peak was identical with that of alanine. The remaining mother liquor was combined with the aqueous washings and extracted three times with 20 ml. portions of ether. The ether layer was washed with water, dried over sodium sulfate and evaporated to dryness to yield additional 7 mg. of crude Va. The water layer was evaporated to dryness under reduced pressure to yield crystalline solid. The solid was dissolved in 0.75 ml. of water containing 60 mg. of sodium bicarbonate and to this was added a solution of 60 mg. of 2,4-dinitrofluorobenzene in 1.5 ml. of ethanol. The mixture was stirred for 2 hours at room temperature and extracted with ether to remove excess 2,4-dinitrofluorobenzene. The aqueous layer was diluted with a small volume of water. The yellowish crystals which precipitated on acidifying with dilute hydrochloric acid were filtered, washed with water and dried to yield 24 mg. of crude N-2,4-dinitrophenyl-L-alanine. Recrystallization from ethyl acetate–petroleum ether gave 9 mg. of the analytical sample, m.p. 170—172.5°; $[\alpha]_D^{24}$ $+ 155.1 \pm 16.6°$ (c, 0.118 in N-NaOH); λ_{max}^{N-HCl} 266 nm (ε, 8,500) and 354 nm (ε, 15,500); ν_{max}^{Nujol} 3350 (NH), 3088, around 2600 (br), 1720 and 1680 (COOH), 1624 and 1590 (phenyl), and 1524, 1505, and 1361 cm^{-1} (aromatic NO_2). Identity with an authentic sample, prepared from L-alanine according to Sanger's method (1945), was determined by mixed m.p., $[\alpha]_D$, λ_{max}^{N-HCl}, and infrared comparisons.

Anal. Calcd. for $C_9H_9O_6N_3$: C, 42.36; H, 3.56; N, 16.47. Found: C, 42.75; H, 3.52; N, 16.35.

Rao et al. reported m.p. 177° and $[\alpha]_D$ + 144° (in N-NaOH) for this compound.

N-[4α-(2-Carboxyethyl)-5-oxo-7aβ,γ(R)-dimethyl-3aα-hexahydroindan-1β-butyryl]-L-alanine (X). a) From 3-Oxo-6aβ,γ(R)-dimethyl-2,3,4,6,6aβ,7,8,9,9aα,9bβ-decahydro-1H-cyclopenta[f]-quinoline-7β-butyric Acid (VIa). — 2.433 g. of the enamine lactam carboxylic acid VIa were dissolved in 150 ml. of anhydrous dioxane containing 2 ml. of tri-n-butylamine by gentle heating. The solution was cooled to 10° and 0.91 ml. of ethyl-chlorocarbonate was added. After 15 minutes at this temperature a solution of 784 mg. of L-alanine in 8.8 ml. of N-NaOH was added at once and the mixture was rapidly mixed. After 4 hours the reaction mixture was evaporated to near dryness under reduced pressure. The syrup was suspended in 50 ml. of dilute hydrochloric acid and extracted with ethyl acetate. The organic layer was washed with water, dried over sodium sulfate and evaporated to dryness to yield 2.958 g. of semicrystalline solid. The solid was dissolved in about 5 ml. of water by the aid of a minimum volume of 5% sodium bicarbonate solution. The solution was over-layered with ether, acidified with dilute hydrochloric acid to a pH of about 2 and thoroughly mixed. The white precipitate which separated on standing overnight in the ice-box was filtered, washed with water and dried to yield 2.054 g. of powder. The mother liquor was combined with the washings and extracted with ethyl acetate. The organic layer was washed with water, dried over sodium sulfate and evaporated to dryness to yield 904 mg. of pale yellow solid. 2.054 g. of the above powder were recrystallized from methanol–ethyl acetate and then acetone to yield

380 mg. of VId, m.p. 181—184°. The mother liquor was evaporated to dryness and combined with 904 mg. of the above solid. The total solids (2.570 g.) were chromatographed on 114 g. of silicic acid (Mallinckrodt) containing 6 g. of Celite and eluted with chloroform with progressively increasing amount of acetone. From eluate with 20% acetone in chloroform 809 mg. of the crude starting material were recovered. 863 mg. of eluate with 30% acetone in chloroform were recrystallized from acetone and gave 351 mg. of N-(3-oxo-6aβ,γ(R)-dimethyl-2,3,4,-6,6aβ,7,8,9,9aα,9bβ-decahydro-1H-cyclopenta[f]quinoline-7β-butyryl)-L-alanine (VId), m.p. 181—184°. Further recrystallization from acetone gave the analytical sample, m. p. 185.5 to 186.5°; $[\alpha]_D^{25}$ + 111.5 ± 1.7° (c, 0.889 in ethanol); λ_{max}^{alc} 231 nm (ε, 13,100); ν_{max}^{Nujol} 3487, 3323, and 3191 (NH), 2535, 2319, and 1719 (COOH), and 1652, 1645, and 1538 cm^{-1} (CONH and enamine lactam).

Anal. Calcd. for $C_{21}H_{32}O_4N_2$: C, 66.99; H, 8.57; N, 7.44. Found: C, 66.94; H, 8.71; N, 7.36.

A solution of 200 mg. of VId in 4 ml. of dioxane containing 0,5 ml. of 0.1N-HCl was heated at 100° for 8 hours. The reaction mixture was evaporated to dryness under reduced pressure and extracted with ethyl acetate. The ethyl acetate extract was washed with water, dried over sodium sulfate and evaporated to dryness to yield 198 mg. of solid. The residue was dissolved in a small volume of methanol and the solution was applied to 6 plates (20 × 20 cm, 0.75 mm thickness) of silica gel G. Separation of at least three distinct bands, as detected by a quite dilute iodine vapor, was achieved by developing with the solvent system 2. The band just below that of the starting material VId was scraped from the plates and extracted with a mixture of chloroform, methanol, and 1/100 N-HCl (10:10:1 v/v). The extract was evaporated to dryness with nitrogen stream at room temperature to yield 137 mg. of crystalline solid. Recrystallization from acetone gave pure X, m.p. 161—163°; $[\alpha]_D^{24}$ + 10.6 ± 0,8° (c, 0.659 in ethanol); ν_{max}^{KBr} 3352 and 3188 (NH), 2520 and 1738 (COOH), 1698 (5-ketone), and 1620 and 1553 cm^{-1} (CONH). Identity with the product X, obtained from the incubation mixture, was determined by mixed m.p., $[\alpha]_D$, and infrared comparisons.

Anal. Calcd. for $C_{21}H_{33}O_6N$: C, 63.77; H, 8.41; N, 3.54. Found: C, 63.92; H, 8.42; N, 3.55.

b) *From 4α-(3-Hydroxypropyl)-5β-hydroxy-6aβ,γ(R)-dimethyl-3aα-hexahydroindan-1β-butyric Acid (XIa).* — 892 mg. of the dihydroxycarboxylic acid XIa were dissolved in 14 ml. of anhydrous dioxane containing 0.71 ml. of tri-n-butylamine. The solution was cooled to 10° and 0.29 ml. of ethyl chlorocarbonate was added. After 15 minutes at this temperature a solution of 280 mg. of L-alanine in 3.1 ml. of N-NaOH was added at once and the mixture was rapidly stirred. After 3 hours the reaction mixture was evaporated to near dryness under reduced pressure. The syrup was dissolved in 10 ml. of water, acidified with dilute hydrochloric acid and extracted with ethyl acetate. The organic layer was washed with water, dried over sodium sulfate and evaporated to dryness to yield 1.010 g. of residue. The residue was chromatographed on 50 g. of silicic acid (Mallinckrodt) and eluted with chloroform with progressively increasing amount of acetone. Eluate with 60% acetone in chloroform gave 754 mg. of almost pure N-[4α-(3-hydroxypropyl)-5β-hydroxy-7aβ,γ(R)-dimethyl-3aα-hexahydroindan-1β-butyryl]-L-alanine (XId). The material was subjected to the chromic acid oxidation without further purification. To a stirred solution of 50 mg. of the above XId in 0.5 ml. of acetone was added dropwise 0.12 ml. of Jones reagent (a solution of 26.7 g. of chromic acid in 23 ml. of concentrated sulfuric acid and 40 ml. of water was made up to 100 ml. with water (BOWERS et al., 1953) under cooling at 0°. After the stirring was continued for 30 minutes at this temperature, a small amount of sodium bisulfite was added to destroy an excess chromic acid. The mixture was extracted with ethyl acetate. The organic layer was washed with saturated sodium chloride solution, dried over sodium sulfate and evaporated to dryness to yield 40 mg. of residue. The residue was dissolved in a minimum volume of 5% sodium bicarbonate solution, over-layered with a small volume of ether and acidified with hydrochloric acid. The white crystals which separated on cooling were filtered, washed with water and dried to yield 10 mg. of X. Recrystallization from methanol–water gave needles, m.p. 160—162°. Identity with a sample obtained from VId was determined by mixed m.p. and infrared comparison.

Acknowledgement

The authors express their thanks to the members of the analytical and the physico-chemical departments of this laboratory for analytical and optical data, and to Dr. M. Ebata in this laboratory for his help in tracing the amino acid by the amino acid autoanalyzer and for his valuable criticism.

References

[1] Bowers, A., T. G. Halsall, E. R. E. Jones, and A. J. Jemin, (1953), J. Chem. Soc., 2548.
[2] Chang, F. N., and C. J. Sih (1964), Biochemistry, **3**, 1551.
[3] Coombe, R. G., Y. Y. Tsong, P. B. Hamilton, and C. J. Sih (1966). J. Biol. Chem., **241**, 1587.
[4] Doorenbos, N. J., C. L. Huang, C. C. Tamorria, and M. T. Wu (1961a), J. Org. Chem., **26**, 2546.
[5] Doorenbos, N. J., and C. L. Huang (1961b), J. Org. Chem., **26**, 4106 and 4548.
[6] Gibson, D. T., K. C. Wang, C. J., Sih, and H. W. Whitlock, Jr. (1966), J. Biol. Chem., **241**, 551.
[7] Hayakawa, S. (1954), Proc. Japan Acad., **30**, 128.
[8] Hayakawa, S., T. Fujii, Y. Saburi, and T. Eguchi (1957), Nature, **179**, 537.
[9] Hayakawa, S., Y. Saburi, and K. Tamaki (1958a), J. Biochem., **45**, 419.
[10] Hayakawa, S., Y. Saburi, and H. Hoshijima (1958b), J. Biochem., **45**, 465.
[11] Hayakawa, S., Y. Saburi, K. Tamaki, and H. Hoshijima (1958c), Nature, **181**, 906.
[12] Hayakawa, S., and B. Samuelsson (1964), J. Biol. Chem., **239**, **94**.
[13] Hayakawa, S., Y. Kanematsu, and T. Fujiwara (1967), Nature, **214**, 520.
[14] Lee, S. S., and C. J. Sih (1964), Biochemistry, **3**, 1267.
[15] Lee, S. S., and C. J. Sih (1967), Biochemistry, **6**, 1395.
[16] Norman, A. (1955), Arkiv Kemi, **8**, 331.
[17] Perelman, M., E. Farkas, E. J. Fornefeld, R. J. Kraay, and R. T. Rapala (1960), J. Am. Chem. Soc., **82**, 2402.
[18] Perelman, M., and E. Farkas (1963), U.S. Patent, 3,086,027.
[19] Poos, G. I., G. E. Arth, R. E. Beyler, and L. H. Sarett (1953), J. Am. Chem. Soc., **75**, 422.
[20] Rao, K. R., and H. A. Sober (1954), J. Am. Chem. Soc., **76**, 1328.
[21] Saburi, Y., S. Hayakawa, and T. Fujii (1955), Japan. J. Exp. Med., **25**, 151.
[22] Sanger, F. (1945), Biochem. J., **39**, 507.
[23] Schubert, K., K.-H. Böhme, and C. Hörhold (1961), Zeitschr. physiol. Chem., **325**, 260.
[24] Schubert, K., K.-H. Böhme, and C. Hörhold (1964), Steroids, **4**, 581.
[25] Schubert, K., K.-H. Böhme, and C. Hörhold (1965), Biochim. Biophys. Acta **111**, 529.
[26] Schubert, K., K.-H. Böhme, and C. Hörhold (1966), Abstract Book, IUPAC 4th International Symposium on the Natural Products, at Stockholm, p. 137.
[27] Shaw, D. A., L. F. Borkenhagen, and P. Talalay (1965), Proc. Nat. Acad. Sci., **54**, 837.
[28] Sih, C. J. (1962), Biochem. Biophys. Acta, **62**, 541.
[29] Sih, C. J., and K. C. Wang (1963a), J. Am. Chem. Soc., **85**, 2135.
[30] Sih, C. J., and A. M. Rahim (1963b), J. Pharm. Sci., **52**, 1075.
[31] Sih, C. J., S. S. Lee, Y. Y. Tsong, and K. C. Wang (1965a), J. Am. Chem. Soc., **87**, 1385.
[32] Sih, C. J., K. C. Wang, D. T. Gibson, and H. W. Whitlock, Jr. (1965b), J. Am. Chem. Soc., **87**, 1386.
[33] Sih, C. J., S. S. Lee, Y. Y. Tsong, and K. C. Wang (1966), J. Biol. Chem., **241**, 540.
[34] Sjövall, J. (1964), Methods of Biochemical Analysis XII, 97.
[35] Tsong, Y. Y., K. C. Wang, and C. J. Sih (1964), Biochim. Biophys. Acta, **93**, 398.
[36] Velluz, L., G. Nominé, G. Amiard, V. Torelli, and J. Cérède (1963), Compt. rend., **257**, 3086.
[37] Wallen, L. L., F. H. Stodola, and R. W. Jackson (1959), Type Reactions in Fermentation Chemistry, Agricultural Research Service, United States Department of Agriculture.
[38] Wang, K. C., and C. J. Sih (1963), Biochemistry, **2**, 1238.
[39] Williams, R. T. (1963), in P. Bernfeld (Editor), Biogenesis of Natural Compounds, Pergamon Press Ltd., Oxford, p. 427.

Microbiological Degradation of Ring C and D of the Steroid Skeleton

K. SCHUBERT, K.-H. BÖHME and C. HÖRHOLD

Institut für Mikrobiologie und experimentelle Therapie, Jena,
der Deutschen Akademie der Wissenschaften zu Berlin, DDR

Summary

In investigations concerning the transformation of progesterone by mycobacteria we found the first degradation product of the sterane skeleton: 7a-methyl-1-acetyl-perhydroindane-5-one-4β-(3-propionic acid) (δ-keto acid). By fermentation of the δ-keto acid with cultures of *Nocardia opaca* we succeeded in isolating α-ketoglutaric acid and succinic acid. A degradation pathway will be discussed.

In studies concerning the mechanism of progesterone transformation by mycobacteria we found the first degradation product of a steroid (1).

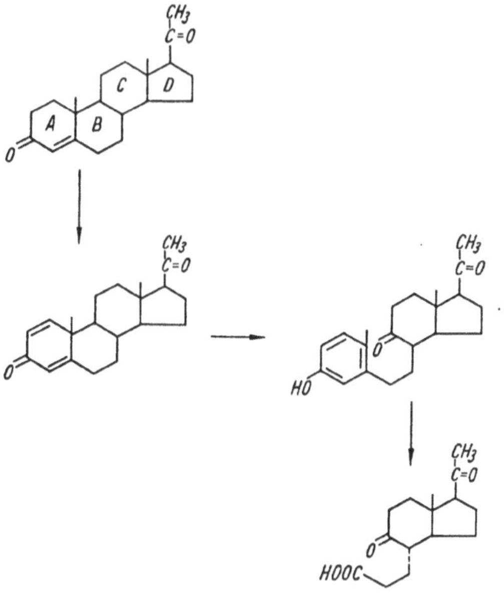

Fig. 1. Microbial formation of a δ-keto acid from progesterone

At first progesterone was dehydrogenated in position one. The simultaneous introduction of a hydroxy-group in position nine and aromatization of ring A led to a seco-compound by opening of ring B. After splitting off a C_6-fragment a δ-keto acid remained, which consists of the rings C and D of the steroid and is a compound of the perhydroindane-type. While Dr. TALALAY made his excellent studies on the degradation of the C_6-fragment (2), we set to investigating the further degradation of the product with the remaining steroid rings C and D (3,4).

Steroids which differ in their substituents on C_{17} were degraded to the corresponding δ-keto acids. The degradation of 17α-hydroxy progesterone gave a δ-keto acid with the remaining

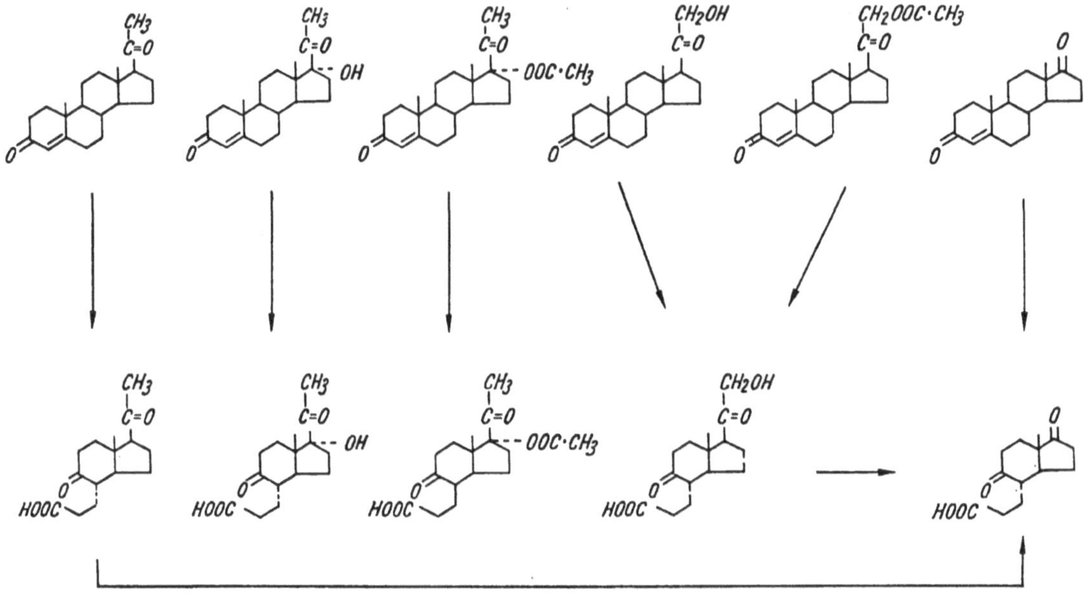

Fig. 2. Microbial formation of δ-keto acids from C_{19}- and C_{21}-steroids

17α-hydroxy-group. 17α-Acetoxy-progesterone will not be saponified by *Nocardia opaca* and we obtained the analogous acetylated δ-keto-acid. 11-Desoxy-corticosterone and 11-desoxy-corticosterone-acetate gave one analogous δ-keto acid. SIH and WANG described the δ-keto-acid which starts from Δ^4-androstene-3,17-dione (5). We obtained the same δ-keto-acid by splitting off the side-chain of the other δ-keto-acids described by us.

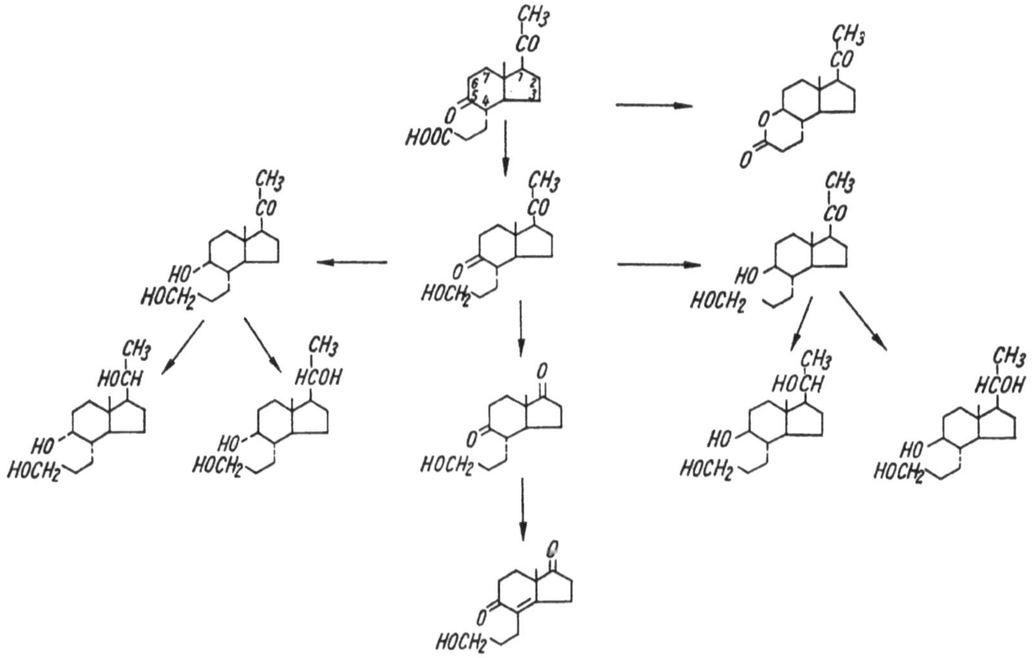

Fig. 3. Hydrogenation of the δ-keto acid from progesterone

Mycobacterium smegmatis is a bacterium, which possesses both dehydrogenating and hydrogenating properties. The reduction of the δ-keto-acid of progeterone is especially favoured under acid conditions. At first we isolated a δ-lacton and further a δ-ketol (6). The splitting off of the side chain yielded another δ-ketol which was further dehydrogenated in 4 position to an α,β-unsaturated ketol. Starting from the ketol with the side chain, the reduction of the keto-group in 5-position yielded 2 stereiosomeric diols, and, finally the reduction of the keto-group in the side chain yielded 4 stereoisomeric triols.

The hydrogenation reactions are side reactions. We also found hydrogenation starting from progesterone and $\Delta^{1,4}$-androstadiene-3,17-dione, leading to the corresponding $5\alpha/3\beta$-OH-compounds. The hydrogenation products obtained are indifferent substances compared to the δ-keto-acid. They accumulated in the fermentation broth and the further degradation of the δ-keto-acid must occur in another way.

Under changed fermentation conditions the δ-keto-acids will be degraded as sole source of carbon to CO_2. At first we did not succeed in isolating any intermediates.

Bacteria species of the genus: *Pseudomonas*, *Corynebacterium*, *Microbacterium*, *Mycobacterium*, *Nocardia* and *Streptomyces* utilized the δ-keto-acids in nutrient solutions as sources of carbon.

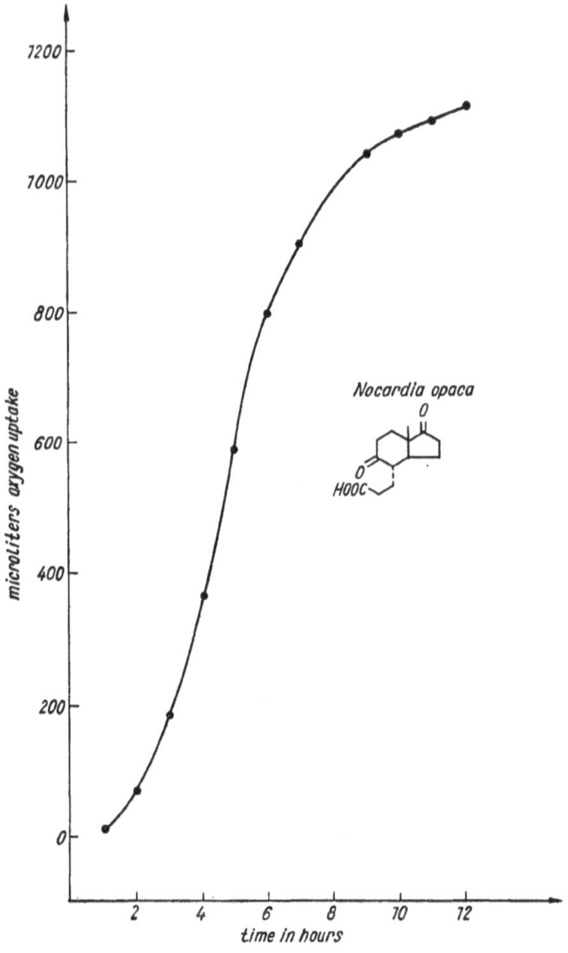

Fig. 4. Respiration of *Nocardia opaca* with δ-keto acids as source of carbon

By means of the WARBURG technique we estimated the oxygen consumption of resting cells of Nocardia opaca in phosphate buffer solution, pH 7, with addition of 0.8 mg δ-keto acid of C_{19}-steroids to 1 ml cell suspension (Fig. 4). The oxygen uptake is plotted as the difference to the endogenous respiration. After 8 hours the δ-keto-acid was utilized.

Keto acid additionally given to a nutrient solution with glycerol as source of carbon, was attacked as second source of carbon (Fig. 5). The glycerol was utilized by *Mycobacterium smegmatis* within 4 hours. After that the degradation of the keto-acid started.

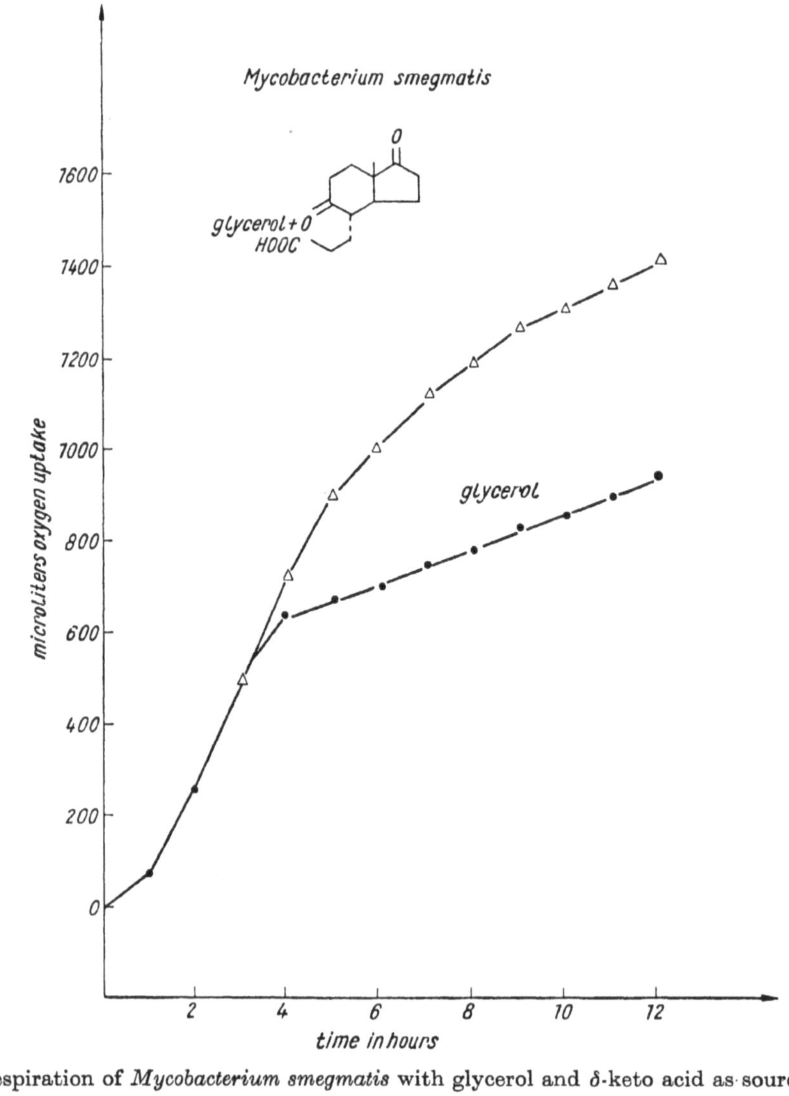

Fig. 5. Respiration of *Mycobacterium smegmatis* with glycerol and δ-keto acid as sources of carbon

The utilization of the keto-acid was inhibited by $\Delta^{1,4}$-androstadiene-3,17-dione, which is a starting material for the microbial production of this δ-keto-acid. (Fig. 6).

In spite of a large screening of different bacteria and additional variation of the culture and fermentation technique we did not succeed in isolating any intermediates. The application of various inhibitors had only the effect to reduce the degradation rate (7).

Without inhibitors almost 50% of the δ-keto-acid were degraded. 1 mM of sodium arsenite inhibited the degradation: 83% keto-acid could be recovered. 1 mM sublimate reduced the degradation to 86%, 2,5 mM 8-hydroxyquinoline to 96% and 0.1 mM EDTA to 75%. 1 mM potassium cyanide showed no effect, for the difference of 6% as compared to the control is not significant. Remarkable is the effect of 10 mM succinic acid. The degradation was promoted: only 10% of keto-acid were detectable in comparison to the control of 44%.

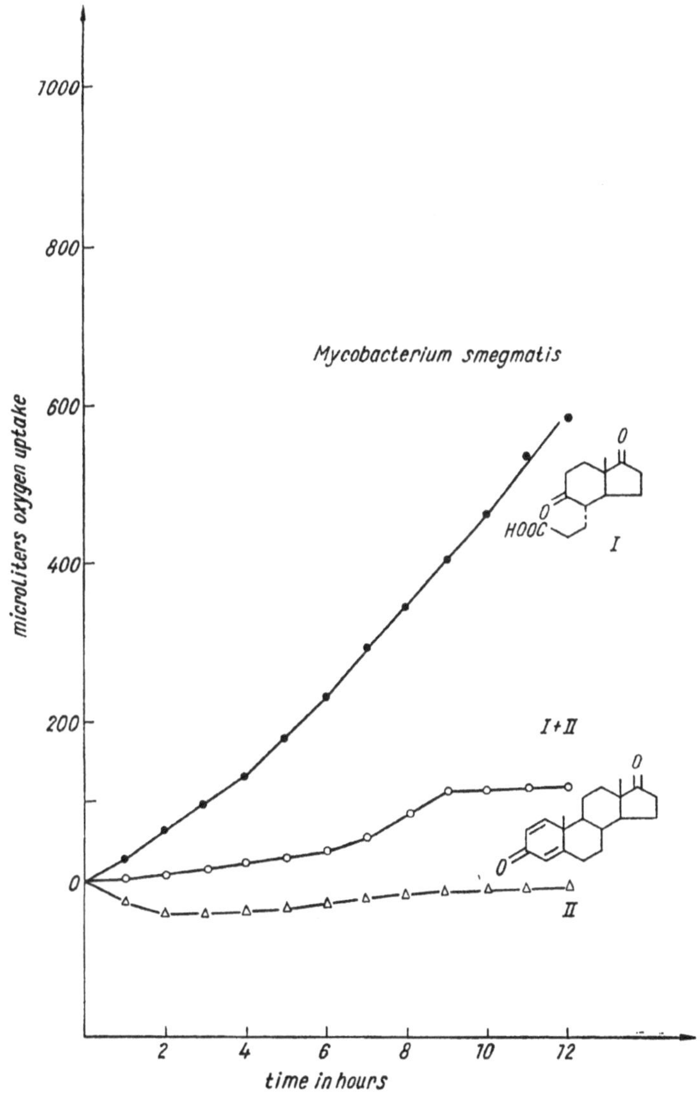

Fig. 6. Inhibition of respiration by $\Delta^{1,4}$-Androstadiene-3,17-dione

Table 1

Influence of additives on the degradation of δ-keto-acid by *Nocardia opaca*

additives	concentration mM	recovered δ-keto acid %
without additive	—	44
Na_3 ASO_3	1,0	83
$HgCl_2$	1,0	86
8-hydoxyquinoline	2,5	96
EDTA	0,1	75
KCN	1,0	38
succinic acid	10,0	10

In further investigations we used the δ-keto-acid of progesterone. By simultaneously increasing the δ-keto-acid concentration and the cell concentration in the fermentation medium we discovered 2 degradation products. 33.25 mg δ-keto-acid in a cell suspension of *Nocardia opaca* with 1 g wet cells in 10 ml phosphate buffer solution of pH 7 were fermented by shaking for 24 hours at 28°. By application of ion exchanger and subsequent column chromatography with silica gel it was possible to isolate 2 degradation acids identified as α-keto-glutaric acid and succinic acid. α-Keto-glutaric acid may be directly separated from the fermentation medium in the form of its 2.4-dinitrophenylhydrazone. In 12 experiments the yields of α-keto-glutaric acid ranged from 8 to 25% and the yields of succinic acid ranged from 1 to 3%. In controls without δ-keto acid the amounts of detectable α-keto-glutaric acid were only one twenty fifth. For quantitative comparison the amount of succinic acid was too small.

For further identification and for the evidence of the formation of α-keto-glutaric acid and succinic acid from the δ-keto acid of progesterone, the δ-keto acid was tritiated by the method of Wilzbach and incubated under the same conditions with *Nocardia opaca*. The results of the isotope dilution indicated the presence of labeled α-keto-glutaric acid-2,4-dinitro-phenylhydrazone and of labeled succinic acid.

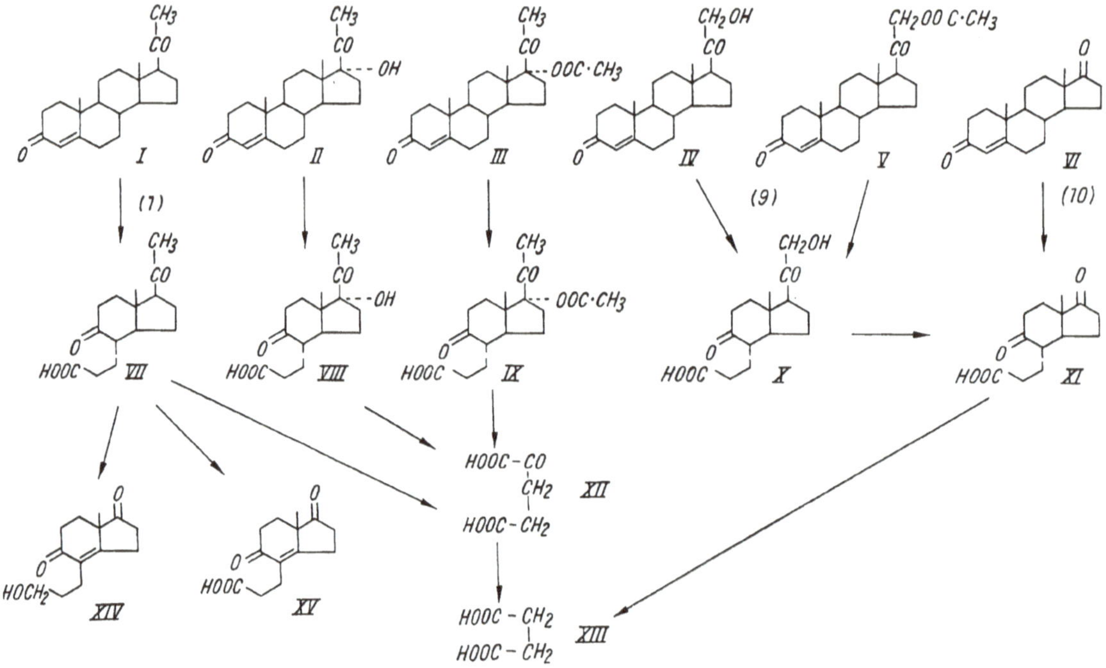

Fig. 7. Microbial formation of δ-keto acids from C_{19}- and C_{21}-steroids and the further degradation to α-keto-glutaric acid and succinic acid

The corresponding δ-keto-acids of progesterone, 17α-hydroxy-progesterone, 17α-acetoxy-progesterone, 11-desoxycorticosterone-acetate and Δ^4-androstene-3,17-dione were studied for comparison in order to detect α-keto-glutaric acid and succinic acid and establish different degradation rates. The δ-keto acids of Δ^4-androstene-3,17-dione, of desoxycorticosterone and of its acetate were degraded fastest of all, and yielded only succinic acid. The degradation of the δ-keto acid of desoxycorticosterone proceeds via the keto acid of the Δ^4-androstene-3,17-dione as an intermediate. The δ-keto acid of the 17α-hydroxy-progesterone was degraded more slowly than the δ-keto acid of the desoxycorticosterone, but faster than the δ-keto acid of progesterone. In both cases α-keto glutaric acid and succinic acid were formed.

The microbial degradation of steroids proceeds in the C_{27}-, C_{21}- and C_{19}-series via analogous δ-keto-acids. Up to now further degradation products have not been detected. The formation of α-ketoglutaric acid and succinic acid led to the formulation of a degradation scheme. We suppose that the degradation of the rings C and D proceeds in an analogous manner as the degradation of rings A and B.

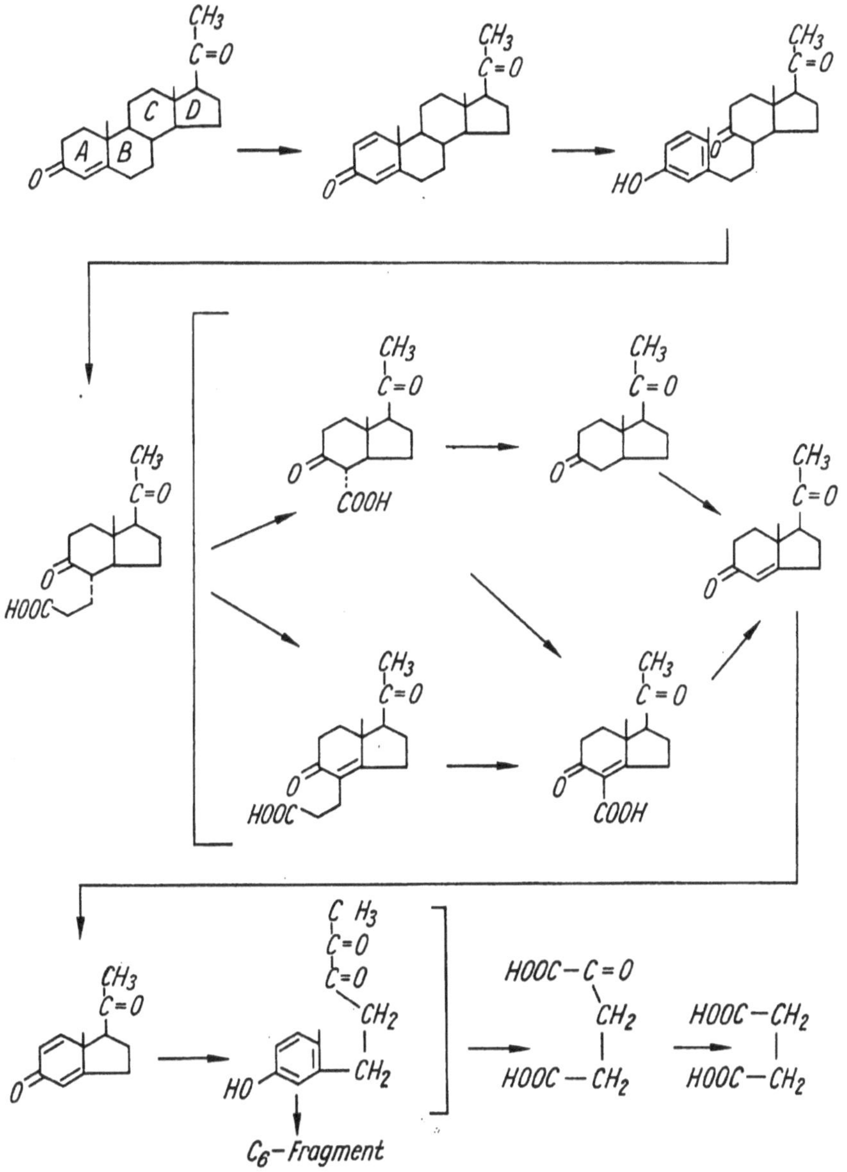

Fig. 8. Degradation scheme of progesterone

Our conception concerning the β-oxydation on the propionic acid side chain have recently been supported by investigations of Lee and Sih with synthetically prepared analogous perhydroindane compounds. The possibility of an introduction of a double bond before the β-oxydation could be proved by the isolation of the α,β-unsaturated keto alcohol 7a-methyl-5,6,7,7a-tetrahydro-indane-1,5-dione-4β(3-propionic-alcohol) and the α,β-unsaturated keto acid 7a-methyl-5,6,7,7a-tetrahydro-indane-1,5-dione-4β-(3-propionic acid) (Fig. 9).

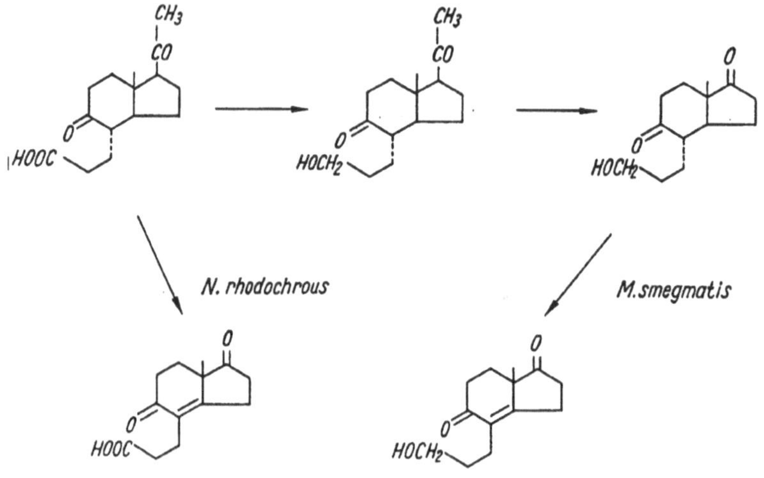

Fig. 9. Microbial introduction of a double bond in position 4 of perhydroindane compounds

It may well be that the keto-group of the sixmembered ring plays the same role for the introduction of 2 double bonds and the following aromatization by dienone-phenole-rearrangement as the oxygen function in 3-position of steroids. After splitting off a C_6-fragment, analogous to the fission of ring A of steroids, α-keto-glutaric acid and from this succinic acid might be obtained from the fragment of the five-membered ring of progesterone after elimination of the methyl-group in the side chain. According to this conception we obtained α-keto-glutaric acid from the keto acids with the side chain of progesterone and 17α-hydroxyprogesterone. The degradation of the δ-keto acids without side chain on the fivemembered ring yielded only succinic acid as expected. According to the supported degradation pathway the yields of α-keto-glutaric acid were always larger than those of succinic acid. Furthermore, succinic acid could be obtained only from α-keto-glutaric acid and not vice versa.

The results show that the microbial degradation of products with rings C and D of steroids proceeds to compounds of the citric acid cycle.

References

[1] Schubert, K., K.-H. Böhme, and C. Hörhold, Hoppe-Seyler's Z. physiol. Chem. **325**, 260 (1961).
[2] Shaw, D., L. F. Borkenhagen, and P. Talalay, Proc. nat. Acad. Sci. USA, **4**, 837 (1965).
[3] Schubert, K., K.-H. Böhme, and C. Hörhold, Acta biol. med. german. **18**, 295 (1967).
[4] Schubert, K., K.-H. Böhme, and C. Hörhold, Acta biol. med. german. **18**, 291 (1967).
[5] Sih, Ch. J., and K. C. Wang, J. Am. Chem. Soc. **85**, 2135 (1963).
[6] Schubert, K., K.-H. Böhme, and C. Hörhold, Steroids **4**, 581 (1964).
[7] Schubert, K., K.-H. Böhme, and C. Hörhold, Biochim. Biophys. Acta **152**, 401 (1968).

Zusammenhänge zwischen Steroidumwandlung und Eigenmetabolismus der Mikroorganismen

O. Hanč, und A. Čižinská

Forschungsinstitut für Pharmazie und Biochemie, Prag, ČSSR

Zusammenfassung

Wir befaßten uns in letzter Zeit mit der Untersuchung des Einflusses von Steroiden auf den Eigenmetabolismus der Mikroben. Bei einigen ausgewählten Bakterien- und Pilzstämmen, die verschiedene Typen von Steroidumwandlungen bewerkstelligen, untersuchten wir den Einfluß der Ausgangssteroide und ihrer Metaboliten auf Wachstum, Zucker- und Sauerstoffverbrauch und stellten fest, daß Steroide, die einer Umwandlung durch mikrobielle Enzyme einer bestimmten Mikrobenart unterliegen, durchweg das Mikrobenwachstum, den Zucker- und Sauerstoffverbrauch hemmen, während die entstehenden Steroidmetaboliten diese Eigenschaft nicht mehr besitzen. Aus der Ermittlung der Wasserlöslichkeit der Ausgangssteroide und ihrer Endmetaboliten ging hervor, daß durch die Umwandlung durchweg in Wasser besser lösliche Steroide entstehen. Vom Standpunkt der Mikrobenzelle handelt es sich um Reaktionen, die schematisch als „Detoxikationsreaktionen" bezeichnet werden können. Schließlich wurden einige Erkenntnisse der Steroidumwandlung und des Metabolismus der Zellen höherer Wirbeltiere mit den bei Mikroorganismen ermittelten Befunden verglichen und ihre Bedeutung für die vergleichende Biochemie und Physiologie hervorgehoben.

Die Mikrobenzellen sind ein sehr geeignetes Objekt für die Untersuchung des Metabolismus der verschiedensten Substanzen, und zwar auch solcher, die kein physiologisches Substrat für Mikroorganismen vorstellen. Bereits hinreichend bekannt ist die breite Adaptabilität dieser Zellen gegenüber verschiedenen Medien, auch wenn die einzelnen Mikrobenarten ursprünglich die verschiedensten Ansprüche, entweder hinreichend spezifisch oder im Gegenteil sehr unspezifisch, stellten.

Schematisch können wir drei Gruppen von metabolischer Einwirkung der Mikrobenzellen auf organische Substanzen verzeichnen. Entweder wird eine solche Substanz als Kohlenstoff- und Energiequelle für das Wachstum der Mikrobenzellen verwertet, sie wird fortlaufend in physiologische Zwichenprodukte überführt, die das Enzymsystem der Zellen auf dem üblichen Wege weiter zu verwerten vermag, oder sie wird durch Redoxreaktionen, häufig kombiniert mit der Spaltung des ursprünglichen Moleküls, in nicht weiter verwertbare Metaboliten überführt. Drittens findet man eine Indifferenz der Mikrobenzellen gegenüber der untersuchten Substanz, die nach Zusatz zum Medium auch nach sehr langer Inkubationszeit unverändert bleibt.

Bei diesen Versuchen arbeiten wir am häufigsten mit einer vorgezüchteten Mikrobenzellmasse, die steril in Puffermedium übertragen wird, wo man den Verlauf des Metabolismus der untersuchten Substanz und die Wirkung zusätzlicher Faktoren, wie stimulierende oder hemmende Zusätze, genau analytisch verfolgen kann. Die vorangehende Wachstumsphase des Mikroben in seinem obligaten Medium ist allerdings nicht ohne Einfluß auf die nachfolgenden metabolischen Leistungen der vorgezüchteten Mikrobenzellen. Sämtliche Befunde über den Metabolismus der verschiedensten mit Mikroben inkubierten organischen Substanzen sind das Ergebnis reiner Empirie, ohne daß es möglich wäre, auch nur qualitativ das Ergebnis vorauszusehen.

Der erste Forscher, der systematisch die metabolischen Leistungen der Mikroben an einer bestimmten Klasse organischer, für die Zellen unphysiologischer Substanzen untersuchte, war vor ungefähr 50 Jahren Carl Neuberg (Neuberg, 1914). Er unterwarf eine Reihe carbonylhaltiger Substanzen der Reduktion mit Hefezellen und stellte nach Reduktion verschiedener Aldehyde und Ketone Metaboliten mit Alkoholfunktion fest. Bei der Reduktion von Benzaldehyd ermittelte er neben Benzylalkohol die Bildung von optisch aktivem Phenylacetylcarbinol (Neuberg, 1921), womit er die Grundlage zur technischen Synthese des optisch aktiven Ephedrins schuf. In der Steroidreihe führte die erste mikrobielle Ketonreduktion vor 30 Jahren der italienische Forscher L. Mamoli durch (Mamoli, 1937). Jedoch erst während der letzten 15 Jahre, nachdem der Amerikaner D. H. Peterson die grundlegende Entdeckung gemacht hatte, daß mittels verschiedener Schimmelpilze die Pregnensteroide in hoher Ausbeute in 11-Stellung hydroxyliert werden können (Peterson, 1952), entwickelte sich die Vielzahl der Verfahren zur mikrobiellen Steroidumwandlung. Diese Arbeiten, die von einer ganzen Reihe von Arbeitskreisen in Amerika, Europa und auch in Japan mit bedeutender Intensität durchgeführt wurden, trugen bekanntlich vor allem zur Entfaltung der Corticoidproduktion bei, denn hinsichtlich der Ausbeute der Reaktionen waren die mikrobiellen Hydroxylierungen und Dehydrierungen in Stellung 1—2 des Steroidskeletts den rein synthetischen Verfahren überlegen.

Über 1000 verschiedene Mikrobenarten wurden auf ihre Fähigkeit zur Umwandlung der Steroidstruktur an einer ganzen Reihe von Steroiden getestet, über 600 Arbeiten wurden auf diesem Gebiet veröffentlicht und an die 800 Patente auf Verfahren von mikrobiellen Steroidumwandlungen erteilt. Heute gibt es bereits 45 monographieartige Übersichten über dieses Thema und drei Monographien in Buchform (Achrem, 1965, Čapek et al., 1966, Charney et al., 1967).

Die Umwandlungsreaktionen am Steroidskelett unter katalytischer Wirkung der mikrobiellen Enzyme haben zumeist Redoxcharakter. Neben Veresterungen, Verseifungen und Isomerisierungen wurden insgesamt 12 Oxydations- und 5 Reduktionsreaktionen in verschiedenen Stellungen am Steroidskelett festgestellt, von denen einige für das Enzymsystem des Mikroorganismus spezifisch sind. Einige Umwandlungen der mikrobiellen Enzyme sind sehr häufig, oft innerhalb einer ganzen Mikrobengattung, andere treten wiederum sehr selten auf und sind auf einige Arten oder Varianten beschränkt.

Eine sehr häufig vorkommende Reaktion ist z. B. die Oxydation der sekundären Alkoholgruppe in 3β-Stellung zu einer Ketogruppe. Eine verhältnismäßig seltene Reaktion ist die Hydrierung der Doppelbindung unter Bildung eines gesättigten Steroidderivates. Große Unterschiede beobachtet man hinsichtlich der Geschwindigkeit von Umwandlungen des Steroidsubstrates und bezüglich der Ausbeute des Endmetaboliten. Am häufigsten treten Gleichgewichtszustände zwischen dem Ausgangssteroid und dem Metaboliten ein. Oft werden bei der Einwirkung des enzymatischen Systems von Mikroben auf ein bestimmtes Steroid mehrere Umwandlungstypen kombiniert, so daß wir eine Reihe von Metaboliten erhalten. Bei der Oxydation von 3-Hydroxy-Δ^5-Steroiden erfolgt gleichzeitig Isomerisierung mit Verschiebung der Doppelbindung, so daß Δ^4-3-Keto-Steroide entstehen. Die Einführung einer Doppelbindung in Stellung 1—2 wird mit der Hydroxylierung in Stellung 9 kombiniert, worauf Spaltung des Steroidskelettes im Ring B folgt. Die Spaltung der Seitenkette der Pregnansteroide zwischen Kohlenstoffatom 17 und 20 wird oft von einer Spaltung des Ringes D begleitet, wobei Testololacton oder dessen Derivate entstehen.

Die Mikrobenenzyme, welche die Umwandlung der Steroide katalysieren, weisen oft eine bedeutende Spezifität in bezug auf die Struktur des Ausgangssteroidsubstrates auf (Sih, 1962, Talalay, 1956, Tamm, 1963, Tweit, 1962, Urech, 1960). Diese Tatsache wurde schon zur enzymatischen Analyse der Funktionsgruppen am Steroidkern verwertet. Bei der Umwandlung analoger Gruppen spielen oft die Substituenten in anderen Stellungen des Steroidskelettes, sowie auch die Länge der Seitenkette in Stellung 17 eine Rolle. Demgegenüber entstehen z. B. durch Oxydation der sekundären Alkoholgruppe in 3β-Stellung sowohl in der Androstan-, Pregnan- als auch in der Cholestanreihe Ketone.

Wie bei der oxydativen Spaltung des Progesterons zu Testosteron (HANČ, 1957) durch *Penicillium notatum* ergaben auch andere Pregnanderivate Verbindungen der Androstanreihe mit einer Hydroxygruppe in 17β-Stellung soweit die Ausgangssteroide nicht durch eine Hydroxygruppe in 17α-Stellung substituiert waren. Demgegenüber ist bei der Hydroxylierung des Cortexolons in 11β-Stellung durch *Absidia orchidis* die Hydroxygruppe in 17α-Stellung eine Voraussetzung (ČAPEK, 1961). Bei der Umwandlung von Cortexon entsteht nur 11α-Epicorticosteron und aus Progesteron das 11α-Hydroxyderivat.

Mit der Substratspezifität bei mikrobiellen Umwandlungen ist die Stellungsspezifität der erfolgten Umwandlungen eng verbunden. Einige Mykobakterien (ČAPEK, 1962) reduzieren beispielsweise die Ketogruppe in Stellung 20 des Pregnanskelettes, während die Ketogruppen in Stellung 3 und 11 unberührt bleiben. Eine Hydroxygruppe wird durch Mikroben vorzugsweise in eine oder mehrere Stellungen des Steroidskelettes eingeführt. Auch bei Verseifung der Steroidacetate zu Steroidalkoholen durch Einwirkung von Mikroben ist die Stellungsspezifität ersichtlich (TAMM, 1959). Durch Esterase des *Fusarium lini* wird die Acetoxygruppe der Herzglykoside in Stellung 3 und 16 hydrolysiert, durch das Enzym von *Giberella saubinetti* lediglich die in Stellung 3.

Außerdem ist die Stereospezifität bei mikrobiellen Steroidumwandlungen auffallend. Einige Streptomyceten und Mykobakterien reduzieren die 20-Ketogruppe der Pregnanderivate zu 20β-Hydroxysteroiden, wogegen die enzymatischen *Rhodotorula*-Systeme diese zu 20α-Hydroxyverbindungen reduzieren. In unseren Versuchen über die Oxydation von epimeren Androstendiolen (HANČ, 1958) wurde 3β-, 17α-Diol vermittels *Penicillium notatum* oder durch *Actinomyces globisporus* in Epitestosteron überführt, wogegen das 17β-Epimere durch den ersten angeführten Mikroben in Testosteron und durch den zweiten in Androstendion umgewandelt wurde. VISCHER und WETTSTEIN (VISCHER, 1956) beobachteten als erste, daß bei der mikrobiellen Umwandlung eines Steroides, sofern ein neues Asymmetriezentrum entsteht, gewöhnlich die Bildung eines einzigen der beiden möglichen Epimeren bevorzugt ist. Wirken die mikrobiellen Enzyme auf das razemische Steroidsubstrat ein, dann wird nur die d-Form (das natürliche Enantiomere) transformiert, wogegen die l-Form von den Enzymen unberührt bleibt.

Mit der Frage der Induktion von Enzymen, welche an mikrobiellen Umwandlungen der Steroide beteiligt sind, beschäftigte sich vor allem TALALAY mit seiner Schule (TALALAY, 1963). Als erster isolierte er Enzyme in reinem Zustand und beschrieb ihre Eigenschaften. Es handelte sich um 3α-Hydroxysteroid-Dehydrogenase, die die reversible Umwandlung von gesättigten 3α-Hydroxy- und 3-Ketosteroiden katalysiert. Δ^4-3-Ketosteroide wurden durch dieses Enzym nicht reduziert. Weiter isolierte er die Dehydrogenasen der 5α- und 5β-Steroide, welche die Einführung der Doppelbindung in den Ring A des Androstanskelettes katalysieren und endlich die Isomerase von Δ^5-3-Ketosteroiden, die die irreversible Umwandlung zu Δ^4-3-Ketosteroiden mit konjugierten Doppelbindungen in der Androstan- und der Pregnanreihe katalysiert. Bei reversiblen Oxydoreduktionen der Steroide verläuft die Wasserstoffübertragung über Diphosphopyridinnukleotid, ähnlich wie z. B. beim System der Alkohol-Dehydrogenase.

Eine zweite Gruppe von enzymologischen Arbeiten, welche sich mit der Erforschung des Dehydrierungsmechanismus in Stellung 1—2 und mit der Steroidspaltung befaßt, stammt aus den Arbeitsgruppen von Prof. SIH in den USA (SIH et al., 1962) und Prof. Kurt SCHUBERT in der DDR (SCHUBERT et al., 1961). Auch die Arbeiten über die 20β-Hydroxysteroid-Dehydrogenase bei *Streptomyces hydrogenans*, veröffentlicht von der Arbeitsgruppe Dr. HÜBENERS und Prof. Schmidt-Thomés in der DBR, müssen hier erwähnt werden (Hübener et al., 1958, Schmidt-Thomé et al., 1962).

Obwohl das Interesse bei der Mehrzahl der Arbeiten, die sich mit Mikroben und Steroiden befassen, auf die enzymatischen Leistungen verschiedener Mikroben an Steroidsubstraten gerichtet ist, fehlen auch Arbeiten hinsichtlich des umgekehrten Gesichtspunktes nicht, nämlich wie Steroide auf Mikroorganismen einwirken, insbesondere auf ihr Wachstum. Dieses Thema wurde monographisch in einer Übersichtsarbeit im Jahre 1964 zusammengefaßt (Buetow et al., 1964). Einige Steroide können bei bestimmten Mikroben als Wachstumsfaktoren oder Stimulatoren wirken, jedoch wurde öfter an einer Reihe von Steroiden beobachtet, daß sie in geeigneter Konzentration als Wachstumsinhibitoren der Mikroben wirken. Nicht nur die eigentlichen Steroidhormone wurden in diesem Sinne an verschiedenen Mikrobenarten getestet, sondern auch ihre Analoga und Derivate. Am häufigsten wurde das Wachstum ausgewählter Mikroben unter Zusatz von Steroiden in verschiedenen Konzentrationen verfolgt, ohne jedoch festzustellen, was mit dem zugesetzten Steroid inzwischen erfolgt. Auch fehlte bisher jedwede Erklärung über die Beeinflussung des Mikrobenwachstums durch Steroide, und man mußte sich mit Hypothesen zufrieden geben, wie z. B. die der Beeinflussung der Zellpermeabilität, oder der Steroidwechselwirkungen mit mikrobiellen Enzymen im Sinne einer Kompetition mit normalen Enzymsubstraten.

Wir verfolgten schon bei der Hydroxylierung von Progesteron in 11α-Stellung mittels *Rhizopus nigricans* nach D. H. Peterson (Peterson, 1962) die Bedeutung der Steroide für die Zuckerverwertung durch Mikroben, und wir haben festgestellt, daß Progesteron die Zuckerverwertung der Mikroben blockiert, wogegen der entstehende Metabolit 11α-Hydroxyprogesteron in gleicher Konzentration in dieser Beziehung bereits wirkungslos ist (Čapek et al., 1957). Weiterhin untersuchten wir, welchen Einfluß die beiden Steroide auf den Sauerstoffverbrauch der Mikroorganismen haben. Analoge Messungen wurden auch bei der Reduktion von Androstendion zu Testosteron durch Hefen und bei der Oxydation von Dehydroepiandrosteron zu Androstendion durch *Actinomyces globisporus* durchgeführt (Čapek, 1958). Auch hier hemmte das Ausgangssteroid den Sauerstoffverbrauch des Mikroben, wogegen der entstehende Metabolit in entsprechender Konzentration ohne Einfluß war.

In letzter Zeit befaßten wir uns von neuem mit dieser Thematik, um die vorausgesetzte Gesetzmäßigkeit zu überprüfen und zu vervollständigen. Schon zu Beginn war uns allerdings klar, daß dieses Problem nicht in seiner ganzen Reichweite erfaßt werden kann und daß die Messungen nur an einer begrenzten Auswahl von Mikroorganismen durchgeführt werden können, wobei uns bekannt war, welche metabolischen Vorgänge am Steroidmolekül diese Mikroorganismen bewirken können. Für diesen Zweck wurden 3 Bakterienarten und 4 Schimmelpilzarten gewählt:

a *Bacillus repens*
b *Mycobacterium flavum*
c *Flavobacterium peregrinum*
d *Penicillium notatum*
e *Aspergillus oryzae*
f *Rhizopus nigricans*
g *Absidia orchidis*

welche folgende Steroidumwandlungen durchführen:

a, b Dehydrierung im Ring A unter Bildung einer neuen Doppelbindung
c Verseifung der Steroidester und Oxydation der sekundären Alkohole zu Ketonen
d Spaltung der Pregnansteroide unter Bildung von Androstanderivaten
e Spaltung der Pregnansteroide unter Bildung von Testololacton
f Hydroxylierung der Pregnansteroide in 11α-Stellung
g Hydroxylierung von Pregnansteroiden in 11β-Stellung

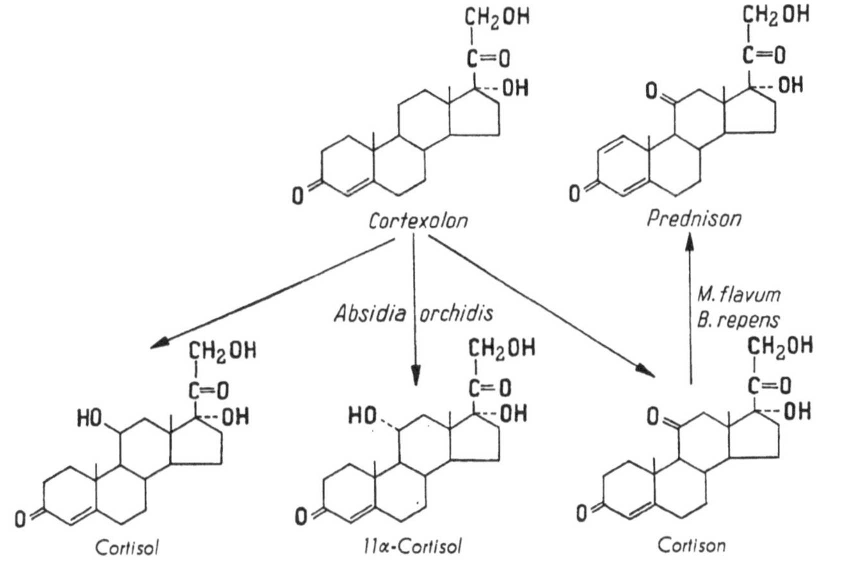

Fig. 1

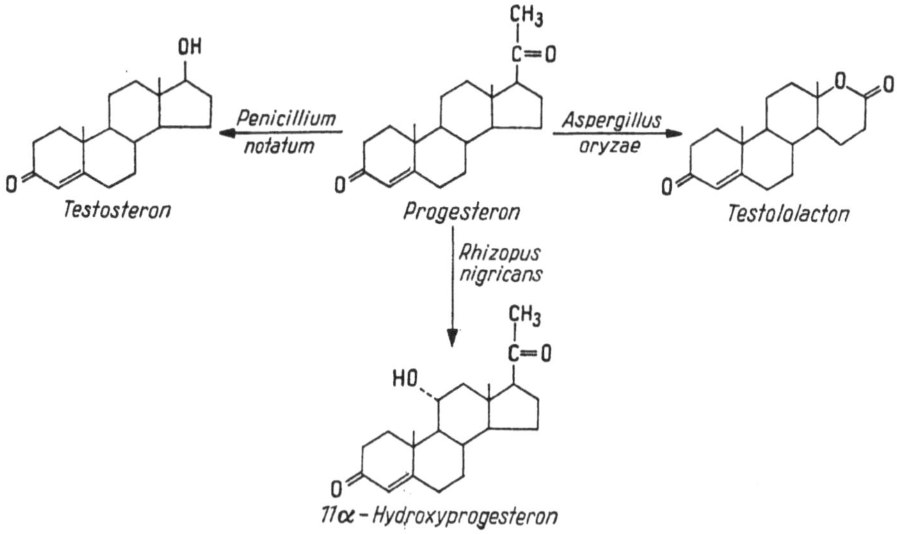

Fig. 2

Für die Messungen wurden insgesamt 12 verschiedene Steroide verwendet, von denen nur 3 Ausgangssteroide im wahren Sinne des Wortes waren, die weiteren 9 stellten schon bekannte Metaboliten dieser ersten 3 Steroide dar (des Progesterons, des Cortexolons und des Acetates von Dehydroepiandrosteron).

Die Umwandlungen der Steroide verliefen bei den Bakterienarten viel schneller als bei den Schimmelpilzen, deshalb wurden die Wachstums-Messungen bei den Bakterien während einer Zeitspanne von 12 resp. 24 Stunden turbidimetrisch verfolgt, dagegen bei den Schimmelpilzen innerhalb von 72 Stunden mit Hilfe der Methode der Trockengewichtsbestimmung des Myzeliums. Gleichzeitig wurde die Zuckerverwertung aus dem Medium unter Einwirkung der zugesetzten Steroide bei den Schimmelpilzen und der Sauerstoffverbrauch bei einer Bakterienart mit Hilfe der Warburgmethode verfolgt. Daneben wurden gleichzeitig die Steroidumwandlungen in den einzelnen Zeitabschnitten mit Hilfe der Papierchromatographie bestimmt.

Alle Versuche wurden dreimal wiederholt und aus den Ergebnissen der Durchschnitt genommen. Die Steroide wurden als 2%ige Äthanollösung dem Medium zugesetzt. Die Serie der Blindversuche ohne Steroide enthielt bei gleichen Bedingungen nur die entsprechende Äthanolmenge. Die Bakterienstämme wurden in Fleisch-Pepton-Bouillon vom pH 7,2, die Schimmelpilzstämme aus Sporendispersion, welche durch ein standardisiertes Verfahren zubereitet wurde, in Maltosemedium gezüchtet. Für die Versuche wurden 100 ml Ansätze in 500 ml Kolben auf der Schüttelmaschine mit 100 Schwingungen pro Minute benutzt.

Turbidimetrische Messungen wurden mit dem Spektrophotometer Prema SF 58 Typ 10 L5 durchgeführt und als Standard sterile Bouillon verwendet. Bei der Genauigkeitsmessung am Wachstum von 5 unabhängigen Kulturen betrug die maximale Abweichung nach 6 Stunden 3 von 100 Teilstrichen der Skala. Die gemessenen Unterschiede wurden daher nur als signifikant betrachtet, wenn sie wenigstens 5 Teilstriche betrugen.

Aus den Ergebnissen stellten wir eine Tabelle zusammen, in welcher die Wachstumsinhibition des Mikroorganismus und die Hemmung der Zuckerverwertung in Prozenten den Blindversuchen ohne Steroide in einzelnen Zeitabständen gegenübergestellt sind.

Tabelle 1

Der Einfluß von Progesteron auf Wachstum und Metabolismus von Schimmelpilzen

Mikroorganismus	Steroid-Umwandlung	Hemmung in %					
		des Myzelwachstums			des Zuckerverbrauchs		
		24	48	72	24	48	72 Std.
Penicillium notatum	Spaltung	46,2	7	2,5	66,7	11,8	1,8
Aspergillus oryzae	Spaltung	59,5	46,5	0	61,8	30,7	3,9
Rhizopus nigricans	Hydroxylierung	36	0	0	55,7	1	0,4

Einfluß von Endmetaboliten

Mikroorganismus	Metabolit	Hemmung in %					
P. notatum	Testosteron	0	0	0	11,5	3,3	3,2
Asp. oryzae	Testololacton	0	0	0	11,4	3,9	1,7
Rh. nigricans	11α-Hydroxy-progesteron	0	0	0	4,9	0,5	0,4

(Konzentration 20 mg/100 ml)

Bei der Untersuchung des Einflusses von Progesteron und Testosteron auf den Schimmelpilz *Penicillium notatum* war die größte Wachstumshemmung nach 24 Stunden bei den Ansätzen mit Progesteron zu beobachten. Die Prozente der Hemmung nehmen mit der Zeit schnell ab ebenso wie die der Hemmung der Zuckerverwertung. Analoge Befunde wurden auch bei *Aspergillus oryzae* festgestellt. Hier jedoch beträgt die Wachstumsinhibition noch nach 48 Stunden Inkubation 46% und die Hemmung der Zuckerverwertung 30%. Bei *Rhizopus nigricans* bewirkt Progesteron nach 24 Stunden eine 36%ige Inhibition des Schimmelpilzwachstums, jedoch nach 48 Stunden ist die Hemmung nicht mehr ersichtlich. Parallel verläuft die Hemmung der Zuckerverwertung. Das Steroid, welches durch das enzymatische System des Mikroorganismus metabolisiert wird, hemmt anfangs stark das Wachstum und die Zuckerverwertung durch den Mikroorganismus. Nach 72 Stunden ist jedoch die Inhibition gleich Null. Wenn wir den Verlauf der Progesteronumwandlung in den einzelnen Intervallen mit Hilfe der Chromatographie vergleichen, zeigt sich, daß dieser ziemlich parallel zu den erhaltenen Hemmungswerten verläuft. Das Endprodukt des Steroidmetabolismus, in allen drei Fällen ein anderes, inhibiert das Mikrobenwachstum schon nicht mehr, obwohl sich noch eine gewisse, wenn auch geringere Hemmung der Zuckerverwertung bemerkbar macht. Zwischenprodukte bei der Umwandlung von Progesteron zu Testololacton mittels *Aspergillus oryzae* weisen eine Hemmung mit fallender Tendenz auf, soweit diese selbst durch enzymatische Katalyse in den Endmetaboliten überführt werden.

Tabelle 2

Der Einfluß von Zwischenprodukten bei der Umwandlung von Progesteron in Testololacton auf *Asp. oryzae*

Zwischenprodukt	weitere Umwandlung	Hemmung in % des Myzelwachstums			Hemmung in % des Zuckerverbrauchs		
		24	48	72	24	48	72 Std.
Androstendion	Spaltung des Rings D	33	23,2	0	30,5	23,4	2,6
Testosteron		26	21	0	26,7	14,6	3,5

(Konzentration 20 mg/100 ml)

Ein ähnliches Ergebnis ist auch aus Messungen des Einflusses von Cortexolon und seiner Metaboliten auf Wachstum und Zuckerverwertung bei *Absidia orchidis* ersichtlich. In einer Konzentration von 20 mg auf 100 ml Medium weist Cortexolon die stärkste Wachstumshemmung nach 24 Stunden auf, nach 72stündiger Inkubationszeit jedoch ist das Wachstum schon wieder übereinstimmend mit den Kontrollansätzen. Auch hier verläuft die Zuckerverwertung parallel. Die Metaboliten des Cortexolons haben keinen Einfluß mehr. Je schneller das Ausgangsteroid umgewandelt wird, um so schneller nimmt die Hemmung ab.

Bei den geprüften Bakterienstämmen wurde die größte Wachstumshemmung nach 6stündiger Inkubation mit dem Steroid, welches der Umwandlung durch das Enzymsystem des gegebenen Mikroben unterliegt, festgestellt.

Tabelle 3

Hemmung des Wachstums von Bakterienstämmen durch das Ausgangssteroid und dessen Metaboliten

Mikroorganismus	Steroid	Wachstumshemmung in % nach Stunden 3	6	9	12	24
B. repens	Cortison	46	55	10	0	0
	Prednison	0	16	0	0	0
M. flavum	Cortison	41	40	27	17,5	0
	Prednison	0	0	– 15	– 13	0
Fl. peregrinum	Acetat von Dehydroepiandrosteron	0	62,5	47	0	0
	Androstendion	0	– 15,6	0	0	0

(20 mg Steroid/100 ml Medium)

In den weiteren Stunden sinkt das Inhibitionsvermögen schnell und nach 12 bis 24 Stunden ist das Wachstum der Kulturen, verglichen mit den Kontrollansätzen, wieder ausgeglichen. Der entstehende Metabolit hemmt das Wachstum des Mikroorganismus in gleicher Konzentration nicht, er wirkt im Gegenteil leicht stimulierend. Diese Wirkung macht sich wieder zwischen der 6ten und 12ten Stunde Inkubationsdauer bemerkbar, dann nimmt sie ab. Ein steter Einfluß auf das Wachstum wurde also bei keinem der geprüften Steroide bemerkt. In unserer Zusammenstellung wurden größtenteils natürliche Steroide geprüft, mit Ausnahme von Prednison und dem Acetat des Dehydroepiandrosterons. Es besteht allerdings die Möglichkeit, daß synthetische Steroide, wie Methyl-, Fluor-, Amino- oder Aza-derivate, auch metabolische Gifte sein können, welche eine dauernde Inhibition des Mikrobenmetabolismus hervorrufen könnten, wie dies schon in einigen Fällen beschrieben wurde (CASAS-CAMPILLO et

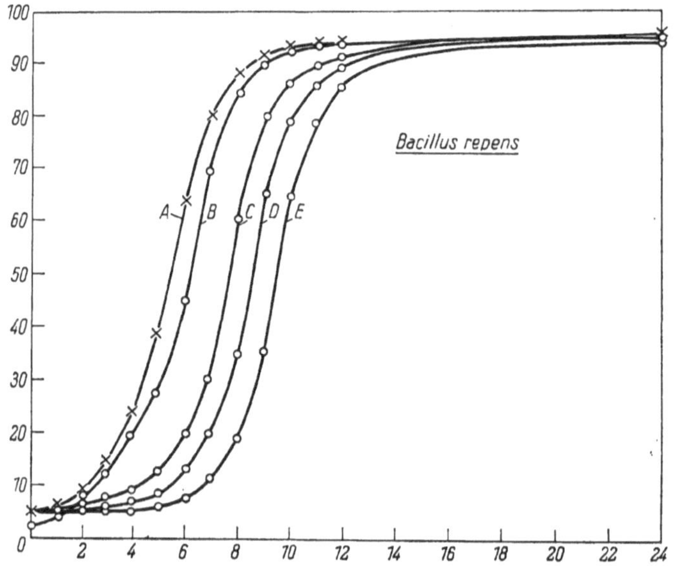

Fig. 3. Wachstumskurven von *Bacillus repens* in Bouillon-Medium pH 7,2 bei 27 °C.
A — ohne Zusätze, B — mit Aethanol, C — 20 mg Cortison, D — 40 mg Cortison, E — 80 mg Cortison.
X-Achse — Zeit in Stunden, Y-Achse — Photometer-Skalenteile als Maß der Trübung

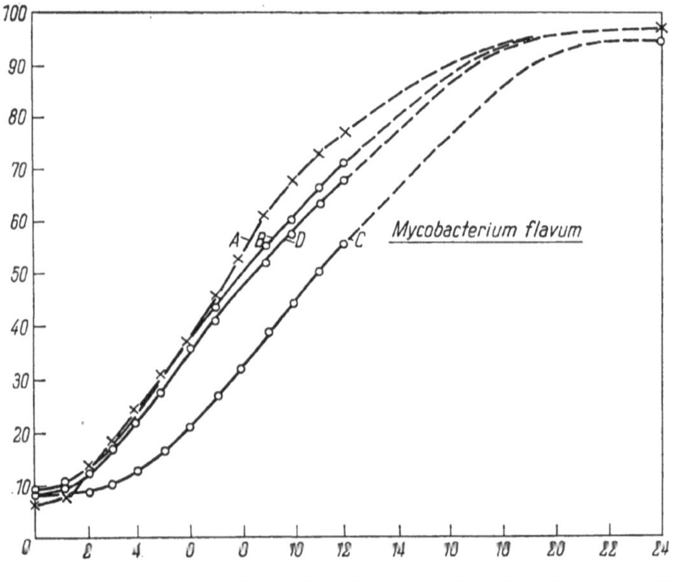

Fig. 4. Wachstumskurven von *Mycobacterium flavum* in Bouillon-Medium pH 7,2 bei 27 °C.
A — ohne Zusätze, B — mit Aethanol, C — 20 mg Cortison, D — 20 mg Prednison.
X-Achse — Zeit in Stunden, Y-Achse — Photometer-Skalenteile als Maß der Trübung

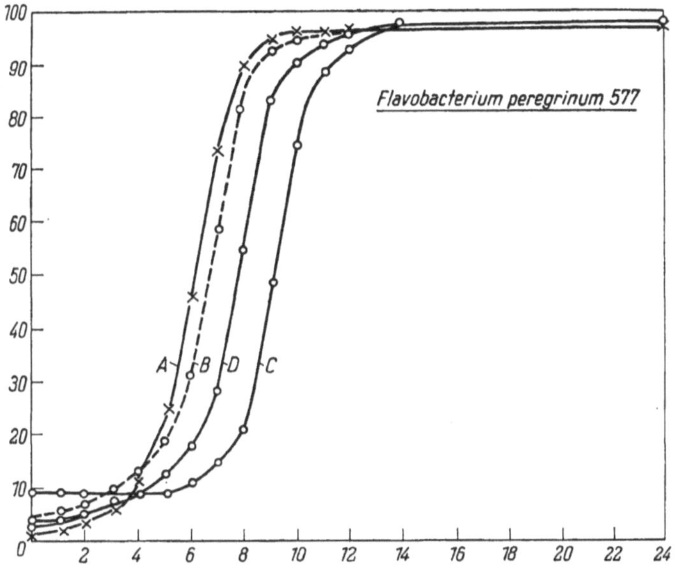

Fig. 5. Wachstumskurven von *Flavobacterium peregrinum* in Bouillon-Medium pH 7,2 bei 27 °C.
A — ohne Zusätze, B — mit Aethanol, C — 20 mg Dehydroepiandrosteron-Acetat,
D — 20 mg Dehydroepiandrosteron.
X-Achse — Zeit in Stunden, Y-Achse — Photometer-Skalenteile als Maß der Trübung

al., 1961, VARICCHIO et al., 1967). Unsere Ergebnisse gelten daher nur für Steroidhormone und deren nahe Metaboliten. Aus den ermittelten Ergebnissen können wir die Folgerung ziehen, daß das Steroid, welches durch das Enzymsystem von Mikroben umgewandelt wird, eine vorübergehende Hemmung des Wachstums und des metabolischen Prozesses der Zellen dieses Mikroorganismus hervorruft, während seine Steroidmetaboliten diese Eigenschaft nicht mehr aufweisen. Eine vorübergehende Inhibition des Mikrobenwachstums, bewirkt durch ein der Umwandlung unterliegendes Steroid, steht im Einklang mit dem Vorgang der Umwandlung dieses Steroids in einen oder mehrere Metabolite. In allen Fällen macht sich dies durch eine Verlängerung der Lag-Phase bemerkbar und durch eine Verzögerung des Einsetzens der Wachstums-Exponentialphase. In jeder anderen Hinsicht behalten die Wachstumskurven ihren Verlauf bei.

Unabhängig wurde in letzter Zeit ein analoges Ergebnis bei Staphylokokken festgestellt (YOTIS et al., 1966), wobei verschiedene natürliche Steroide das Wachstum und die Utilisation von Brenztraubensäure hemmten, und diese Inhibition durchweg während der Inkubationszeit abnahm. YOTIS und STANKE nahmen auf Grund dieser Ergebnisse an, daß die Steroide hierbei einer Umwandlung unterliegen, obgleich sie sich nicht eingehender damit befaßten.

Bei Versuchen zur Ermittlung des Einflusses von Cortison in verschiedenen Konzentrationen auf den Sauerstoffverbrauch bei *Bacillus repens*, der Cortison in Prednison überführt, wurde nach Intervallen von 30 Minuten beinahe konstante Hemmung festgestellt, und zwar bei der Cortisonkonzentration von 50 µg/ml 15—17%, bei 100 µg/ml 44—46%. Prednison bewirkt in derselben Konzentration nur 4—11% Hemmung mit dem Maximum nach 60 Minuten Inkubation. Dieses Ergebnis steht mit der bekannten Tatsache in Übereinstimmung, daß Prednison bei dieser Umwandlung von Cortison nicht den Endmetaboliten darstellt, sondern einmal weiter in 20-Stellung reduziert und zum anderen oxydativ gespalten wird.

Wir untersuchten weiter, wie die Löslichkeit der Steroide im Medium bzw. in Wasser, mit ihrer mikrobiellen Umwandlung zusammenhängt. Die Steroidumwandlungen erfolgten gewöhnlich in Dispersionen von 0,2—0,8‰ Steroid. Das Verhältnis des Steroids zur Trockensubstanz der Mikrobenmasse betrugt ungefähr 1:10. In vereinzelten Fällen kam jedoch eine bedeutend höhere Steroidkonzentration zur Anwendung. Kondo und Masuo (Kondo et al., 1961) beschreiben z. B. eine Dehydrierung von Cortisol zu Prednisolon auch bei einer Konzentration von einigen Prozenten des Ausgangssteroids, also weit über die Löslichkeit der zu metabolisierenden Substanz hinaus, und dennoch wurden hohe Ausbeuten an Umwandlungsprodukt erhalten. Die japanischen Autoren beschrieben dieses Phänomen als ,,Pseudokristallfermentation" des Steroids. Das Ausgangssteroid geht offenbar allmählich in Lösung, so wie es mit der entsprechenden Dehydrogenase des Mikroben reagiert, und das gebildete Prednisolon kristallisiert nach Diffusion aus den Zellen wieder aus, wenn es die entsprechende, durch seine Löslichkeit im Medium gegebene Konzentration erreicht hat. Gewöhnlich jedoch hemmt die hohe Konzentration des Steroids in der Dispersion im Medium die sonst glatt verlaufende Steroidumwandlung und die Ausbeuten an den gewünschten Metaboliten sind bedeutend niedriger als bei den normalerweise angewandten Steroidkonzentrationen.

Die Wasserlöslichkeit der Steroide ist zumeist sehr niedrig, es wurden für einige Substanzen die Werte 0,4—285 µg/ml im Temperaturbereich 25—37 °C angegeben (Kabasakalian et al., 1966). Die Wasserlöslichkeit der Steroidester, z. B. der Acetate, ist gegenüber den Steroidalkoholen ca. 10 × geringer. Für den Solvatationseffekt ist offenbar die freie Hydroxylgruppe in Stellung 17β, 21 oder 3 des Steroidskeletts sehr wichtig, wie auch die Sauerstofffunktion in 11-Stellung des Rings C. Der Hydratationsprozeß geht von der Möglichkeit der Bildung von Wasserstoffbrücken bei sauerstoffhaltigen Substituenten des Steroidmoleküls aus.

Die Löslichkeit einiger in unserer Arbeit verwendeten Steroide bestimmten wir in Wasser und in 2%igem Alkohol nach der üblichen bekannten Technik der Darstellung von wäßrigen Steroidsuspensionen im Überschuß und durch Bestimmung des gelösten Anteils nach 24-stündiger Gleichgewichtseinstellung bei 27 °C spektrophotometrisch bei 246 mµ gegenüber Standardproben. Dehydroepiandrosteron und sein Acetat wurden kolorimetrisch nach Reaktion mit Schwefelsäure bei 405 mµ Wellenlänge gegenüber der Reagenzlösung als Vergleichslösung gemessen.

Tabelle 4

Löslichkeit der Steroide in destilliertem Wasser und in 2%igem Aethanol bei 27 °C

Steroid	in dest. Wasser µg/ml	in 2%igem Aethanol µg/ml
Dehydroepiandrosteron	28	35
— Acetat	3	3,5
Androstendion	50	86
Testosteron	30	38
Testololacton	20	24
Progesteron	15	20
11α-Hydroxyprogesteron	78	113
Cortexolon	240	280
Cortisol	350	450
Epicortisol	360	470
Cortison	280	365
Prednison	150	188

Bei den meisten untersuchten mikrobiellen Steroidumwandlungen entstehen aus dem ursprünglichen Ausgangssteroid wasserlöslichere Metaboliten. Die Wasserlöslichkeit allein ist jedoch nicht für die eigentliche Steroidumwandlung entscheidend, denn auch Steroide mit sehr geringer Wasserlöslichkeit werden metabolisch umgewandelt.

Die Beeinflussung der Oberflächenspannung des Mediums durch Steroide ist nach den von uns gemessenen Werten nur sehr geringfügig. Sie kommt auch durch keine deduzierbare Gesetzmäßigkeit im Hinblick auf die Steroidumwandlung durch Mikroorganismen oder die Wirkung der Steroide auf das Mikrobenwachstum zum Ausdruck.

Um zur schematischen Einteilung der Beziehung der Mikroorganismen zu organischen Substanzen, die normalerweise keine physiologischen Substrate für deren Enzyme darstellen, zurückzukehren, kann festgestellt werden, daß bei den Steroiden sämtliche drei zu Beginn dieser Mitteilung angeführten Möglichkeiten eintreten können. Ein Teil der Steroide wird von manchen Mikrobenarten auch nach langer Inkubation nicht umgewandelt, andere Steroide werden zu Derivaten umgewandelt, und schließlich vermögen manche Mikrobenarten die Steroide als Kohlenstoffquelle zu verwerten und weitgehend zu spalten, oft bis zu CO_2 und H_2O. Die bei derartigen Steroidumwandlungen beteiligten mikrobiellen Enzyme haben oft induktiven Charakter.

Vom Gesichtspunkt der Mikrobenzelle aus kann man die Steroidumwandlung gewissermaßen als „Detoxikations"prozeß betrachten, bei dem das Ausgangssteroid, welches Wachstum, Atmung und Zuckerverwertung durch den Mikroorganismus hemmt, von einem induzierbaren Enzymsystem in einen Metaboliten überführt wird, der diese Eigenschaften nicht mehr besitzt und wasserlöslicher als die Ausgangssubstanz ist. In dieser Weise verhält sich die Mikrobenzelle auch gegenüber anderen organischen Substanzen. Eingangs erwähnten wir bereits die Versuche Neubergs, bei denen Benzaldehyd durch Hefe in Benzylalkohol und Phenylacetylcarbinol überführt wurde. Auch hier hemmt unseren Messungen zufolge Benzaldehyd die Atmung und Zuckerverwertung durch Hefe, während seine Metaboliten diese Eigenschaften nicht besitzen. Wie man vor kurzem feststellte, wurden auch z. B. bestimmte zur Holzkonservierung benützte organische Arsenderivate durch Holzpilze in andere Arsenderivate überführt, die das Wachstum dieser Pilze nicht mehr hemmten (Duncan et al., 1964). Diese Fragen hängen eng mit der „Resistenz" von Mikroben und auch lebenden Zellen z. B. gegenüber Heilmitteln zusammen.

Die Steroide stellen also hinsichtlich ihrer Beziehung zur Mikrobenzelle keine Ausnahme dar. Bei der Untersuchung des Einflusses geringer Konzentrationen von Steroidhormonen auf den Metabolismus der Mitochondrien von Zellen des tierischen Organismus (Vallejos et al., 1967) wurde Hemmung der Respiration und der oxidativen Phosphorylierung festgestellt. Progesteron bewirkt eine stärkere Hemmung der Energieübertragung als seine Derivate, die bei seinem Metabolismus in den Zellen von Lebewesen entstehen. Es ist möglich, die bei Mikrobenzellen ermittelten Ergebnisse bis zu einem gewissen Grad mit der metabolischen Wirkung der Steroidhormone in den Zellen von Lebewesen zu vergleichen.

Für die eigentliche physiologische Funktion im lebenden Organismus, d. h. für die „Hormonwirkung", die sich zumeist nur morphologisch am Makroorganismus beobachten läßt, bringen allerdings derartige biochemische Befunde bisher keine direkte Erklärung. Es wurde z. B. auch der Einfluß der verschiedenen Steroidhormone auf die Dehydrogenasensysteme der Zellen verschiedener Gewebe von Lebewesen untersucht und ihr bedeutender Anteil bei Transhydrierungen ermittelt (Talalay, 1960). Wie diese wenig spezifische Aktivität mit der eigentlichen Hormonwirkung der Steroide zusammenhängt, ist vorläufig nicht ganz klar.

Auch über das Verhältnis der biologischen Aktivität der eigentlichen Steroidhormone und ihrer Metaboliten fehlt uns bisher eine klare Vorstellung. Durch verschiedene biologische Teste wurde z. B. „androgene" Aktivität der verschiedensten Androstanderivate ermittelt. Es wurde angegeben, daß Androsteron, Dehydroepiandrosteron, Androstendion und andere physiologische Metaboliten oder Vorstufen von Testosteron so und so viel Prozente seiner Wirkung haben, ohne daß wir bis heute beweisen können, ob diese Substanzen an sich überhaupt „androgene" Aktivität auf die einzelnen Akzeptororgane oder -zellen haben oder ob sie durch den Organismus — mit verschiedener Geschwindigkeit — in Testosteron als einzigem tatsächlich physiologischem „Androgen" überführt werden. Eine definitive Antwort auf diese Fragen kann die sich entwickelnde Molekularbiologie geben, die Kenntnis der enzymatischen Schlüsselreaktionen, die für diesen oder jenen biologischen Effekt verantwortlich sind, der sich nur am Komplex der Zellen und Gewebe des lebenden Organismus messen läßt. Zur Überbrückung unserer derzeitigen Unkenntnis können ohne Zweifel die biochemischen Befunde über Metabolismus und Einwirkung der Steroide bei Mikroben eine beachtliche Bedeutung als Modellversuche zu den „Schlüsselreaktionen" auch der tierischen Zellen haben und so nicht nur der vergleichenden Biochemie, sondern vielleicht auch der Wachstumsphysiologie dienen.

Literatur

[1] Achrem, A. A., J. A. Titow, Mikrobiologische Umwandlungen von Steroiden, (in russ. Sprache), 504 S., Moskau (1965).
[2] Buetow, B. E., B. H. Levedahl, Ann. Rev. Microbiol. **18**, 167 (1964).
[3] Čapek, A., O. Hanč, E. Malikova, Čs. Mikrobiol. **2**, 282 (1957).
[4] Čapek, A., H. Pavlů, O. Hanč, Naturwiss. **45**, 89 (1958); Folia biol. **4**, 337 (1958).
[5] Čapek, A., O. Hanč, Folia microbiol. **6**, 237 (1961).
[6] Čapek, A., O. Hanč, Folia microbiol. **7**, 181 (1962).
[7] Čapek, A., O. Hanč, M. Tadra, Microbial transformations of steroids, 239 S., Prag (1966).
[8] Casas-Camillo, C. et al., J. Bacteriol. **81**, 366 (1961).
[9] Charney, W., H. Herzog, Microbiological transformation of steroids. A Handbook, 670 S., New York (1967).
[10] Duncan, C. G., F. J. Deverall, Appl. Microbiol. **12**, 57 (1964).
[11] Hanč, O. et al., Arzneimittelforsch. **7**, 175 (1957).
[12] Hanč, O. et al., Coll. czech. chem. Comm. **23**, 982 (1958).
[13] Hübener, J. et al., Z. physiol. Chem. **313**, 124 (1958); Biochim. Biophys. Acta **35**, 270 (1959); Biochem. Z. **333**, 95 (1960).
[14] Kabasakalian, P. et al., J. Pharm. Sci. **55**, 642 (1966).
[15] Kondo, E., E. Masuo, J. gen. appl. Microbiol. (Japan) **7**, 113 (1961).
[16] Mamoli, L., A. Vercellone, Ber. **70**, 470 (1937).
[17] Neuberg, C., E. Welde, Biochem. Z. **62**, 477 (1914).
[18] Neuberg, C., J. Hirsch, Biochem. Z. **115**, 282 (1921); **121**, 311 (1921); **127**, 327 (1922); **128**, 610 (1922).
[19] Peterson, D. H., H. C. Murray, J. am. chem. Soc. **74**, 1871, 5933 (1952).
[20] Schmidt-Thomé, J. et al., Biochem. Z. **336**, 322 (1962).
[21] Schubert, K. et al., Z. Naturforsch. **15b**, 584 (1960); **16b**, 595 (1961); **18b**, 988 (1963); Ber. dtsch. Akad. Wissenschaft. **2**, 308 (1960); Z. physiol. Chem. **325**, 260 (1961).
[22] Sih, Ch. J. et al., Biochim. biophys. Acta **56**, 584 (1962); **62**, 541 (1962); **64**, 409 (1962).
[23] Talalay, P., P. I. Marcus, J. biol. Chem. **218**, 675 (1956).
[24] Talalay, P., H. G. Williams-Ashman, Rec. Progr. Hormone Res. **16**, 1 (1960).
[25] Talalay, P., in Boyer, P. D., The enzymes **7**, 177 (1963).
[26] Tamm, Ch., A. Gubler, Helv. chim. Acta **42**, 239 (1959).
[27] Tamm, Ch. et al., Helv. chim. Acta **46**, 889 (1963).
[28] Tweit, R. C. et al., J. org. Chem. **27**, 3654 (1962).
[29] Urech, J. et al., Helv. chim. Acta **43**, 1077 (1960).
[30] Vallejos, R. H., A. O. M. Stoppani, Biochim. biophys. Acta **131**, 295 (1967).
[31] Varicchio, F. et al., J. Bacteriol. **93**, 627 (1967).
[32] Vischer, E. et al., Experientia **12**, 50 (1956).
[33] Yotis, W., R. Stanke, J. Bacteriol. **92**, 1285 (1966).

Microbiological Conversion of Cholesterol into Androsta-1,4-diene-3,17-dione

W. F. van der Waard, J. Doodewaard, J. de Flines and S. van der Weele
Royal Netherlands Fermentation Industries Ltd., Delft, Holland

Summary

Strains of *Mycobacterium* capable of complete breakdown of cholesterol were isolated from soil samples. The degradation of the ring system was found to be inhibited by certain inorganic ions such as Ni^{++}, Co^{++}, Pb^{++} and SeO_3^{--}, androsta-1,4-diene-3,17-dione (ADD) being the main product. Based on these facts a fermentation process for the production of ADD in shaken flasks was developed.

The possible role of iron in the breakdown process of cholesterol is shortly discussed.

Degradation of cholesterol is not a new object of investigation for the microbiologist. One can go back as far as 1913 to find that Soehngen had already microorganisms growing in a medium containing this steroid as the sole source of carbon. In the course of time other investigators have studied the breakdown of cholesterol, e.g. Tak, Turfitt, Horvath and Kramli, Arnaudi, Whitmarsh, Schubert, Sih. But direction and interest have changed with time. Soehngen contented himself with observing the microbes multiply in the medium with cholesterol as the sole source of carbon and from this fact concluded that the steroid was consumed by the bacteria.

Tak (1942) tried to prove that cholesterol really disappeared and observed a 90% breakdown of the added steroid within 28 days of incubation. Later on microbiologists became interested in the way in which cholesterol was broken down in those processes. Formation of cholest-4-ene-3-one was readily proved by Turfitt (1946). Demonstration of products which were further degraded was difficult and in most cases only very small amounts of such intermediates could be detected.

Until the fifties the interest of the microbiologist in cholesterol was more or less of academic nature. Chemists had already discovered that choleserol could serve as a raw material for the preparation of steroid hormones. Oxidation of the side chain by means of chromic acid with formation of androstane derivatives found application in steroid industry, although yields were rather low, due to the formation of many byproducts. However, the discovery of the steroid-hydroxylating and -dehydrogenating properties of many microorganisms, followed by the application of steroid conversion with the aid of moulds and bacteria on a large industrial scale, renewed the interest of microbiologists in cholesterol degradation. At this time, when the raw materials for the preparation of steroid hormones were rather scarce, the cholesterol molecule formed an irresistable challenge. Also in our laboratarory we have tried to alter this molecule in order to obtain some useful product. At that time we ran a screening program in which several steroids were added to cultures of all kinds of microorganisms. Addition of cholesterol, however, did not yield any interesting product. Sometimes we observed a complete breakdown of the substrate. From this fact we concluded that it would be essential to block the breakdown of the ring system in some way in order to limit the degradation to the removal of the aliphatic side chain of the molecule.

SIH et al. (1963) solved this problem by making use of the fact that some microorganisms very slowly degrade steroids with an aromatic A-ring in comparison with normal steroids. Using 19-hydroxycholestenone as a substrate they obtained oestron in good yield.

In our work we for some time used 5α-chlorcholestanol, prepared by simple addition of HCl, as a substrate. Again we got no interesting products, propably due to the backformation of cholesterol in the cultures. In this way the ring system was insufficiently protected. It was then decided to try to limit degradation by addition of enzyme inhibitors. The screening program we set up contained three phases:

1. Isolation of microorganisms able to break down cholesterol rapidly and completely. Literature data on the quantities of cholesterol which could be utilized by microorganisms, were not interesting from the technical point of view.

2. Optimalization of this breakdown process. Our minimum goal was the degradation of 1 g cholesterol/l in 24 hours using a low cost medium.

3. Addition of enzyme inhibitors in different concentrations and following the degradation process by means of thin layer chromatography.

For accumulation of the microorganisms we used a mineral medium with cholesterol as the sole source of carbon. Both shaken flasks and Petri dishes with an agar medium were used. No difficulty was encountered in isolating a great variety of microorganisms able to grow on cholesterol in the primary culture. However, once they were isolated on a medium such as peptoneagar, we did not succeed in making them grow again in a medium with cholesterol and mineral salts, not even when the medium was fortified with soil extract or with small amounts of yeast extract.

This problem was solved by growing the organisms in a complete medium like peptone or yeast extract and adding the cholesterol afterwards. It was therefore possible to select certain bacteria which after growth in an inexpensive medium like 0.5% corn steep liquor, were able to degrade at least 1 g cholesterol/l in 24 hours after addition of the steroid. Using these conditions we started screening enzyme inhibitors. It was found that after adding salts of certain heavy metals with the steroid a new compound accumulated in the medium. This compound could readily be isolated and identified as androsta-1,4-diene-3,17-dione (ADD).

In preliminary experiments best results were obtained with addition of nickelsulphate. However, also cobaltchloride, cadmiumsulphate and to a lesser extent leadacetate and sodiumselenite caused similar accumulation of the new product.

Microorganisms capable of forming ADD under these conditions were shown to belong to the genus *Mycobacterium*. Some other Mycobacteria from our collection showed the same effect; provided they possessed the property to utilize cholesterol to some extent. Among them were *Mycobacterium butyricum ATCC 11.314* and also the organism used by TAK in 1942, named *Mycobacterium cholesterolicum.*

For the development of a suitable fermentation process we selected nickelsulphate as enzyme inhibitor and one of our own isolates, identified as *Mycobacterium sp. strain KNGSF 70*. This strain is deposited at the Centraal Bureau voor Schimmelcultures at Baarn (Holland) and listed under number 495.66.

Further investigations showed that similar results could be obtained using substrates as beta-sitosterol, campesterol and stigmasterol, although the conversion rate was low compared with that of cholesterol. Also cholest-4-ene-3-one may serve as a substrate. Occasionally we observed this compound in the culture shortly after addition of cholesterol.

However, the rate in which cholesterol is transformed into cholestenone is not limiting for the whole process as was shown by adding cholestenone instead of cholesterol as the substrate. In this case there was no increase in the rate of formation of ADD.

If the concentration of the inhibitor and other conditions are carefully selected, the end product of this process will be ADD. In addition small quantities of androst-4-ene-3,17-dione (AD) can be detected. During the fermentation the latter compound is further dehydrogenated to ADD. Traces of 17α-hydroxy-androsta-1,4-diene-3-one and testosterone are sometimes found, mainly in the mother liquor after crystallization of ADD. How those products were formed has not been investigated.

The strain is maintained on broth agar and grows very well in 0,5% corn steep liquor medium, prepared with tap water. As an inoculum a culture in 0,5% peptone or yeast extract is used which gave excellent results. Addition of 0,25% Tween 80 to all media improves growth and homogeneity of the cultures as is known to be the case with many strains of *Mycobacterium*.

Small amounts of phosphate e.g. 0,125% disodiumphosphate improved the reproducibility of the fermentation.

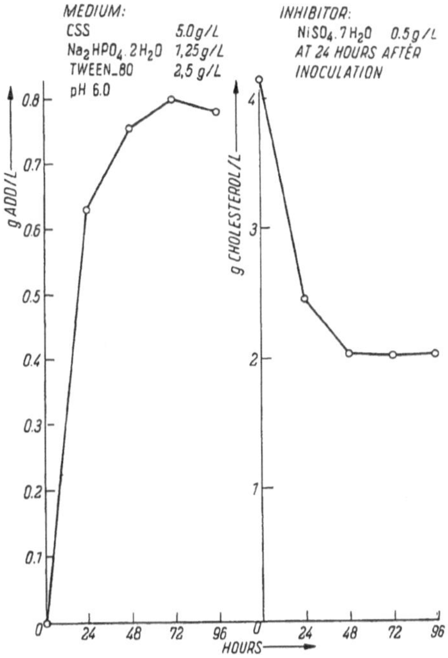

Fig. 1

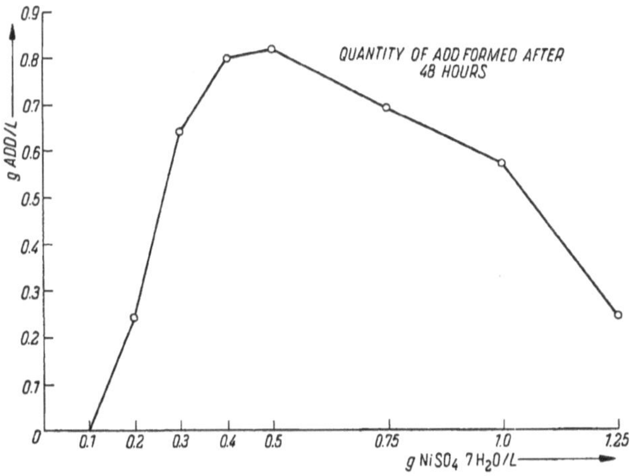

Fig. 2

Fig. 1 shows the formation of ADD under the above described conditions. The concentration of the inhibitor was 0.500 g $NiSO_4 \cdot 7\ H_2O$/l (104 mg Ni/l). This is the optimal concentration in shaken flasks but there is a rather broad optimum as is shown in fig. 2. Too little inhibitor does not prevent the breakdown of the ring system, cholesterol is quickly utilized and yields of ADD are poor. Too much nickel gives opposite results: slow formation of ADD, but high conversion yields. Under the conditions, optimal for the formation of ADD, the conversion yield is generally 65%. This figure is bound to fluctuate due to changes in the fermentation conditions beyond control so far.

Experiments were run at 30 °C; variation of temperature did not greatly affect the results.

As was shown in fig. 1, it is possible to obtain about 800 mg of ADD/l in shaken flasks in 72 hours.

Whitmarsh (1964) isolated small amounts of ADD and AD from a culture of *Nocardia* when 8-hydroxychinoline was added with the cholesterol.

Investigators of the Noda Research Institute at Japan have developed such a process starting with cholestenone and producing ADD using sequestering agents such as 8-hydroxy chinoline and α,α'-dipyridyl as enzyme inhibitors. Our strain, *Mycobacterium 70* formed only traces of ADD upon addition of 8-hydroxychinoline but using ethylenediaminedi-(-ortho-hydroxyphenyl acetic acid), known as CHEL 138 we obtained ADD in fair amounts.

This indicates that sometimes sequestering agents may cause accumulation of ADD and that on the other hand the same result may be obtained by addition of nickel ions. Is there a connection between these two phenomena? It is generally assumed that the microbial breakdown of the ring system is initiated by hydroxylation at carbon 9 (Sih et al., 1962). We found that iron strongly counteracts the effect of nickelsulphate. Only a few mg of ferrosulphate completely undo the effect of 500 mg nickelsulphate/l. This negative influence of iron can be checked to some extent by increasing the nickel concentration in the culture as is shown in fig. 3.

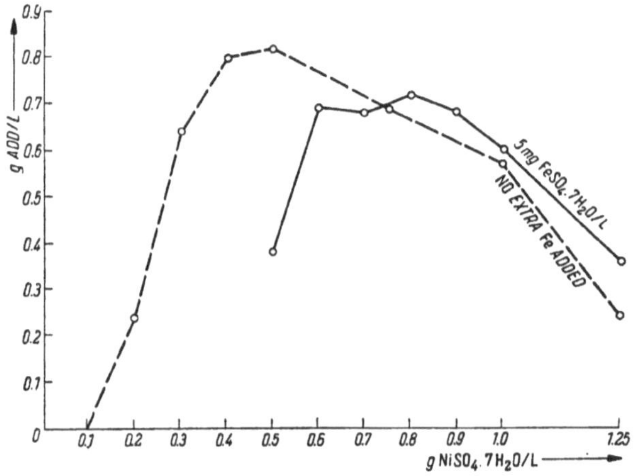

Fig. 3

It seems therefore that iron has something to do with the hydroxylation at carbon 9 and that in certain cases the metal is bound by chelating agents, or, like in the fermentation with *Mycobacterium 70*, is superseded by another ion like nickel. In both cases the formation of ADD may be the result.

If this were true it should be possible to accumulate ADD in a medium that is completely free from iron, assuming that growth and cholesterol degradation would take place under these conditions. When we cultivated *Mycobacterium 70* in a medium poor in iron (distilled water, casaminoacids) we did observe the formation of a small amount of ADD from cholesterol. After prolonged incubation on the shaker the product disappeared. So nickel may not be really necessary to obtain ADD from cholesterol. When cadmiumsulphate was used as an inhibitor ADD was not always obtained. We did not check the optimal conditions in this case but sometimes we found 3-hydroxy-9,10-seco-androsta-1,3,5(10)-triene-9,17-dione on the chromatograms. This compound could also be isolated and identified. The formation of this steroid indicates that with *Mycobacterium* the degradation of the ring system may follow the normal course.

About the fate of the side chain there is not much to be said. We did not succeed in detecting methylheptanone in the culture in spite of many gaschromatographic analyses. Horvath and Kramli (1947) have reported the isolation of this compound from cultures of *Azotobacter*.

If the ketone was formed in our cultures, it propably was utilized by the organism for if methylheptanone was added in small amounts to our shaken flasks it completely disappeared during prolonged fermentation.

From the recent work of SIH et al. (1967) with *Nocardia restrictus* we know that the degradation of the side chain may take place stepwise. Until now we did not succeed in isolating acid intermediates during the fermentation.

Summarizing we may say that strains of *Mycobacterium* which are able to degrade cholesterol, form androstadienedione in the culture after addition of the ions of some heavy metals like nickel and cobalt with the steroid substrate.

With strain KNGSF 70 and 500 mg nickelsulphate/l about 800 mg of androstadienedione was formed in 72 hours.

References

[1] HORVATH, J., c. s. Nature **160**, 639 (1947).
[2] NODA RES. INST., Neth. Pat. Appl. 65.02883.
[3] SIH, C. J. et al., Biochim. Biophys. Acta **62**, 541 (1962); J. Am. Chem. Soc. **85**, 2135 (1964); J. Am. Chem. Soc. **89**, 1956 (1967).
[4] SOEHNGEN, N. L., Zentr. Bakteriol. Parasitenk. Abt. II **37** (1913) 595.
[5] TAK, J. D., Antonie van Leeuwenhoek, J. Microbiol. Serol. **8** (1942) 32.
[6] TURFITT, G. E., Biochem. J. **40** (1946) 79.
[7] WHITMARSH, J. M., Biochem. J. **90** (1964) 23P.

Hydroxylation with Ferro-Ascorbate as a Completion or Substitute of the Microbiological Transformation of Steroids

B. MATKOVICS and G. GÖNDÖS[1]

Department of Organic Chemistry, "József Attila" University of Szeged, Szeged, Hungary

Summary

Ferro-ascorbate hydroxylation was studied on different models, namely aromatic, aliphatic, heterocyclic and more complicated compounds i.e. deoxycholic acid, codein, brasiline. In the case of deoxycholic acid we succeeded to determine the stereochemistry of hydroxylation too.

We have been busy for years with the problem of ferro-ascorbate hydroxylation, described first by UDENFRIEND et al. (1954) with the aim to clear up the chemical possibilities of extension and usage of the method on various model compounds (MATKOVICS et al. 1967).

First according to the original conceptions the ferro-ascorbate hydroxylation of different aromatic compounds was examined. The examinations of MASON et al. (1956) carried out with ^{18}O elucidated that the oxygen introduced into aromatic compounds (into salicylate), arises from molecular oxygen and parallelly the Fenton-hydroxylation the reaction occurs exclusively at electronegative sites of the aromatic rings (contrary to free radical hydroxylation). Accordingly the hydroxylating agent of the ferro-ascorbate hydroxylation is a cation containing chelate, deriving from the air oxygen, namely the ferryl ion, $Fe(OH)^{3+} \rightleftharpoons FeO^{2+} + H^+$, formation of which requires molecular oxygen as an electron source. In this respect a similarity can be assumed with the oxygen-dependent monohydroxylating oxidases.

In our experiments the same ferro-ascorbate system was used as described by UDENFRIEND et al. (1954) consisting of ascorbic acid, $FeSO_4 \cdot 7\,H_2O$ (ferro-sulphate), buffer and ethylendiamin tetraacetic acid (EDTA) aerated by O_2 at 37 °C. Employing H_2O_2 (hydrogen peroxide) instead of aeration the optimal ratio of stock-solution and H_2O_2 was 1:1,5 mol/l. (discussed later).

A mixture of 0,3 mol $Na_2HPO_4 \cdot 2\,H_2O$ and 0,3 mol KH_2PO_4 was used as buffer.

For steroid hydroxylation the system of KESTON et al. (1954) was used in the same concentration. If some change occurred in the concentrations this will be mentioned.

Of the aromatic compounds phenylalanine (Phe.), tyrozine (Tyr.), dioxyphenylalanine (DOPA), salicylic acid (S. a.) and adrenaline (Adr.) had been hydroxylated by ferro-ascorbate and as it is shown in fig. 1 the identification of the products has been mostly carried out. The formation of overhydroxylated products is well visible in the case of Phe. and other aromatics but their identification has not been so far carried out due to lack of comparative compounds and low concentration. But the small paper-chromatographic Rf values indicate an overhydroxylation (Fig. 1).

The transformation of proline to hydroxyproline has been examined in vivo and in vitro (also with ferro-ascorbate) (CHVAPIL et al. 1959); (JUVA et al. 1966). The same examination was made by us (MATKOVICS et al. 1964) and a hydroxylation of the heterocyclic methylene group took place (Fig. 2). Similary occurs the formation of hydroxylysine (PROCKOP et al. 1966).

---

[1] Vortrag nicht gehalten.

Using our experiences experiments were made in order to see whether the methyl-group at the end of a carbohydrate chain can be turned into an oxymethyl-group with the ferro-ascorbate system. DL-alanine, an amino acid, was chosen as a model and under the earlier conditions the formation of serine could be demonstrated by paper chromatography (Fig. 3) (Matkovics et al. 1964). Fig. 4 represents the serine detection by the Bender and Hobein automatic amino acid analyser.

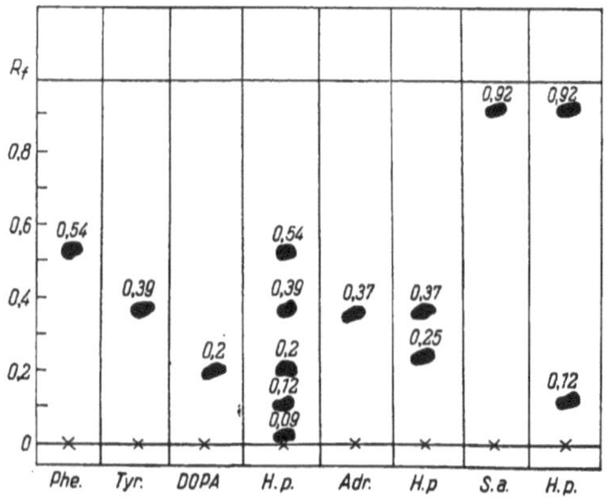

Fig. 1

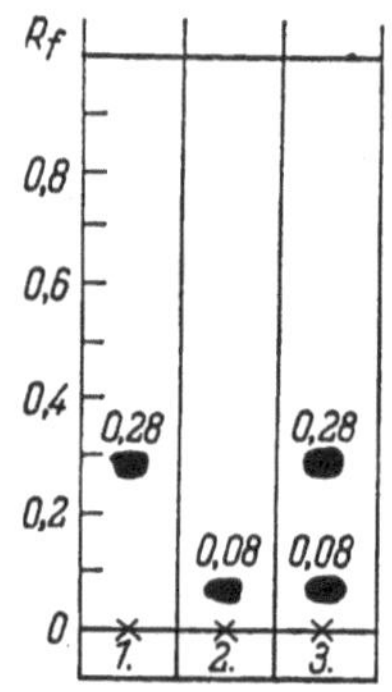

Fig. 2

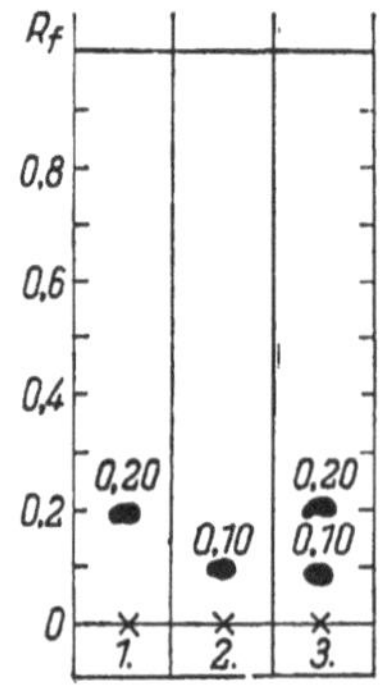

Fig. 3

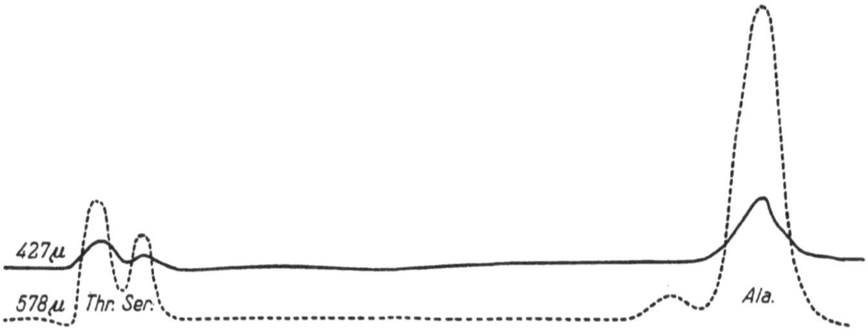

Fig. 4

Having extended the hydroxylation method on the methyl-group too, we studied the hydroxylation of larger molecules, namely, brasiline into haematoxyline (Gábor et al. 1964), deoxycholic acid into cholic acid (Matkovics et al. 1965), codeine into l-hydroxy-codeine (Selmeci et al. 1967). From these reactions we would emphasize the ferro-ascorbate transformation of deoxycholic acid into cholic acid as the parameters has been most extensively studied and it fits in the profile of the Symposium.

From steroids the ferro-ascorbate hydroxylation of 11-deoxy-17α-hydroxycorticosterone, corticosterone, ^{14}C-deoxycorticosterone (Keston et al. 1964) and deoxycorticosterone (Cier et al. 1958) was tried, however, in no case the reaction products were isolated and identified. Only the fact of hydroxylation or other transformations was demonstrated by chromatography.

Efforts were made to elucidate all the parameters, as mentioned before, in the case of the ferro-ascorbate hydroxylation of deoxy-cholic acid. Fig. 5 shows the time dependence of the samples taken, the spot of cholic acid appeared at pH 6.8 after 18 hours oxygenation. On the other hand at the optimal pH, as seen in fig. 6, the cholic acid appears after 3 hours oxygenation and after 8 hours the ratio of deoxycholic acid and cholic acid is 38.8 to 61.2% in the solution. If the oxygenation is substituted by H_2O_2, the optimal ratio of stock-solution and peroxide is 1:1,5 mol/l, then the cholic acid: deoxy-cholic acid ratio is 58.4 to 41.6% (Fig. 7).

In the last figure (Fig. 8) is well visible the disappearance of deoxycholic acid (1), the increase of cholic acid (2) and later that of a third (3) unknown material. The cholic acid appears here very early in the third hour.

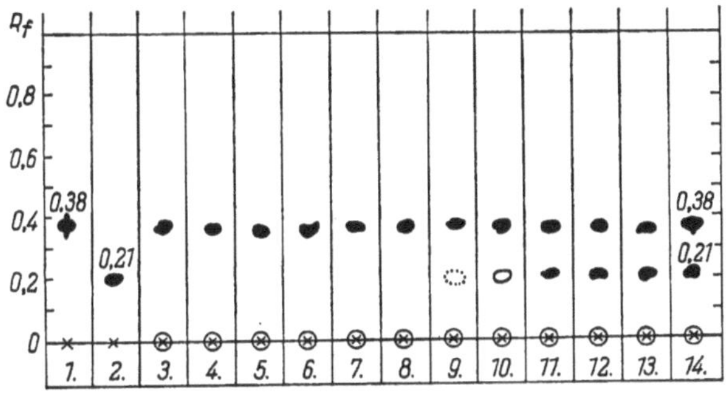

Fig. 5

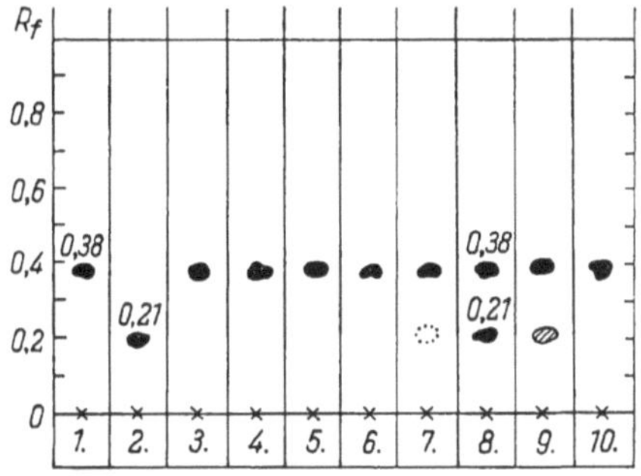

Fig. 6

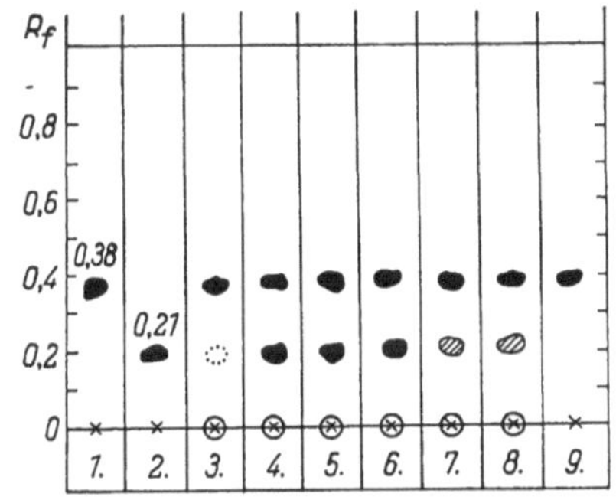

Fig. 7

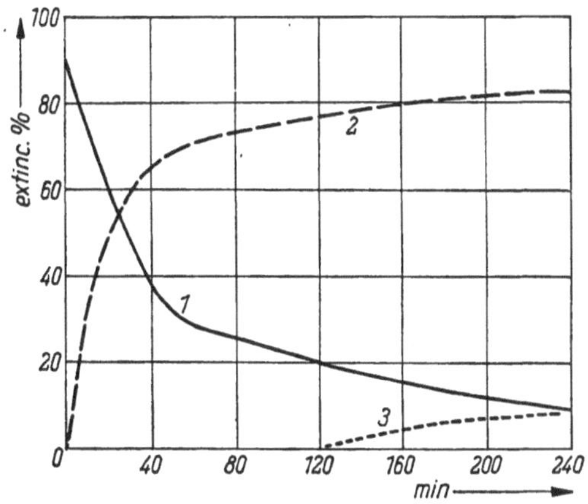

Fig. 8

Consequently the ferro-ascorbate transformation of deoxy-cholic acid into cholic acid shows a very adequate example to throw light not only on the hydroxylation of a C-7 methylene group but also it clears up the stereochemical mechanism of hydroxylation.

Our ferro-ascorbate hydroxylation failed in some cases e.g. estrone, estradiol, Girard T derivative of estrone and Reichstein S (17α-hydroxy-11-deoxycorticosterone).

Despite the failures in our opinion the ferro-ascorbate system, as well as the "hydroxylase" isolated from B. megaterium (Wilson et al. 1965), the Fenton-reaction and irradiation can equally be used to hydroxylate or transform the steroids (Wheeler et al. 1967), but the different transformations must be investigated and tried.

References

[1] Cier, A., C. Nofre, A. Revol, Compt. rend., **274**, 542 (1958).
[2] Chvapil, M., J. Hurych, Nature (London), **184**, 1145 (1959).
[3] Gábor, M., B. Matkovics, Gy. Göndös, Planta Medica, **12**, 419 (1964).
[4] Hecker, E., G. Nowoczek, Z. f. Naturforsch., **21** 153 (1966).
[5] Juva, K., D. J. Prockop, A. W. Cooper, I. W. Lash, Science, **152**, 92 (1966).

[6] Keston, A. S., M. Carsiotis, Arch. Biochem. Biophys. **52**, 282 (1954).
[7] Mason, H. S., I. Onoprienko, Fed. Proc., **15**, 310 (1956).
[8] Matkovics, B., Gy. Filvig, Gy. Göndös, Biochim. Biophys, Acta, **86**, 180 (1964).
[9] Matkovics, B., P. Pénzes, Gy. Göndös, Steroids, **5**, 451 (1965).
[10] Matkovics, B., Gy. Göndös, Biol. Közl. (under publication).
[11] Prockop, D. J., E. Weinstein, T. Mulveny, Biochem. Biophys. Res. Comm., **22**, 124 (1966).
[12] Selmeci, Gy., B. Matkovics, Gy. Göndös, K. Kovács, Pharmazie, **22**, 292 (1967).
[13] Udenfriend, S., C. T. Clark, J. Axelrod, B. B. Brodie, J. Biol. Chem., **208**, 731 (1954).
[14] Wheeler, O. H., R. Montalvo, J. Appl. Rad. Isotopes, **18**, 127 (1967).
[15] Wilson, J. E., C. S. Vestling, Arch. Biochem. Biophys., **110**, 401 (1965).

Preparation of Steroid Metabolites by Microbial Reactions

J. SCHLEGEL

Institut für Mikrobiologie und experimentelle Therapie, Jena,
der Deutschen Akademie der Wissenschaften zu Berlin, DDR

Summary

For investigations of the steroid metabolism in mammalians steroid metabolites are needed both for comparison and, in larger amounts, for testing their biological activity. Microbiological reactions have proved to be favourable for preparing labeled and unlabeled steroid metabolites. Besides hydroxylation reactions the stereospecific hydrogenation of the Δ^4-3-ketosteroids to 5α- and 5β-compounds is of importance. One main route of the mammalian steroid metabolism is the hydrogenation of the ring A unsaturated steroid hormones and steroid drugs to metabolites of the $3\alpha/5\beta$ and $3\beta/5\alpha$ type. *Clostridium* spec. reduce stereospecifically Δ^4-, $\Delta^{1,4}$-, $\Delta^{4,6}$ and $\Delta^{1,4,6}$-3-ketosteroids to the $3\alpha/5\beta$ type. Using preparative thin-layer chromatography a number of 3α-hydroxy-5β-steroids has been prepared at gram scale.

In investigations concerning the steroid metabolism difficulties will arise from the lack of steroid metabolites needed for comparison. It is not so easy to synthesize steroid metabolites in a chemical way. Often the yields are very low. This especially happens if no stereospecific reaction is available for the preparation of the required compounds. A number of separation procedures are necessary to isolate the desired isomer from the reaction mixture.

This point of view is of the greatest importance if labeled steroid metabolites are needed for further investigations of the steroid metabolism or for experiments on the mechanism of action. Only reactions with a high stereospecificity and therefore good yields without losses of labeled substance can be used for such purposes. Occasionally steroid metabolites were prepared by stereospecific biochemical reactions with in vitro experiments of mammalian organs. However, only small amounts of the needed samples were obtained in such a way.

A few micrograms are sufficient for the purpose of comparison. But it is also necessary to test steroid metabolites for their biological activity. There are many evidences that metabolites of steroid hormones cannot be considered as physiologically inactive compounds. The same applies to metabolites of steroid drugs, and the reason for some actions or undesired actions of steroid drugs respectively is likely to be seen in the physiological activity of the metabolites.

Gram-amounts of steroid metabolites are required in order to test their unknown biological activity. Microbiological reactions can favourably be used for the preparation of labeled and unlabeled steroid metabolites. Cultures of microorganisms are easily obtainable. The size of the fermentation cultures can be chosen as one likes and the fermentation can be repeated.

Examples for the microbiological preparation of labeled steroid metabolites are 4-^{14}C- and 7-^{3}H-15- and 16-hydroxy-steroids. GIANNOPOULOS and SOLOMON (1) isolated 15α-hydroxyprogesterone from human pregnancy urine. This metabolite is capable of antagonizing the actions of corticosteroids. For further investigations of the metabolism of this substance they used specifically tritium-labeled 15α-hydroxy-progesterone which was prepared in the Squibb Laboratories from 7-^{3}H-progesterone with a 15α-hydroxylating strain of microorganisms.

For the investigation of the in vivo metabolism of 16-hydroxy metabolites (2) Neher from the CIBA Laboratories prepared (4-^{14}C)-16α-hydroxy-progesterone, Fried from the Squibb Laboratories (7-^{3}H)-16α-hydroxy-progesterone and Vezina from the Ayerst Laboratories (7-^{3}H)-16α-hydroxy-dehydroepiandrosterone with different strains of 16α-hydroxylating microorganisms. A 16α-hydroxy group partially inhibits the Δ^5-isomerase action and seems to influence the reduction of ring A and B.

Usually, microbial reactions are stereospecific. The variety of the known reactions at the steroid molecule in mammalians can be obtained by the different strains of microorganisms.

Vice versa in an important special case, in the field of steroid degradation, many microbial degradation reactions are being studied which have hitherto been unknown in mammalians, although evidence exists that steroids are metabolized to carbondioxide (3,4). By means of microorganisms a possible pathway of steroid degradation has been elaborated by the groups of Talalay, Schubert and Sih.

The isolated lower molecular products of the steroids will be of great importance, if the steroid degradation in mammalians is under investigation.

For the preparation of other steroid metabolites we especially studied a hydrogenation reaction of *Clostridium paraputrificum*. (5).

As is well known one main route of the mammalian steroid metabolism is the hydrogenation of the unsaturated ring A and the 3-ketogroup of steroid hormones and steroid drugs. In this way metabolites of different types are obtained. The formation rate of these metabolites varies. Corticosteroid hormones for example are preferably hydrogenated to the 5β/3α-hydroxycompounds. We observed that the anaerob bacteria isolated from the intestine transformed androstenedione to etiocholanolone with yields of more than 85%. This steroid is a metabolite of the 5β/3α type.

Then the stereospecificity of the microbiological reaction was examined with regard to 25 unsaturated 3-ketosteroids of the C-19-, C-21, C-24 and C-27 series with double bonds in position Δ^{4}-, $\Delta^{1,4}$-, $\Delta^{4,6}$- and $\Delta^{1,4,6}$. All Δ^4-3-ketosteroids of the C-19 and C-21 series but the l-form of 19-nortestosterone were hydrogenated to the 5β/3α tetrahydroderivatives. Examples are given in table 1.

Table 1

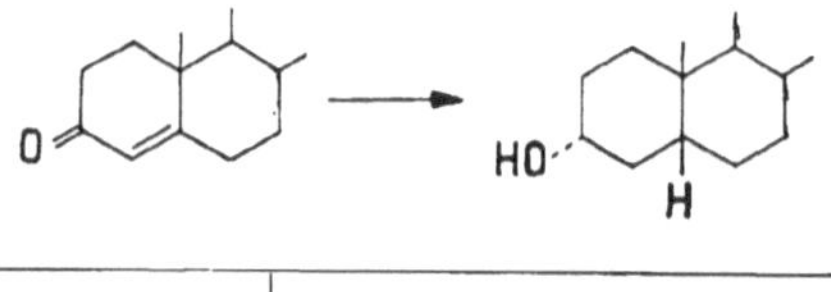

Substrate	Hydrogenation product
Testosteron	5β-Androstan-3α,17β-diol
Progesteron	Pregnan-3α-ol-20-on
Desoxycorticosteron	Pregnan-3α,21-diol-20-on
Cortison	Pregnan-3α,17α,21-triol-11,20-dion
Desoxycorticosteron-21-Acetat	Pregnan-3α,21-diol-20-on-21-Acetat
Cortison-21-Acetat	Pregnan-3α,17α,21-triol-11,20-dion-21-Acetat

21-acetates of corticosteroids were saponified only in small amounts.

Δ^4-double bonds of $\Delta^{1,4}$-3-ketosteroids were saturated selectively. Thus under certain conditions $\Delta^{1,4}$-3-ketosteroids may be transformed to the 5β-Δ^1-3-ketoderivatives with yields up to 80%. Examples are given in table 2.

Table 2

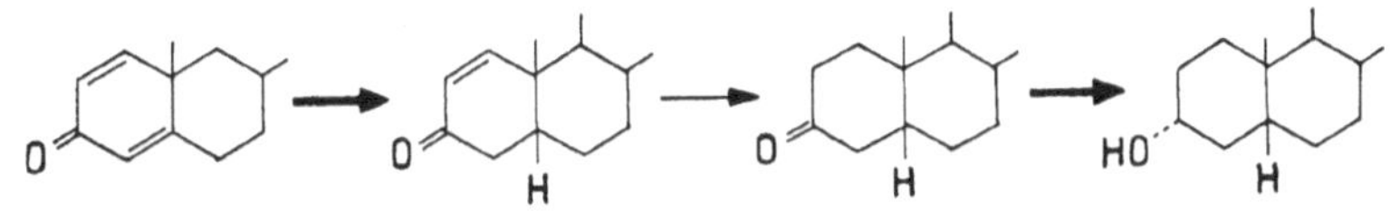

Substrate	Hydrogenation product
$\Delta^{1,4}$-Androstadien-3,17-dion	Δ^{1}-5β-Androsten-3,17-dion 5β-Androstan-3,17-dion 5β-Androstan-3α-ol-17-on
$\Delta^{1,4}$-Pregnadien-17α,21-diol-3,11,20-trion (Prednison)	Δ^{1}-Pregnen-17α,21-diol-3,11,20-trion Pregnan-3α,17α,21-triol-11,20-dion
17a-Oxa-17-oxo-D-Homo-$\Delta^{1,4}$-androstadien-3-on (Testololacton)	17a-Oxa-17-oxo-D-homo-Δ^{1}-5β-androsten-3-on 17a-Oxa-17-oxo-D-Homo-5β-androstan-3α-ol

Δ^{6}-double bonds of $\Delta^{1,4,6}$-5β-$\Delta^{1,6}$- and $\Delta^{4,6}$-3-ketosteroids were not attacked.

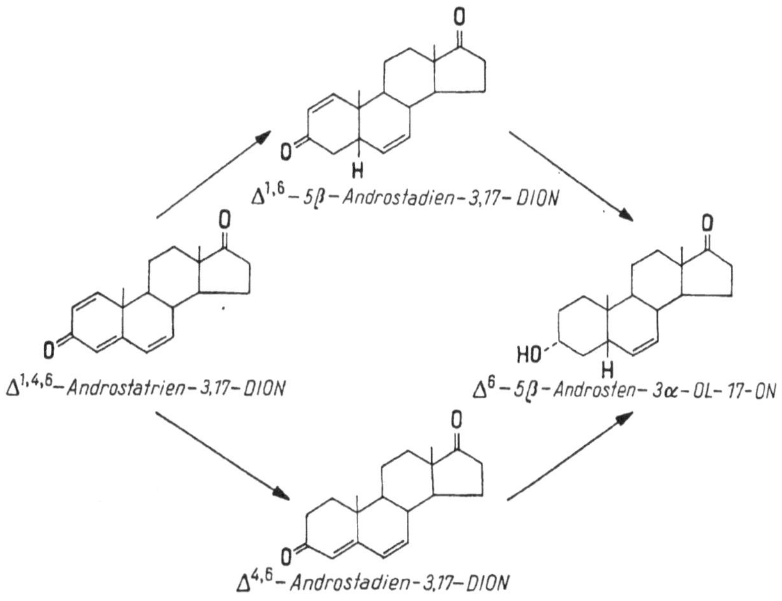

Hydrogenation of $\Delta^{1,4,6}$-Androstatriene-3,17-dione, 5β-$\Delta^{1,6}$- and $\Delta^{4,6}$-Androstadiene-3,17-dione by Cl. paraputrificum

Functional groups like hydroxyl, carbonyl, methyl and lactone in position 6 (6), 11 and 17 do not influence the stereospecific hydrogenation. Only a chlorine atom in position 4 of $\Delta^{1,4}$-3-ketosteroids inhibits the reduction of the Δ^{1}-double bond.

By this hydrogenation some compounds were obtained which had not been described before.

The results showed that it is possible to stereospecifically transform 3-keto steroid hormones and steroid drugs to metabolites of the 5β/3α type by a microbial reaction. For investigations of new steroid drugs the corresponding 5β/3α metabolites are easy to prepare. As already mentioned it is necessary to have steroid metabolites in g-amounts for testing their biological activity. It is especially on the 5β-steroids that interesting results have been reported lately. For preparing 5β-metabolites in gram amounts, 1 g of the substrates were incubated in 3 liter cultures of *Cl. paraputrificum.* The tetrahydroderivatives of some Δ^{4}-3-ketosteroids were obtained only by crystallization of the extracts. In other cases small amounts of unchanged substrate and dihydroproduct had to be separated.

After incubation of $\Delta^{1,4}$-3-ketosteroids the 5β-Δ^{1}-3-ketoderivatives as main products had to be isolated from hexahydroderivatives. Especially the preparative thin layer chromatography with silicagel and alumina as adsorbent was used successfully for the purification and separation of the hydrogenation products. Their R_s-values related to the respective substrates were sufficiently smaller or larger than 1.

On 20 by 80 cm glass plates and 1 mm layers of the adsorbent 0,5—1 g of steroid mixture was chromatographed.

Without difficulty g-amounts of $5\beta/3\alpha$ hydroxy and 5β-Δ^{1}-3-ketometabolites were obtained in high purity. The yields ranged from 50—80%.

These results encourage us to prepare other metabolites of the steroid metabolism in mammalians by microbial reactions.

References

[1] Giannopoulos, G. and S. Solomon, Biochemistry **6**, 1226 (1967).

[2] Ruse, J. L. and S. Solomon, Biochemistry **5**, 1065 (1966).

[3] Heard, R. D. et al., Recent Progress in Hormone Research **IX**, 383 (1954).

[4] Brusca, D. R. and A. M. Gawienowski, Steroids **312**, 209 (1964).

[5] Schubert, K., J. Schlegel, C. Hörhold, Steroids, Suppl. I, 175 (1965).

[6] Schubert, K., J. Schlegel, K.-H. Böhme, C. Hörhold, Biochim. Biophys. Acta **144**, 132 (1967).

Transformations of Cardenolides by Microorganisms of the Intestine

I. Herrmann and K. Repke

Institut für Biochemie, Berlin-Buch, der Deutschen Akademie der Wissenschaften zu Berlin, DDR

Summary

The metabolic fate of lanatoside C in the animal organism is analysed. The results of both *in vivo*- and *in vitro*-experiments prove that in rats the removal of the glucose residue, the hydrolysis of the acetic ester group and the saturation of the C,C-double bond in the lactone ring (order of preferred reaction sequence) are effected by microorganisms of the gut. The same appears also to be true for man. Several pieces of evidence favour the conclusion that the oral activity of lanatoside C in man may require the deglucosylation of the molecule and may individually vary in dependence on the rate of its detoxification by reduction of the lactone ring.

In the course of studies on the metabolic fate of cardioactive steroids in the animal organism (for a review see Repke, 1963), we have found that the cardenolide glycoside lanatoside C undergoes profound metabolic transformations which appear to be due to microorganisms of the flora of the alimentary tract. Figure 1 shows the structural formula of lanatoside C

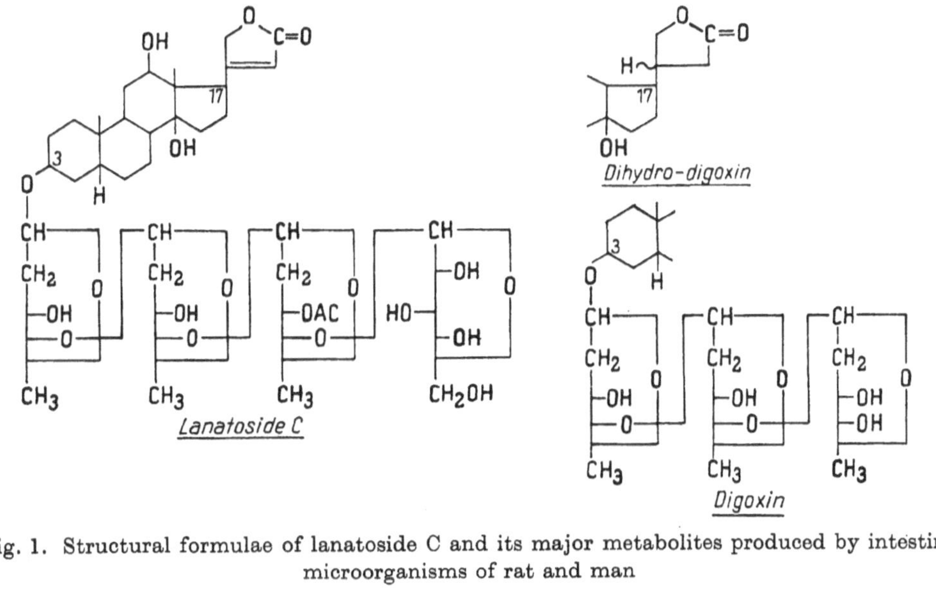

Fig. 1. Structural formulae of lanatoside C and its major metabolites produced by intestinal microorganisms of rat and man

and the partial formulae of its major metabolites digoxin and dihydro-digoxin found in our studies. Compared to lanatoside C, in digoxin the terminal glucose residue and the acetic acid ester group are removed. In dihydro-digoxin, the C,C-double bond of the butenolide ring at C-17 is saturated.

Whereas the cleavage of the glucosidic linkage in various cardenolide glycosides has been already described (Brown et al., 1955; Engler et al., 1958; Lauterbach et al., 1960b) the reduction of the double bond in the butenolide ring was not observed as yet. The present paper aims to evidence the microbial origin of these transformations, to clarify the reaction sequences and the possible biological significance of the metabolic reactions.

The identity of the metabolites was checked by comparison with authentic samples of digoxin and dihydro-digoxin with regard to various characteristics. To this end, the paper chromatographic mobilities, the reactions in specific colour tests and the spectroscopic properties were determined. — In several solvent systems, the chromatographic mobilities of lanatoside C, digoxin and dihydro-digoxin are highly different and very characteristic. The presence of the digitoxose chain is evidenced by the reaction with xanthydrol which is a specific reagent for 2-deoxysugars. This reagent is also used for the quantitative determination of the glycosides. The preservation of the butenolide ring was checked by the application of the meta-dinitrobenzene or picric acid test which is negative in the case of saturation of the C,C-double bond. Finally, also the spectra in sulfuric acid enable us to differentiate between digoxin and dihydro-digoxin and to identify the ring saturated metabolite (for methodic details see Repke, 1958, 1959, 1960).

In the first step of our studies, we explored the *in-vivo*-metabolism of lanatoside C in the rat. Table 1 shows the appearance of excretion products in urine and feces after intravenous injection of lanatoside C. Within the first 5 hours unchanged lanatoside C is excreted by urine in considerable amounts, but metabolites cannot be traced. Within 5—24 hours after injection, when no further lanatoside C is excreted, small amounts of the metabolites digoxin and dihydro-digoxin are appearing in urine. The feces, however, contain besides small

Table 1

Excretion products in urine and feces after *intravenous* injection of lanatoside C (5 n moles/g). Amounts in mole % of the dose

excretion products	urine		feces
	0–5 hours	5–24 hours	0–24 hours
lanatoside C	19	0	0
digoxin	0	5	4
digoxigenin- bis -digitoxoside	0	1	0
dihydro-digoxin	0	0–2	21

amounts of digoxin as much as 21 mole % of the dose in form of the metabolite dihydro-digoxin. The lag phase for the appearance of metabolites in urine and the abundance of dihydrodigoxin in feces might be explained by the assumption that lanatoside C is firstly excreted by bile and transported into the lower parts of the gut and that it is metabolized only there by intestinal microorganisms. Although the temporal appearance of the metabolites and their distribution between urine and feces thus strongly point to a microbial origin of the transformations, we found it necessary to check the possibility that animal enzymes are also involved.

To exclude transformations by intestinal microorganisms, these studies were performed *in vitro* using different, enzymatically highly active preparations of rat tissues. When incubated with homogenates of liver and small intestine or with liver slices, lanatoside C is recovered unchanged under all conditions studied. Since it had been shown earlier that liver and intestine are the major tissues in effecting other transformations of cardenolides (Lauterbach et al., 1960a, 1960b), this negative result appears to exclude the possibility that animal enzymes are involved in the removal of glucose and acetic acid residues as well as in the reduction of the butenolide ring.

Table 2

Transformations of lanatoside C by the contents of rat caecum
0,2 μmole of lanatoside C was incubated with 1 g caecum contents (mixed of 3 rats) in 10 ml liver-broth under anaerobic conditions (stream of 5% CO_2 in N_2) at 37 °C for 24 hours. Amounts in mole % of the substrate

Experiment No.	Compounds found after incubation		
	lanatoside C	digoxin	dihydro-digoxin
1	0	53	38
2	0	79	20
3	0	105	0

To check the conclusion that these metabolic transformations are effected by microorganisms of the gut, the transformations of lanatoside C by the contents of rat caecum were studied. Table 2 shows that under appropriate conditions lanatoside C is completely transformed to digoxin and in most experiments also to dihydro-digoxin.

It should be noted that the sum of the molar amounts of the metabolites digoxin and dihydro-digoxin recovered essentially corresponds with the added amount of lanatoside C. This means, that significant further transformations such as the hydrolysis of the butenolide ring or the removal of the digitoxose chain do not occur.

Whereas the degradation of lanatoside C to digoxin was found in all experiments, the production of dihydro-digoxin was absent with some caecum contents. The lack of parallelism between the two transformations favours the conclusion, that the removal of glucose and acetic acid residues on the one side and the saturation of the butenolide ring on the other side are effected by different microorganisms. This conclusion is further supported by the following observations. When lanatoside C is incubated with caecum contents in the presence of oxygen instead of nitrogen, the cleavage of the glucoside is somewhat reduced only, but the saturation of the butenolide ring is completely suppressed.

The partial or complete dependence of the transformations upon the absence of oxygen seems to exclude the possibility that digestive enzymes formed by the intestine are involved. This conclusion is confirmed by the observation that lanatoside C is not metabolized by sterile filtrates of caecum contents suspensions.

Table 3

Transformation of lanatoside C by human feces
0,2 μmole of lanatoside C was incubated with 1 g feces from different individuals under same conditions as observed with caecum contents of rats

Individual No.	Compounds found after incubation		
	lanatoside C	digoxin	dihydro-digoxin
1	0	60	32
2	0	96	0
3	0	0	92

The above-described transformations of lanatoside C are also obtained by incubation with human feces. As shown in Table 3, human feces of different individuals effectively degradate lanatoside C to the glucose-free metabolites digoxin or dihydro-digoxin in all experiments, but the production of dihydro-digoxin exhibits again considerable individual variations. It should be added here, that the feces of dog and cat also deglucosylate lanatoside C (Table 4), but interestingly enough they are unable to hydrogenate the double bond in the butenolide ring.

These observations anew suggest that the two transformations are performed by different microorganisms.

Table 4

Transformations of lanatoside C by feces of various species

0,2 μmole of lanatoside C was incubated with 1 g feces in 10 ml liver-broth under anaerobic conditions at 37 °C for 2 hours. Amounts in mole % of the substrate

Species	Compounds found after incubation			
	lanatoside C	deacetyl-lanatoside C	acetyl-digoxin	digoxin
rat	0	0	5–19	71–98
dog	48–49	0	25–37	22–27
man	59–73	0	30–38	15–26
cat	87–103	0	0–8	0–35

As to the nature of the microorganisms involved, the following may be said. According to earlier studies carried out with other cardenolide glucosides (LAUTERBACH et al., 1960b), the cleavage of the beta-glucosidic linkage in lanatoside C may be effected by cellulolytic bacteria known to be present in the intestinal flora. Concerning the saturation of the butenolide ring, however, we have no knowledge about the nature of the microorganisms involved. Certain strains of *Clostridium paraputrificum* and *Clostridium tertium* isolated from human feces were found to be unable to saturate the butenolide ring. Under the same incubation conditions, these strains effectively reduce the C,C-double bond in rings A and B of hormone type steroids (SCHUBERT et al., 1963). Further studies are required to decide this question.

We were interested to learn more about the reactions sequence which leads to the transformation of lanatoside C into digoxin. In order to trace the possible intermediary products, short-term incubations of lanatoside C with feces of different species were carried out. Table 4 shows the transformations of lanatoside C by feces of various species when incubated under anaerobic conditions for 2 hours only. Firstly, it should be noted that deacetyl-lanatoside C is not produced. Thus, a primary deacetylation of lanatoside C before the removal of the glucose residue does not appear to occur. Possibly, the neighbouring glucosidic linkage protects the ester group from the attack of esterolytic enzymes. As further shown in this table, whenever digoxin is produced also acetyl-digoxin will be formed. So, the removal of the glucose residue yielding acetyl-digoxin is one of the first steps which is followed by esterolysis giving digoxin. Interestingly enough no dihydro-digoxin is formed in these short-term experiments by the feces of rat and man otherwise able to produce this metabolite. This observation favours the conculsion that the saturation of the butenolide ring is not a primary transformation of lanatoside C, but that the metabolite digoxin undergoes this reductive process.

To check this conclusion, the reduction of the butenolide ring of digoxin by the contents of rat caecum and by human feces was studied. Although there are again considerable individual variations, in most experiments dihydro-digoxin is produced. Thus, the transformation of lanatoside C into dihydro-digoxin appears to go via the intermediary product digoxin.

Summing up this part of our paper, the pathway of the transformations of lanatoside C seems to be as follows (fig. 1). The first step is the removal of the glucose residue from the sugar side chain yielding acetyl-digoxin. The second step is the hydrolysis of the ester group giving digoxin. The final step is the saturation of the butenolide ring of digoxin forming dihydro-digoxin.

In the final part of our paper, we wish to touch briefly the possible significance of the various transformations of lanatoside C. As well-known, lanatoside C is a renowned drug for the treatment of cardiac failure. Since it is recommended for oral application, too its microbial transformations could have some significance for the oral activity of the drug. To decide this question, we performed a model experiment in rats of which the enzymatic activities of the intestinal flora were shown to be similar to those of man. Lanatoside C was injected into the duodenum of the rat to study its intestinal absorption. Interestingly enough, we found that no unchanged lanatoside C is absorbed into the blood. Instead of the drug, its metabolite digoxin only appears in the blood.

Figure 2 shows the digoxin level in blood after intraduodenal injection of lanatoside C. The lag phase of 6 hours for the appearance of the metabolite in blood is easily explained by the fact, that it must be transported to the lower parts of the gut to be degradated here to digoxin by microorganisms. If the same is also true for man, which is favoured by our observations on the glucolytic activities of human intestinal microorganisms, it may be concluded that the oral effectivity of lanatoside C depends upon its degradation to digoxin and that this metabolite is the real agent during oral treatment with the drug.

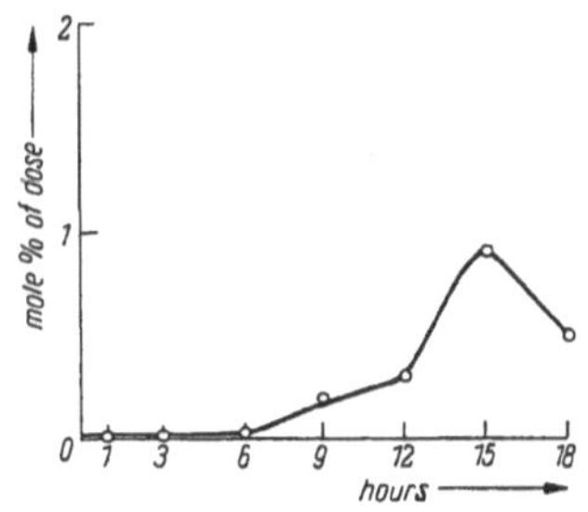

Fig. 2. Digoxin level in blood after intraduodenal injection of lanatoside C (10 nmoles/g) in rats

The significance of the transformation of lanatoside C into dihydro-digoxin also obtained with human intestinal microorganisms is just reverse. The cardiac activity of dihydro-digoxin is about twenty times lower than that of the drug. Thus, the production of dihydro-digoxin means an almost complete detoxification of lanatoside C. The rate of this detoxification reaction showed considerable variations with the intestinal bacteria of different individuals. So, our observations might possibly explain the statement (BATTERMAN et al., 1947) that the oral effectivity of lanatoside C shows considerable individual variations. Of course, the conclusions about the possible therapeutic significance of the microbial transformations of lanatoside C must be checked by experiments in man.

References

[1] BATTERMAN, R. C., and A. C. DE GRAFF, Amer. Heart, J. **34**, 663 (1947).
[2] BROWN, B. T., D. RANGER and S. E. WRIGHT, J. Pharmac. Exp. Ther. **113**, 353 (1955).
[3] ENGLER, R., P. HOLTZ und H. W. RAUDONAT, Naunyn-Schmiedeberg's Arch. Exp. Path. Pharmak. **233**, 393 (1958).
[4] LAUTERBACH, F. und K. REPKE, Naunyn-Schmiedeberg's Arch. Exp. Path. Pharmak. **239**, 196 (1960a).
[5] LAUTERBACH, F. und K. REPKE, Naunyn-Schmiedeberg's Arch. Exp. Path. Pharmak. **240**, 45 (1960b).
[6] REPKE, K., Naunyn-Schmiedeberg's Arch. Exp. Path. Pharmak. **233**, 261 (1958).
[7] REPKE, K., Naunyn-Schmiedeberg's Arch. Exp. Path. Pharmak. **237**, 34 (1959).
[8] REPKE, K., Naunyn-Schmiedeberg's Arch. Exp. Path. Pharmak. **239**, 131 (1960).
[9] REPKE, K., Proc. First Int. Pharmacol. Meeting Stockholm 1961, Vol. 3, p. 47–73, Pergamon Press, London–New York (1963).
[10] SCHUBERT, K., J. SCHLEGEL und C. HÖRHOLD, Hoppe-Seyler's Zschr. Physiol. Chem. **332**, 310 (1963).

Combined Gas Chromatography-Mass Spectrometry in Steroid Research

H. Adlercreutz

Steroid Research Laboratory, Department of Clinical Chemistry,
University of Helsinki, Helsinki, Finland

Summary

The use of combined gas chromatography-mass spectrometry in steroid hormone research is described. A brief description of the instrumentation and technique is given. The method can be used for the evaluation of the specificity of gas chromatographic methods and for the identification of hormonal steroids in biological material. By varying the conditions during mass spectrometry it is possible to obtain specific mass spectra of even structurally closely related steroids. The mass spectrometer can also be used as a highly sensitive and specific detector during gas chromatography. In the case of the trimethylsilyl ether derivative of pregnanediol, it has been possible to quantitate less than 10 pg of the steroid.

The combination of gas liquid chromatography (GLC), using packed columns, and mass spectrometry (MS) in the same instrument is a very recent development, which has created a highly efficient new tool for the analytical chemist (Ryhage, 1962, 1964; Ryhage et al., 1965). The combination of the great separating capacity of gas chromatography with the ability of mass spectrometry to provide immediate structural information for every compound separated is a definite improvement over the rather time-consuming earlier methods in which the components were first isolated separately and then submitted to various analytical procedures. Especially when working with biological extracts containing the compounds under investigation in very low concentrations in the presence of impurities which are difficult to eliminate completely, such a combination of instruments will rapidly give us detailed structural information which cannot be obtained by any other means.

Instrumentation

Our experience is mainly based on the LKB 9000 Gas chromatograph-mass spectrometer developed by Ryhage and coworkers in the laboratory for Mass Spectrometry at the Karolinska Institutet, Stockholm, Sweden. It is the first completely integrated instrument. In this apparatus the gas chromatograph is of the standard type with a flash heater and an oven 14 cm in diameter and 15 cm high containing a coiled glass column. Helium is used as carrier gas. The glass column is connected to a highly efficient molecule separator containing a jet system and two separator chambers. The effluent from the gas chromatograph enters the first of the two chambers through a jet at very high speed. The first chamber is connected to a fore vacuum pump which removes part of the helium gas. The molecules with higher molecular weight proceed directly through a second jet into the second separator chamber, where a highly efficient oil-diffusion pump removes most of the remaining helium gas before the sample enters the ionization chamber, where it is ionized by electron impact.

The electron impact transfers a large amount of energy to the organic molecule, which is decomposed, and great numbers of positively charged fragments are formed. The ion beam is accelerated and focussed before it enters the magnetic field in the analyzer tube. In the magnetic field the groups of ions are deflected according to their individual mass to charge (m/e) ratios. If the magnetic field is kept constant, only a particular group of ions of equal m/e ratio is focussed on the collector slit. To obtain a record of all the ions formed, the magnetic field is changed. In this way the whole mass range of interest can be scanned. The signal from the final anode in the multiplier is connected to a rapid UV-recorder with three galvanometers recording the mass spectrum at three different gains.

Of the total ionization current, about 10 per cent is collected and recorded with an ordinary pen recorder. In this way a portion of the ions produced is used for gas chromatographic recording of the compounds emerging from the column. When the gas chromatographic recording is followed, mass spectra may be taken at any time during the course of elution of the compounds from the GLC column.

Technique

The extracts of the biological samples are first pretested by GLC on the same type of stationary phase and, if the gas chromatograms obtained does not contain too many impurities, the samples may be used as such for mass spectrometry. In practically all studies carried out in this laboratory the trimethylsilyl ether (TMSi) derivatives of the steroids have been used, since the fragmentation pattern of these compounds is very characteristic. Frequently, double derivatives are also used (e.g. TMSi derivatives of methylated steroids). The sample, dissolved in 1—5 μl of n-hexane, is injected into the flash-heater, which is kept at a temperature 20—30° higher than the column. The valve between the molecule separator and the ionization chamber is kept closed when the solvent front emerges from the column. Usually, the electron energy is kept at 20 eV before the scan is started, in order to avoid ionization of the carrier gas. At the moment when the scan starts, the electron energy is automatically switched over to a preselected value, usually 70 eV.

The scan speed and UV-recorder paper speed may be varied. This will depend on the mass range to be scanned and the elution time of the gas chromatographic peak, since, for a small peak, a short scanning time must be chosen if a reasonable sample concentration is to be maintained throughout the mass spectrometric investigation.

When a mass spectrum has been taken, the various peaks are localized and the m/e values noted. A mass spectrum of the background which is produced by bleed from the GLC column, is taken after every investigation. Every liquid phase has its characteristic peaks, which may be used as an internal standard when localizing the peaks of the compound under investigation.

In Fig. 1 the mass spectrum of the background (below) can be seen and also one mass spectrum of the TMSi derivative of pregnanediol. If there are difficulties in locating the ions in the upper region of the spectrum, another mass spectrum with a higher gain is taken. The peaks of the mass spectra of the compound and the background are measured and those of the background subtracted from the peaks obtained with the compound under investigation. With the highest peak taken as 100 per cent, a line diagram is drawn, as seen in Fig. 2. Since the base peak at 117 is very high, the spectrum has not been calculated with this as 100%; in this case the peak at 118 has been used. In this way, some other typical fragments of pregnanediol trimethylsilyl ether derivative can be seen.

Application to steroid research

Investigation of the specificity of gas chromatographic methods

When methods for the estimation of hormonal steroids are developed, it is often of great value to use combined GLC-MS to check the various steps being investigated in the procedures.

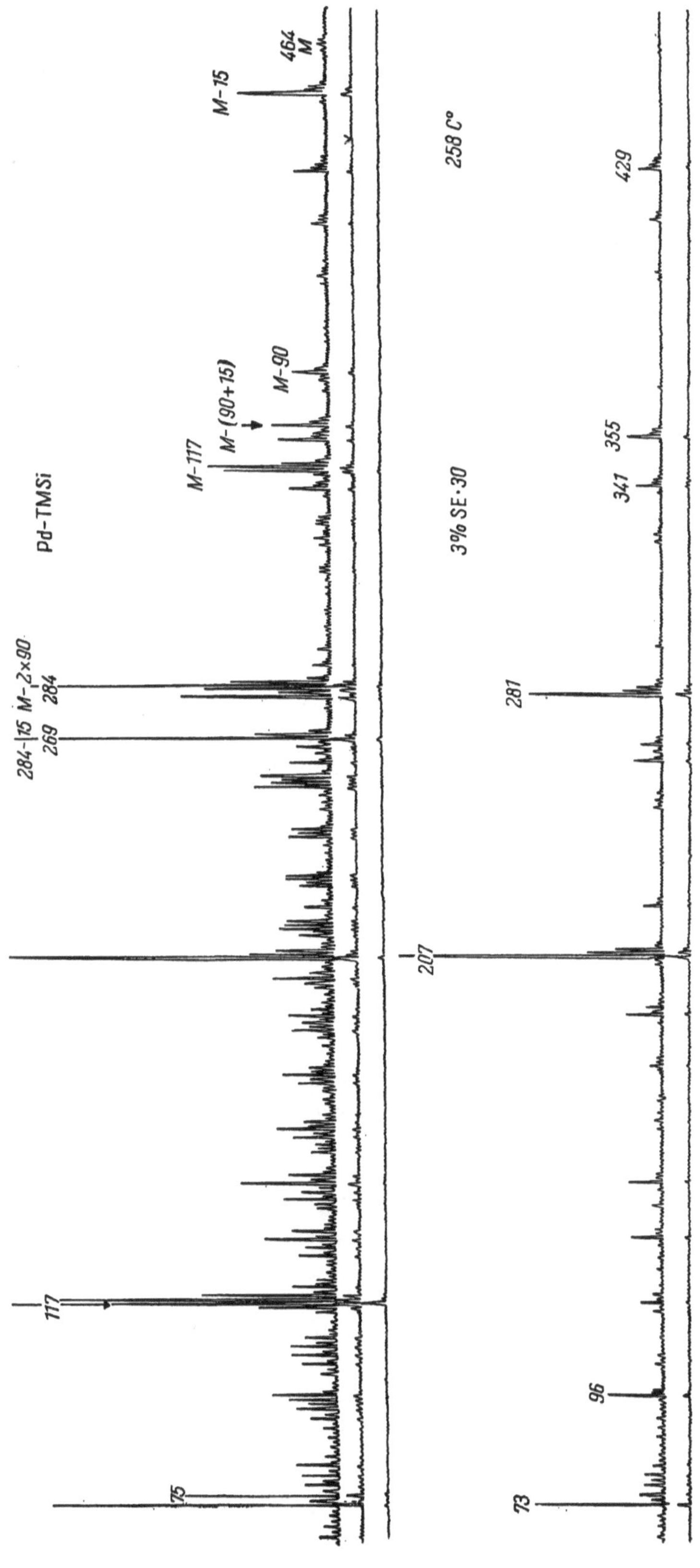

Fig. 1. Mass spectrometric recording of the ditrimethylsilyl ether derivative of pregnanediol reference standard (upper figure) and of the background (lower figure), using the LKB 9000 Gas chromatograph-mass spectrometer and an electron energy of 70 eV. The gas chromatographic column contained 3% SE-30 on 100—120 mesh Gas-Chrom Q. Column temperature 258 °C

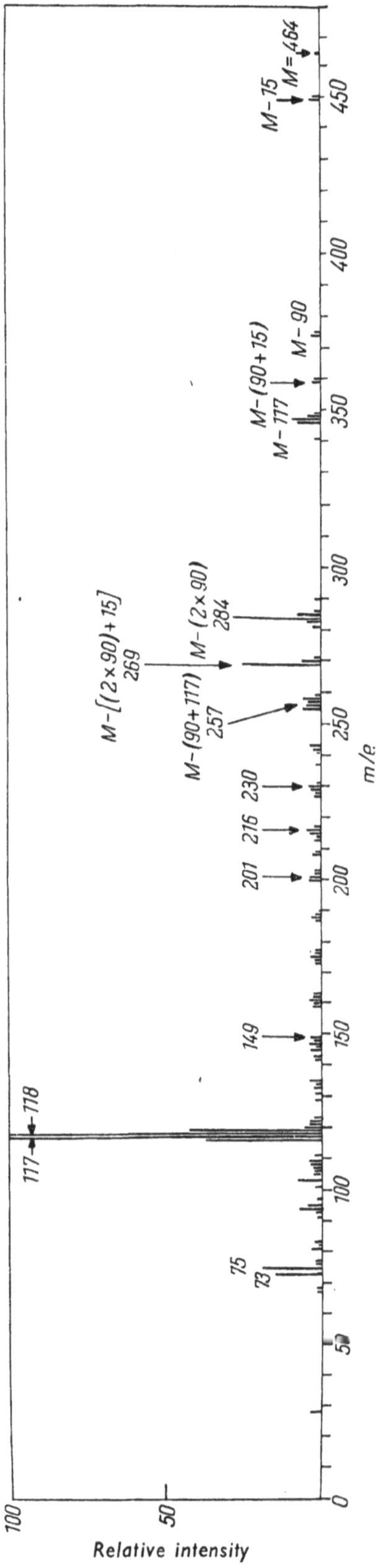

Fig. 2. Mass spectrum of the ditrimethylsilyl ether derivative of the pregnanediol reference standard. LKB 9000 Gas chromatograph — mass spectrometer. The column contained 1% QF-1 on 100–140 mesh Gas-Chrom P

The structure of many compounds is readily changed during the processing of extracts, but with this technique it is possible to obtain rapid information concerning the quantity and quality of these changes. Especially in work with the highly labile oestrogens 16α- and 16β-hydroxy-oestrone, 16-keto-oestradiol-17β and 2-hydroxy-oestrogens, it has been possible to follow structural changes, and thus rapidly to evaluate new modifications of the methods.

Finally, when the method has been developed, the extracts obtained from biological material are investigated and the specificity of the method assessed. When such investigations are carried out, the extracts are first examined on several different gas chromatographic columns to determine which of them will give the most specific result. Thereafter this column, if possible, is used in the GLC-MS instrument for assessing the specificity, and several mass spectra are taken, the peak being "sliced", at the time when the compound is emerging from the column. The spectra are then compared with those of authentic reference compounds and with each other. The relative abundance of each fragment should be approximately the same throughout the whole peak.

The specificity of a number of methods for the estimation of hormonal steroids in urine, blood and bile has been investigated by combined GLC-MS. Many of these methods are now used for research purposes or for routine clinical investigations. One of the main advantages of this technique is that the actual samples used for gas chromatographic estimation of steroid concentrations may be checked with respect to their specificity if the gas chromatographic tracings leave any suspicion that the peak used for quantitation contains some other substance(s). In our experience this is a not uncommon situation in routine clinical work. In many instances the degree of contamination can also be evaluated.

Several methods for the estimation of neutral steroids in biological fluids have been investigated with the GLC-MS technique. SJÖVALL et al. (1966) developed a procedure for serum dehydro*epi*androsterone sulphate and found that the purification procedure used resulted in a very specific estimate of this compound in human serum. VIHKO (1966), investigating some solvolyzable steroids in human plasma by the GLC-MS technique, was able to identify 5 neutral steroids, which could all be estimated. They were: Δ-5-androstane $3\beta,17\beta$-diol, Δ-5-pregnene-$3\beta,20\alpha$-diol, androsterone, dehydro*epi*androsterone and *epi*androsterone. Later, VIHKO et al. (1967) also identified pregn-5-ene-$3\beta,17\alpha,20\alpha$-triol in peripheral plasma after administration of Metopirone.

SALOKANGAS et al. (1967) presented a method for the estimation of urinary androsterone, etiocholanolone and dehydro*epi*androsterone. In order to achieve a high degree of specificity they used double derivatives (TMSi derivatives of the methyloximes of the steroids) for gas chromatography. The mass spectrometric results indicated that the compounds estimated in urine had identical mass spectra with the same derivatives of authentic reference standards. The high specificity with which the 11-deoxy-17-ketosteroids can be estimated has permitted the detection of a significant periodic variation of these steroids during the menstrual cycle, during which the correlation of the excretion of androsterone and etiocholanolone to oestrone is highly significant ($P < 0.001$), and that of dehydro*epi*androsterone to pregnanediol is also significant ($P < 0.01$) (ADLERCREUTZ et al., 1967).

A further development of the above-mentioned procedure was described by SCHAUMAN (1967), who included the estimation of pregnanediol and pregnanetriol in the method and found by GLC-MS that the estimation of these additional steroids was also specific.

A new rapid method for the routine clinical investigations of aldosterone in urine was proposed by SALOKANGAS and ADLERCREUTZ (1967a, b). Conversion of the aldosterone to its γ-lactone according to RAPP and EIK-NES (1966), yields a derivative of this labile compound which is stable in gas chromatography. By taking several mass spectra of the urinary compound emerging from a gas chromatographic column it was possible to show that, despite the rather short purification procedure, the peak used for estimation gives a reliable estimate of urinary aldosterone.

Identification of unknown steroids

Combined GLC-MS is at present the most sensitive tool available for the structural analysis of unknown compounds. Very much information may be gained by studies of the retention behaviour of the unknown compound on various selective and nonselective gas chromatographic liquid phases. The use of various derivatives of the compounds, for example trimethylsilyl ether derivatives, acetonides, methyloximes, methyl ethers and double derivatives, considerably increases the amount of structural information which can be obtained by using the gas chromatographic part of the combined instrument alone.

If the unknown compound has the same retention time as a reference standard on several gas chromatographic columns, and the mass spectra are identical, no interpretation of the mass spectrum is usually needed. However, with a completely unknown compound, interpretation of the spectrum is necessary in order to try to find characterizing fragments.

Owing to the very characteristic fragmentation pattern of steroid trimethylsilyl ethers, it is advisable to use this derivative for the mass spectrometric investigation of compounds containing hydroxyl groups. Then there is usually no problem of finding the molecular ion, since the loss of trimethylsilanol is very characteristic and gives a peak in the mass spectrum corresponding to M-90 (ENEROTH et al., 1964). Other typical fragments are M-15 (loss of a methyl group), M-(90 + 15), M-(2 × 90) and M-((2 × 90) + 15). When once the molecular weight of the unknown compound is known, a considerable number of steroids can be excluded. If the compound contains carbonyl groups, steroid methyloximes may be formed and these derivatives have a characteristic fragment at m/e M-30 (FALES and LUUKKAINEN, 1965), which may aid in identification of the molecular ion. Acetates are not suitable for this purpose. The loss of acetic acid on electron impact will frequently occur to such an extent that the molecular ion will hardly be detectable (see BIEMAN, 1962). For some steroids it may be very difficult to obtain information concerning the molecular weight, since they are destroyed during gas chromatography. This is especially true for the corticosteroids, which lose their side-chain and give rise to 17-ketosteroids (VANDEN HEUVEL and HORNING, 1960). However, the side-chain may be stabilized by methyloxime formation followed by conversion of the reacting hydroxyl groups to trimethylsilyl ethers (GARDINER and HORNING, 1966).

Steroids having a side-chain with a secondary alcohol and a methyl group like pregnanediol (Fig. 2) will as its TMSi derivative, gives a very intense peak at m/e 117. This is due to loss of the side-chain at C-17 (ADLERCREUTZ et al., 1966a). Similarly, with pregnanolones and pregnenolones, the side-chain is lost (PETERSON, 1962), in the case of TMSi derivatives together with the trimethylsilanol group at C-3, with formation of fragments at m/e M-(43 + 90). A more general fragmentation process is the elimination of ring D, with a resultant fragment M-(side-chain + 42) which may be used for establishing the size of the side-chain in sterols (RYHAGE and STENHAGEN, 1960).

In steroids with a keto-group in ring D, such as the 17-ketosteroids, ring D is also eliminated, and ions are formed at m/e M-56 and M-57 (BUDZIKIEVICZ and DJERASSI, 1962). Compounds having a double bond at C_5-C_6 and a 3β-hydroxyl group show an intense ion at m/e 129, which represents a group formed by the trimethylsilanol group + carbon atoms 1, 2 and 3 from ring A (DIEKMAN and DJERASSI, 1967). This peak may be the base peak in the mass spectrum of the TMSi derivative of dehydro*epi*androsterone. It is also present in the other naturally occurring 17-ketosteroids but the intensity is not so pronounced. The same fragment is also seen in C_{19}-steroids having a hydroxyl group at C-17. In the latter case, the ion at m/e 129 might represent the trimethylsilanol group attached to C-17 and three carbon and four hydrogen atoms from ring D (VIHKO, 1966). In this connection it must be emphasized that variation of the condition during mass spectrometry may considerably influence the fragmentation pattern of the TMSi derivatives, as also pointed out by DIEKMAN and DJERASSI (1967).

In the case of oestrogens, especially the polyhydroxy compounds, many difficulties arose when these steroids had to be identified by this technique. We gained our first experience when trying to identify 11-dehydro-oestradiol-17α in pregnancy urine (LUUKKAINEN and ADLERCREUTZ, 1965). By taking mass spectra of the TMSi derivative of the compound isolated and its methylation product it was possible to localize the fragments containing rings A and B, since the mass of the fragments was 58 mass units less for the methylated compound. This technique has later been used for the identification of other oestrogens and has proved very useful.

The specific identification of oestrogens having 3 hydroxyl groups is a more difficult problem. The fragmentation pattern of these compounds was completely unknown and the work was started by analyzing reference compounds of a number of polyhydroxy oestrogens. Some of these experiments had to be carried out with high resolution instruments or with the LKB low resolution instrument in combination with the recently developed so-called peak matching device.

The fragmentation pattern of oestriol is seen in Fig. 3. The upper panel shows the spectrum of the authentic oestriol reference standard as its TMSi derivative and the lower figure shows the spectrum of the urinary compound obtained from pregnancy urine. The spectra are practically identical. The proposed structure of the fragments is given below. All the main fragments from 218 to 504 have been elucidated (LUUKKAINEN and ADLERCREUTZ, 1967). The fragmentation pattern is highly interesting and could not have been clarified without the high resolution measurements shown in the upper figure. F represents the figure found and C the calculated theoretical value for the proposed fragment. As can be seen, there is good agreement between these values.

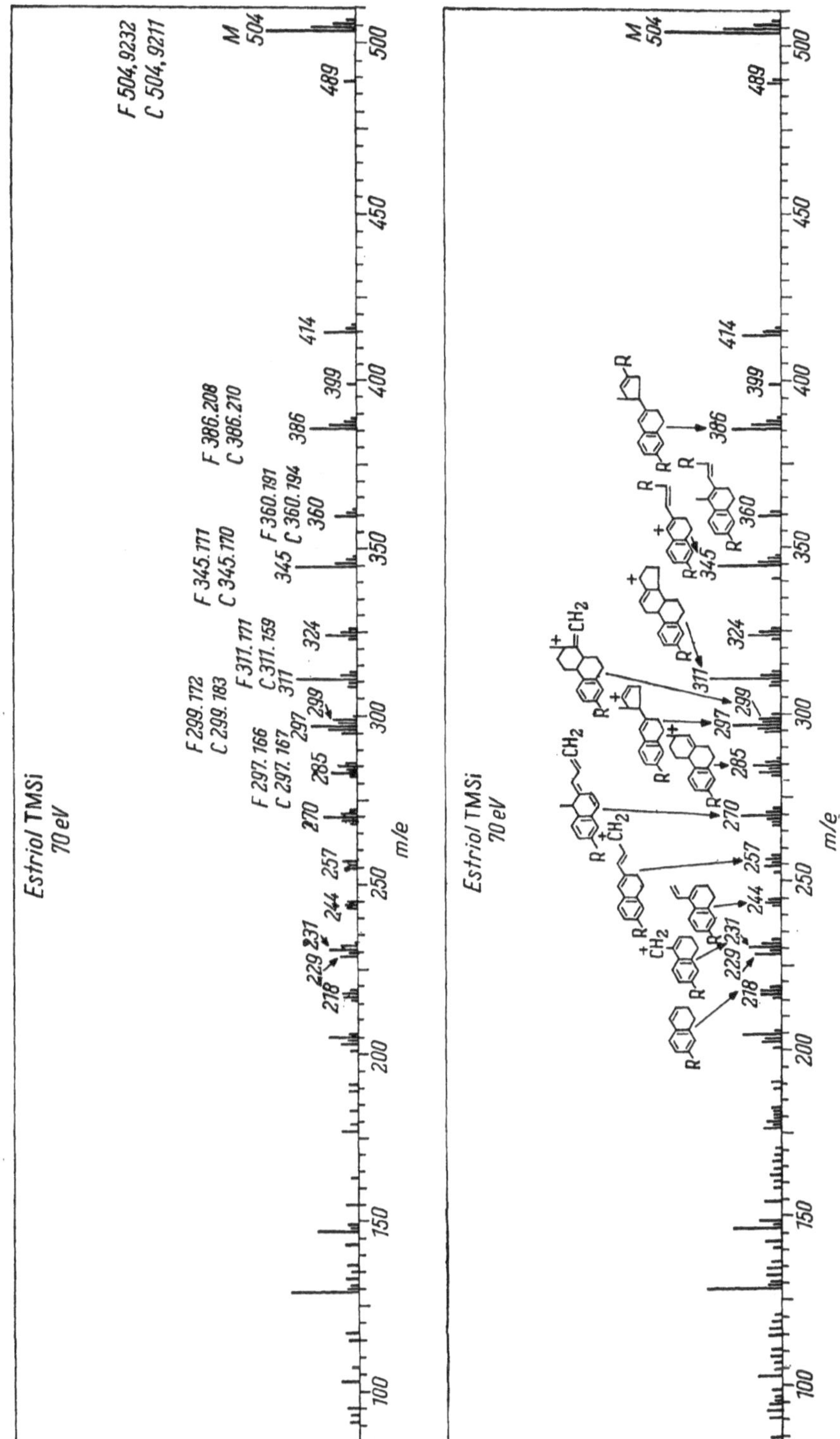

Fig. 3. Mass spectrum of the tritrimethylsilyl ether derivative of oestriol. The spectrum of the authentic reference oestriol is shown above, including the data of the high resolution measurements of some ions. F — found, C — calculated. The spectrum of the oestriol-like compound from urine (below) with the proposed structures of the fragments. According to LUUKKAINEN and ADLERCREUTZ (1967). Published with the permission of the Editor of Ann. Med. exp. Fenn.

In attempts to estimate the four epimeric oestriols substituted at C-16 and C-17, the overlap in gas chromatography of the usually large amounts of oestriol into the region of the other epimeric oestriols is especially liable to interfere with the estimation of 17-*epi*oestrol, which is present in very small amounts. The best way to overcome these difficulties is to convert the cis-glycolic oestrogens into acetonides and separate them from the transglycolic ones (Adlercreutz et al., 1966b). It was found that the acetonides or their TMSi derivatives gave excellent results in gas chromatography. The mass spectra of the TMSi derivatives of the acetonides of 16- and 17-*epi*oestriol differ clearly from each other (Adlercreutz and Luukkainen, 1968). The most striking difference is that the fragment at m/e 342 is missing from the spectrum of 17-*epi*oestriol. This fragment arises by the loss of acetone from the molecular ion of the derivative of 16-*epi*oestriol (Fig. 4) according to the scheme given by De Jongh and Bieman (1964). This fragment has the structure of a steroidal epoxide. Furthermore, in the spectrum of the TMSi derivative of 17-*epi*oestriol acetonide there are fragments at m/e 312 and 155 which are not present in the spectrum of the corresponding derivative of 16-*epi*oestriol. The ion at m/e 155 probably contains the ring D with the acetonide group unfragmented (Adlercreutz and Luukkainen, 1968).

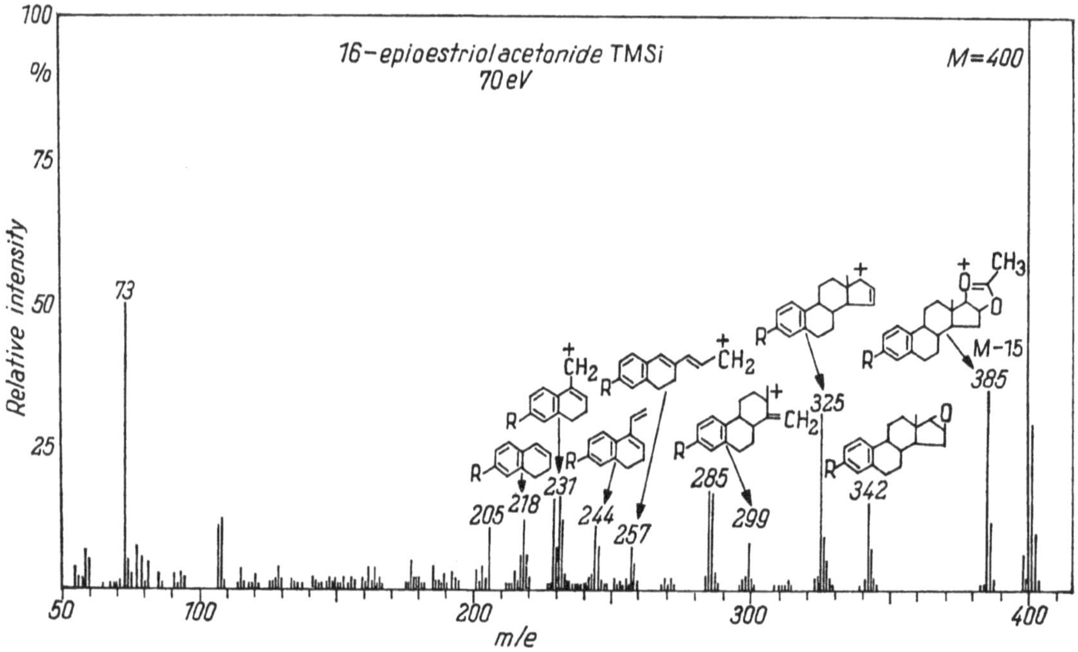

Fig. 4. Mass spectrum of the trimethylsilyl ether derivative of the 16-*epi*oestriol acetonide reference standard. Same conditions as in Fig. 1

The mass spectra of 16,17-*epi*oestriol and oestriol are very similar if the electron energy used is 70 eV. When the energy was lowered to 16 eV, which usually gives clearly different spectra for epimeric compounds, it was found that the spectra were still practically identical. However, when the energy was lowered to 13 eV, the spectra produced differed from each other in regard to the intensity of the ions (Adlercreutz and Luukkainen, 1968). Since oestriol and 16,17-*epi*oestriol have clearly different retention times in gas chromatography, it is now theoretically not difficult to identify these two compounds separately in extracts of biological material. However, owing to the very small amounts of 16,17-*epi*oestriol present it is in practice still a rather difficult problem.

In attempts to separate 6α- from 6β-hydroxyoestradiol by mass spectrometry, it was found that the difference was very small when 70 eV were used for the electron energy. However, with 16 eV, a clear difference could be observed, especially with regard to the ratio of the molecular ion (M) to the fragment (M-15) (ADLERCREUTZ and LUUKKAINEN, 1968). In addition, the retention times of the two compounds differ clearly.

Table 1

Relative intensities of fragments between m/e 311 and 504 for TMSi derivatives of oestrogen triols

m/e	311	324	345	360	386	399	414	489	504
Oestriol	22	10	19	6	15	3	12	4	35
15α-hydroxyoestradiol-17β	—	7	—	—	—	4	4	6	100
7α-hydroxyoestradiol-17β	—	19	—	—	—	4	74	2.5	5
6α-hydroxyoestradiol-17β	—	22.5	—	—	—	7	100	—	—
2-hydroxyoestradiol-17β	—	—	—	—	—	—	4	4	100

Table 1 shows the relative intensities of fragments between 311 and 504 for TMSi derivatives of a number of oestrogen triols. As can be seen, the structural isomers can easily be distinguished from each other, using an electron energy of 70 eV. In this respect the structural isomers clearly differ from the stereoisomers.

The data obtained on the oestrogens show that it is possible to identify these steroids in biological material by combining the gas chromatographic results with those obtained from the mass spectrometer. However, in these studies, it has been necessary to have the compounds at least partially separated before mass spectrometry. Now it is possible to use the mass spectrometer in a new way by including a so-called accelerating voltage alternator, as suggested by SWEELEY et al. (1966). By continuous recording of two mass spectrometric peaks of two compounds, unresolved in gas chromatography, it is possible to separate the compounds by this technique. In this way it is also possible to obtain the ratio of a labelled compound to an unlabelled compound.

The combined gas chromatograph-mass spectrometer may also be used in gas chromatography as a highly sensitive and specific detector which enables us to estimate steroids at the picogram level. In Fig. 5 the mass spectrometric recording of the first part of the mass spectrum of pregnanediol trimethylsilyl ether can be seen. As also shown in Fig. 2, the fragment at m/e 117 including the side-chain of pregnanediol gives an extremely intense ion, which in this case can be estimated only with the most insensitive galvanometer, since the other galvanometer responses go off the chart. If this peak is continuously scanned just as the compound emerges from the gas chromatographic column, it is possible to estimate picogram amounts of it (JAAKONMÄKI, LUUKKAINEN, ADLERCREUTZ, unpublished results). By modifying the instrument

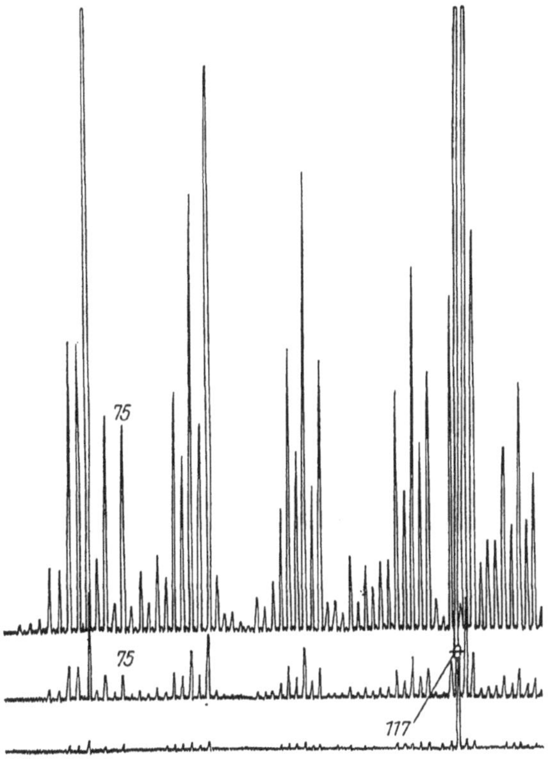

Fig. 5. Mass spectrometric recording of the first part of the spectrum of the ditrimethylsilyl ether derivative of the pregnanediol reference standard. The fragment at m/e 117 has been used for quantitation of the pregnanediol at picogram level

slightly, it seems possible in some instances to measure levels as low as onetenth of a picogram. It is of the greatest interest that such small amounts of steroid pass through the gas chromatographic column unabsorbed and that the response seems to be linear.

It can be concluded that combination of gas chromatography and mass spectrometry in the same instrument provides us with the most sensitive and specific tool we have today for studies on the steroids in biological fluids. We are witnessing the beginning of a very interesting period in steroid research and in the development of instruments of this kind and I think if we meet here in Jena three years from now, many new achievements will have come about through the use of this technique.

References:

[1] Adlercreutz, H., and T. Luukkainen, Proceedings of the Round Table an Gas Chromatography of Hormonal Steroids, applied to biological fluids, held May 23–25, 1967. ed. Scholler, R., and Jayle, M. F. 93 (1968)
[2] Adlercreutz, H., T. Luukkainen and W. Taylor, European J. Steroids **1**, 117 (1966a).
[3] Adlercreutz, H., S. Laiho, and T. Luukkainen, Proc. of the Meeting on gaschromatographic Determination of Hormonal Steroids, Rome 14 (1966b).
[4] Adlercreutz, H., T. Luukkainen, and A. Svanborg, Ann. Med. exp. Fenn. **45**, 277 (1967).
[5] Bieman, K., Mass spectrometry. Organic chemical applications, McGraw-Hill Series in Advanced Chemistry, New York, San Francisco, Toronto, London (1962).
[6] Budzikiewicz, H., and C. Djerassi, J. Amer. chem. Soc. **84**, 1430 (1962).
[7] De Jong, D. C., and K. Bieman, J. Amer. chem. Soc. **86**, 67 (1964).
[8] Diekman, J., and C. Djerassi, J. org. Chem. **32**, 1005 (1967).
[9] Eneroth, P., K. Hellström, and R. Ryhage, J. Lipid. Res. **5**, 245 (1964).
[10] Fales, H. M., and T. Luukkainen, Anal. Chem. **37**, 955 (1965).
[11] Gardiner, W. L., and E. C. Horning, Biochem. biophys. Acta **115**, 524 (1966).
[12] Luukkainen, T., and H. Adlercreutz, Biochim. biophys. Acta **107**, 579 (1965).
[13] Luukkainen, T., and H. Adlercreutz , Ann. Med. exp. Fenn. **45**, 264 (1967).
[14] Peterson, L., Anal Chem. **34**, 1781 (1962).
[15] Rapp, J. P., and K. B. Eik-Nes, Anal Biochem. **15**, 386 (1966).
[16] Ryhage, R., Arkiv Kemi **20**, 185 (1962).
[17] Ryhage, R., Anal Chem. **36**, 759 (1964).
[18] Ryhage, R., and E. Stenhagen, J. Lipid. Res. **1**, 361 (1960).
[19] Ryhage, R., S. Wikström and G. R. Waller, Anal Chem. **37**, 435 (1965).
[20] Salokangas, A., and H. Adlercreutz, Acta endocr. (Kbh.) Suppl. **119**, 97 (1967a).
[21] Salokangas, A., and H. Adlercreutz, Scand. J. clin. Lab. Invest. **19**, Suppl. **100**, 134 (1967b).
[22] Salokangas, A., K.-O. Schauman, H. Adlercreutz, and T. Luukkainen, Scand. J. clin. Lab. Invest. **19**, Suppl. **95**, 59 (1967).
[23] Schauman, K.-O., Acta endocr. (Kbh.) Suppl. **119**, 117 (1967).
[24] Sjövall, K., J. Sjövall, K. Maddock, and E. C. Horning, Anal. Biochem. **14**, 337 (1966).
[25] Sweeley, C. C., L. H. Elliott, I. Fries, and R. Ryhage, Anal. Chem. **38**, 1549 (1966).
[26] VandenHeuvel, W. J. A., and E. C. Horning, Biochim. Biophys. Research Comm. **3**, 356 (1960).
[27] Vihko, R., Acta endocr. (Kbh.) Suppl. **109**, (1966).
[28] Vihko, R., O. Jänne, and T. Luukkainen, Acta endocr. (Kbh.) Suppl. **119**, 39 (1967).

Anomalous Steroids and Fungi

Ž. Procházka

Institute of Organic Chemistry and Biochemistry,
Czechoslovak Academy of Sciences, Prague, ČSSR

Summary

Data are presented how *Rhizopus nigricans* hydroxylates various steroids, especially those with an anomalous skeleton. The question is also discussed if, and how, the quantitative determination of all products of hydroxylation of one compound with one microorganism can contribute to the study of the anatomy of hydroxylating enzymes. The substances submitted to hydroxylation were the following ones: Reichstein's S, 14β,18-cycloprogesterone, dehydroepiandrosterone, B-nor analog of the latter, and 3α,5α-cyclo-6β-hydroxy-androstan-17-one.

It is a well established fact that microbial hydroxylations of steroids usually do not lead to a single product, but that a mixture of isomeric products is formed. The question arose, whether this was caused by the presence of several different specific enzymes in the organism, each of them hydroxylating at one position only, or by the non-specificity of a single enzyme.

From the available data in the literature (Tamm, 1962) can be deduced that the second alternative is true. If so, the study of the course of hydroxylations of structurally different steroids and the determination of the yields of all isomeric products formed can represent a way how to study the anatomy of an enzyme. The expression "anatomy" was used probably for the first time by Ringold, and it seems that this word should be synonymous with the expression "general structure", which includes the primary, secondary and tertiary structure of an enzyme. For organic chemists this line of research is very attractive, because certain biochemical conclusions may be attained by purely chemical means.

We have studied, therefore, in the past few years, the structures of as many transformation products as we were able to isolate from a fermentation medium, where one steroid was submitted to the transformation with one microorganism. We dealt predominantly with steroids having anomalous skeletons, which can contribute to a further small refinement of the study of the anatomy of hydroxylating enzymes. In all experiments presented here, *Rhizopus nigricans* was used for oxygenations.

Rhizopus nigricans is generally considered to be an excellent 11α-hydroxylator. Our results show that it is so indeed. However, this does not mean, that it is specific for 11α-hydroxylation in all instances. This will be illustrated by our experimental results.

It is well known that progesterone is hydroxylated in a very high yield to 11α-hydroxyprogesterone. One is tempted to say that it is transformed so specifically. However, the same microorganism transforms another similar pregnene derivative, Reichstein's substance S, in a rather different manner. We were able to isolate from the reaction mixture (fermentation medium) five isomeric products of mono-hydroxylation. Again, 11α-hydroxylation was predominant. 6β-hydroxy-derivative was on the second place in yield, 7β-hydroxy-compound on the third, 6α- on the fourth and 11β-hydroxy-derivative (hydrocortisone) was formed only in trace amounts. Their ratio was very roughly: 23:13:5:5:0.2 (expressed as yields). No simple stereochemical considerations can help us to understand this sequence. The only conclusion we can draw from this result is, that the enzyme attacks the molecule in its central part, far from the polar ends of the molecule. But why 6β-derivative, although more sterically hindered, is formed in pretty higher yield than the 6α- or 7β-derivative, we cannot explain as yet.

When a slightly modified progesterone, i.e. 14β,18-cycloprogesterone was hydroxylated, the result was as expected (Procházka et al., 1965). The change in the shape of the molecule is so negligible in this compound, that the course of hydroxylation was practically the same as with normal progesterone. Cycloprogesterone was hydroxylated also in high yield at the position 11α.

However, when we have used an androstane derivative with a double bond shifted to the position 5—6, i.e. dehydroepiandrosterone, *Rhizopus nigricans* hydroxylated it almost exclusively into position 7α and 7β. No 11α hydroxylation was observed. It is easy to explain, why position 7 was attacked. It is an allylic position, attacked even by molecular oxygen. However, why exactly *Rhizopus nigricans* leaves the 11α-position untouched is beyond our understanding.

In another experimental series we have submitted to hydroxylation with *Rhizopus nigricans* the B-nor-analog of dehydroepiandrosterone (Procházka et al., 1961). In this case the position 7 is no more present in the molecule. The only hydrogen atom at the allylic position of the five-membered ring is tertiary. Obviously, *Rhizopus nigricans* is not able to introduce a tertiary hydroxyl and so it does two other transformations. It hydroxylates the compound at the position 11α, i.e. it does its usual job, but it also attacks the molecule in the neighbourhood of the former allylic position of the normal steroid, i.e. it adds the oxygen atom directly to the double bond, giving rise to the 5,6-epoxide.

To prove definitely and by direct correlation that the hydroxy-group was indeed at C-11 in the B-nor steroid we needed 11α-hydroxylated dehydroepiandrosterone. As we were unable to screen out a microorganism capable to hydroxylate this compound directly at C-11, we decided to hydroxylate another compound with an anomalous skeleton, which can easily be transformed to dehydroepiandrosterone, i.e. 3α,5α-cyclo-6β-hydroxyandrostan-17-one, which we shall call i-dehydroepiandrosterone.

The hydroxylation of this compound has shown again that *Rhizopus nigricans* is not as specific for 11α-hydroxylation as might be concluded on the basis of the hydroxylation of progesterone. We were able to isolate three different isomeric products of monohydroxylation from the reaction mixture. The structure of these compounds was proved as follows. The main product was again 11α-hydroxy-derivative. Its structure was proved by transforming it first to the normal derivative, i.e. 11α-hydroxylated dehydroepiandrosterone, which was then oxidized to the known 11α-hydroxyandrost-4-ene-3,17-dione. The authentic specimen of the latter compound was obtained by sodium bismutate oxidation of epi-hydrocortisone. Physical constants of both specimens were identical. The structure of the second product was proved by Dr. Samek from our Institute by NMR-spectroscopy. The spectrum showed that the second product had the structure of 1α-hydroxy-derivative. The third product of hydroxylation was formed only in a very small yield. It was shown chromatographically, by transforming it to the normal steroid, that it must possess the structure of 3α,5α-cyclo-6β,7-dihydroxy-androstan-17-one. NMR-spectra fully corroborated this conclusion.

From these experiments we can see again that *Rhizopus nigricans* prefers to hydroxylate the central parts of the molecule, which are most remote from the polar substituents already present in the molecule. However, the absence of a substituent at the position 3, and the presence of the hydroxy-group at C-6 caused that position 1 was also attacked. This position, in this particular case, is also remote from the polar substituents. On the other hand, the tendency of *Rhizopus nigricans* to attack the central parts of the molecule is so strong in this microorganism, that it did one transformation which is rather uncommon for most microorganisms. It hydroxylated a position vicinal to a hydroxy group present already in the molecule, i.e. the position 7.

Just to complete this picture, I would like to mention that *Rhizopus nigricans* transformed the i-dehydroepiandrosterone also in an additional way. The fourth product which we were able to isolate from our fermentation medium was both hydroxylated, and reduced. Although the structure of this product is not yet completely elucidated, we propose for it the structure of 3α,5α-cyclo-6β,11α,17β-trihydroxyandrostane.

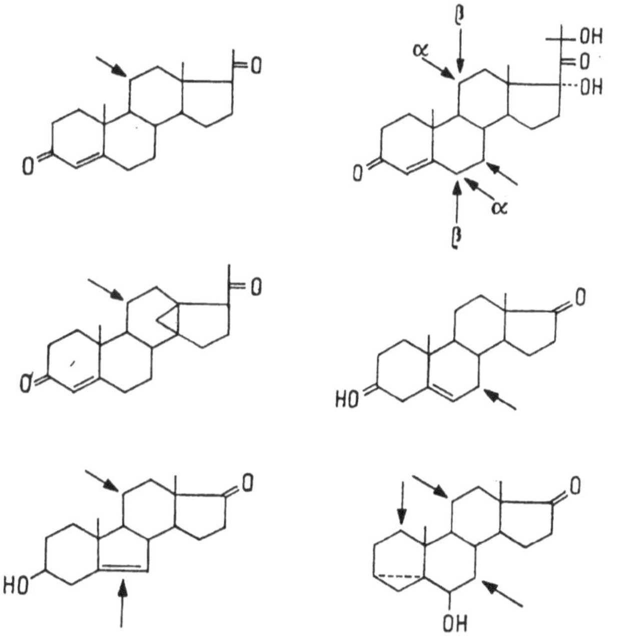

Fig. 1. Sites of hydroxylation of various steroids with *Rhizopus nigricans*

Although our results with *Rhizopus nigricans* are not very numerous, and no very useful conclusions about the anatomy of its hydroxylase can be drawn, we believe that this organical-chemical approach to the study of the anatomy of hydroxylating enzymes is not without merit. A disadvantage of this method consists in the fact that it is experimentally more requiring than the usual experiments with steroid transformations, where the experimenter was interested only in microorganisms able to transform in a specific way, i.e. in high yield, and in the isolation of the main product only. The isolation of all products of hydroxylation from the medium is a rather tedious job, because the products are isomeric, and hence difficult to separate.

At the end I want to express my thanks to Dr. Nina Lestrovaya from Moscow, Mr. Petar Prekajski from Belgrade, and Dr. Zdeněk Samek from Prague, who contributed mostly to obtain some novel results presented in this paper.

References

[1] Procházka, Ž., J. Fajkoš, J. Joska and F. Šorm, Collection Czech. Chem. Commun. **26**, 2068 (1961).
[2] Procházka, Ž., and F. Šorm, Collection Czech. Chem. Commun. **30**, 1874 (1965).
[3] Tamm, C., Angew. Chemie **74**, 225 (1962).

Introduction of Oxygen Functions in 3-Desoxysteroids of the C_{19}-Series by *Corynebacterium*

K. Schubert, W. Schumann and C. Hörhold

Institut für Mikrobiologie und experimentelle Therapie, Jena,
der Deutschen Akademie der Wissenschaften zu Berlin, DDR

Summary

3-Desoxy-C_{19}-steroids were attacked by strains of the genus *Corynebacterium*. The introduction of oxygen-functions in 3-desoxy-C_{19}-compounds occurred instead of degradation products. In the C_{27}- and C_{21}-series no transformation could be observed with analogous compounds. With *Corynebacterium* species SG 985 oxygenation took place in 3-position under isomerization of the double bonds ($\Delta^3 \rightarrow \Delta^4$-3-keto group and $\Delta^{3,5} \rightarrow \Delta^{4,6}$-3-keto group). In the Δ^4-compound an oxygen-function was introduced in the 3-position by the same strain under formation of a Δ^4-3-keto group without isomerization. With the strain *Corynebacterium pseudodiphtheriticum* B 2 the introduction of an oxygen-function in 7- instead of in 3-position was found (Δ^5- $\rightarrow \Delta^5$-7-keto group and $\Delta^{3,5} \rightarrow \Delta^{3,5}$-7-keto group).

Natural occurring steroid hormones like testosterone, progesterone or cortisone possess in 3-position of the steroid skeleton an oxygen-function in conjugation with a double bond in the 4,5-position. Already partial hydrogenation of this α,β-unsaturated keto group leads to a decrease in biological activity. The α,β-unsaturated keto group is also important in the microbiological degradation of steroids. Schubert, Böhme and Hörhold have shown the degradation of progesterone by *Mycobacterium smegmatis*, which occurs through the introduction of an additional double bond in the 1,2-position followed by aromatization and splitting of the ring A (1, 2).

In the last few years 3-desoxysteroids have become of certain importance, since several synthesized compounds of this kind also possess hormone activity. An oxygen-function in 3-position, therefore, is not absolutely necessary for biological activity. Ring A unsaturated 3-desoxy-compounds in the C_{19}-series possess androgenic and anabolic activity, whereas analogous compounds in the C_{21}-series have gestagenic activity.

In search of new microbial degradation mechanisms, excepting the above-mentioned pathway via the oxygen-function in 3-position, 3-desoxy-compounds in the C_{27}-, C_{21}- and C_{19}-series were prepared and incubated with bacteria or fungi respectively.

Degradation products caused by microbial attack to the steroid skeleton were not found in our investigations with 3-desoxysteroids. Only in the presence of a Δ^4-3-keto group in the ring A degradation seems to be possible.

The following investigations led to the detection of some transformation products referring to the microbial introduction of oxygen-functions.

As the first degradation steps occur under dehydrogenation, we started with $\Delta^{3,5}$-cholestadiene as substrate. A total of 200 different microorganisms were investigated under aerobic or anaerobic conditions respectively by variation of the culture medium and the incubation times. Among the investigated strains there were a great number of bacteria and fungi able to utilize cholesterol as well as some so-called unidentified petroleum-bacteria. Thin-layer chromatography of culture extracts did not yield any transformation products, which means that no attack of the inserted 3-desoxysteroid had taken place. The hydrogenated compounds Δ^5-cholestene and 5α-cholestane were neither degradated nor transformed under these conditions. In order to eliminate the influence of the side-chain of a C_{27}-steroid, one 3-desoxysteroid of the C_{21}-series $\Delta^{3,5}$-pregnadiene-20-one was investigated. This compound was not attacked by microorganisms either. Conversion only took place with the analogous compound in the C_{19}-series $\Delta^{3,5}$-androstadiene-17-one by *Corynebacterium* species. The influence of the side-chain is evident. Therefore the following investigations were made with 3-desoxy-C_{19}-steroids only.

In the case of $\Delta^{3,5}$-androstadiene-17-one (I) *Corynebacterium* species SG 985 effected a decrease of the added steroid. After extraction of the culture medium a UV-positive, more polar product was established by thin-layer chromatography. The UV-maximum at 282 nm is corresponding with a $\Delta^{4,6}$-3-keto group. At first the established introduction of an oxygen-function in 3-position of a 3-desoxysteroid was surprising. As to the behaviour in paper chromatography, melting-point, infrared-spectrum and elementary analysis, however, this substance was identical with an authentic sample of $\Delta^{4,6}$-androstadiene-3,17-dione (II).

With that for the first time was established the microbial introduction of an oxygen-function in 3-position of a 3-desoxysteroid.

In a review by PETERSON (3) it is revealed that oxygenation by microbial enzymes takes place till now in 23 different positions as contrasted with 18 positions of oxygenation by mammalian enzymes. Microbial oxygenation at C-atom 3 and C-atom 20 could not be observed so far, because all natural starting-materials possess an oxygen-function in these positions. It is worth noting that the reaction simultaneously is caused both by an oxygenase and an isomerase. The latter is responsible for the double bond shift from $\Delta^{3,5}$ to $\Delta^{4,6}$. When the isolated metabolite $\Delta^{4,6}$-androstadiene-3,17-dione (II) was incubated again with the above-mentioned *Corynebacterium*, $\Delta^{4,6}$-androstadiene-17β-ol-3-one (III) was isolated. The $\Delta^{4,6}$-3-keto group was preserved, only the 17-keto group was reduced to a 17β-hydroxy group.

In this connection the question arose wether double bonds may influence microbial transformations. The saturated compound 5α-androstane-17-one was not attacked by *Corynebacterium* species SG 985.

On the other hand, Δ^3-androstene-17-one (IV) was transformed with formation of a Δ^4-3-keto-group in ring A. The transformation product was identical with authentic Δ^4-androstene-3,17-dione (V). This is another case of introduction of an oxygen-function in 3-position with a double bond shift from Δ^3 to Δ^4.

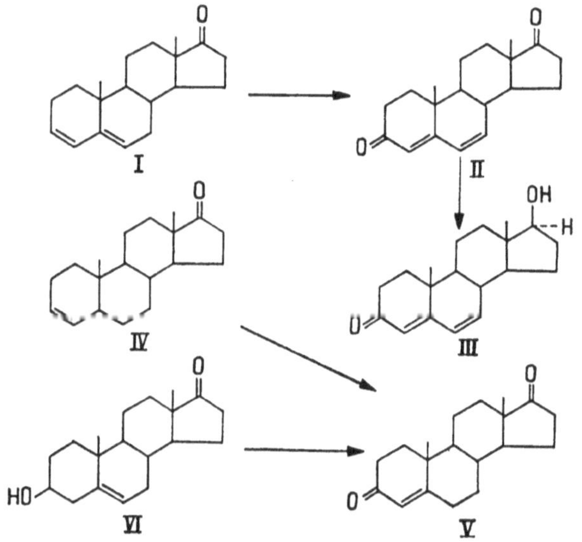

There exists no indication of direct microbial introduction of a keto group in steroids. Transformation products from Δ^{3}-androstene-17-one and $\Delta^{3,5}$-androstadiene-17-one are Δ^{4}-androstene-3,17-dione or $\Delta^{4,6}$-androstadiene-3,17-dione respectively, but the corresponding hydroxy compounds were not found in our investigations.

Compared with this, dehydroepiandrosterone (VI) was converted by the same strain SG 985 into Δ^{4}-androstene-3,17-dione (V). One may assume primary hydroxylation and secondary dehydrogenation of the introduced hydroxy group into a keto group.

The introduction of an oxygen-function by the strain SG 985 in Δ^{4}-androstene-17-one (VII) also led to the formation of an α,β-unsaturated keto group (VIII) with a UV-maximum at 241 nm. Hereby oxygenation may have taken place in 3-position without a double bond shift. Further investigations should be made.

An analogous result was published by a Japanese team. They found an oxygen-function introduced by mammalian enzymes in a Δ^{4}-3-desoxysteroid (IX) yielding a Δ^{4}-3-ketosteroid (X). This reaction shows that a 3-desoxysteroid drug may be oxygenated during metabolism in the 3-position. The above-mentioned microbial system may be useful for investigating this type of reaction.

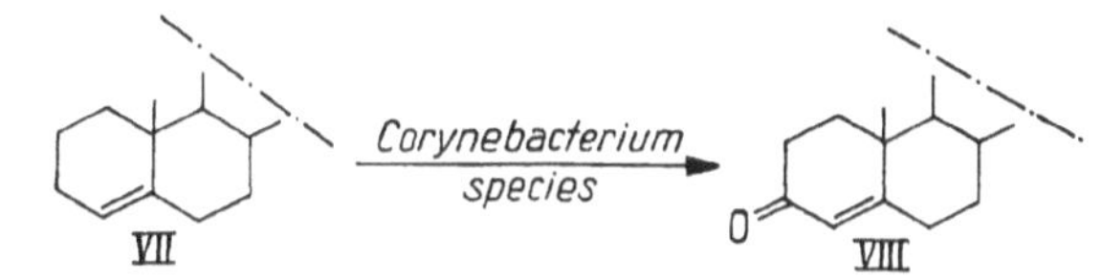

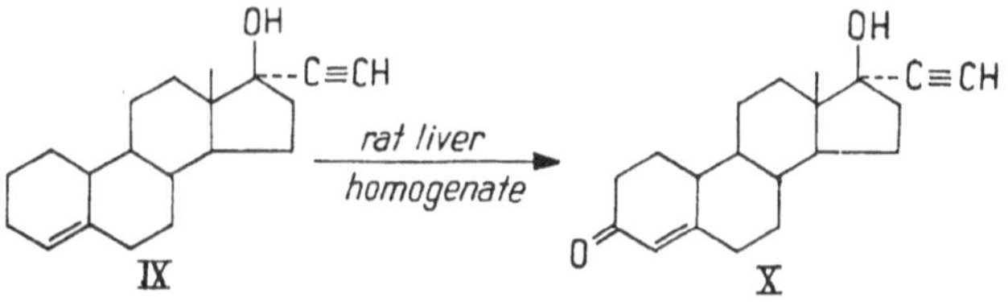

The observed oxygenations with *Corynebebacterium* species occur stereospecifically. The introduction of oxygen takes place either with isomerization of the present double bonds or without a double bond shift under formation of a low-energy conjugated system.

With another strain of the same genus, namely with *Corynebacterium pseudodiphtheriticum* B 2 an oxygen-function in 7- instead of in 3-position was introduced in $\Delta^{3,5}$-androstadiene-17-one (I). Microbial hydroxylations of C 19-steroids in 7-position have been known for rather a long time,' but each time only 7-hydroxy compounds instead of 7-keto compounds were found. In the meantime we have established two UV-positive, more polar products with a UV-maximum at 279 nm, corresponding with a $\Delta^{3,5}$-7-keto group. Characteristical bands in the infrared spectra indicate a hydroxy group in each compound.

A molecular weight of 300 was found for the two products by means of negative ion molecular mass spectrometry. Therefore each of the two compounds must possess one hydroxy group. By paperchromatography product XI was found to be less polar than product XII, that means both hydroxy groups must be located in different positions. The tertiary nature of the introduced hydroxy groups was established through resistance to oxidative attack and to acetylation. Tertiary hydroxy groups are possible only at the C-atoms 8, 9 or 14, since in 5-position there is already a double bond. The two compounds were dehydrated only under ionic conditions according to a transelimination.

Transelimination chiefly occurs with axial hydroxy groups. The hydrogen atoms or substituted groups at the C-atoms 8, 9 or 14 are situated in axial position.

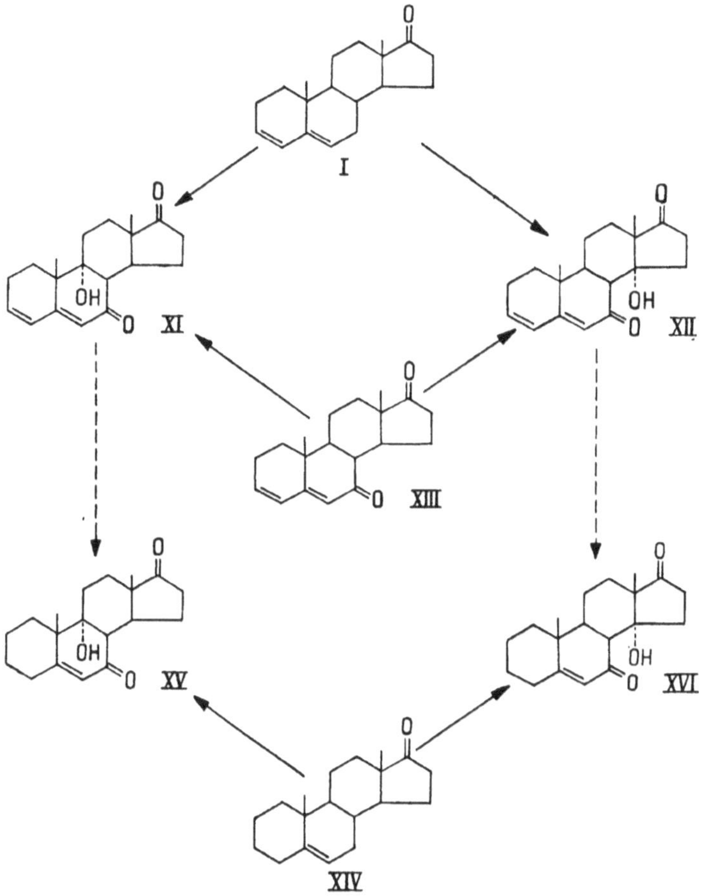

A 9α-hydroxy group is indicated in compound XI. This finding is supported by the so-called "steroid numbers", obtained by gas-chromatography. The position and type of functional groups may be made out by determining the retention time of the steroid proved and of the basic steroid skeletone under the same conditions. The 9α-position of the hydroxy group in XI was also proved by NMR.

In order to localize the position of the hydroxy group in compound XII, only the 8β- and 14α-position was still left. Comparing the NMR-spectra of the free compound with that of the hydroxy compound, the hydroxy group probably is located in 14-position, since an 8β-hydroxy group will cause a pronounced downfield shift of both angular methyl groups.

The same hydroxy compounds were isolated in higher yields from $\Delta^{3,5}$-androstadiene-7,17-dione (XIII) as a metabolite after incubation with *Corynebacterium pseudodiphtheriticum* B 2. The compounds XI and XII do not suffer further transformation when incubated a new.

The introduction of an oxygen-function in 7-position of Δ^{5}-androstene-17-one (XIV) occurred also with the B 2-strain, forming a Δ^{5}-7-keto group and additional hydroxy groups in the same positions (XV) and (XVI) as demonstrated by catalytic hydrogenation of compounds XI and XII.

In the case of Δ^{5}-androstene-17-one and $\Delta^{3,5}$-androstadiene-17-one the formation of a keto-group with the B 2-strain always appeared in conjugation with the present double bond.

Now we know three species-specific mechanisms in microbial oxygenation:

1. Oxygenation occurs in the absence of double bonds.
2. An oxygen-function will be directed through C-C-double bonds in the steroid-skeleton to the neighbouring C-atom.
3. An oxygen-function is introduced under isomerization of the double bond.

References

[1] SCHUBERT, K., K.-H. BÖHME und C. HÖRHOLD, Hoppe-Seyler's Z. physiol. Chemie **325**, 260 (1961).
[2] SCHUBERT, K., K.-H. BÖHME und C. HÖRHOLD, Z. f. Naturforsch. **16b**, 595 (1961).
[3] PETERSON, D. H., in "Biochemistry of industrial microorganisms" Academic Press, New York 1963, S. 537.

Über einige neue Ergebnisse der enzymatischen Umwandlung von C_{18}-Steroiden durch verschiedene Mikroorganismen

A. Schubert, F. W. Kunstmann und D. Onken
Wissenschaftliche Laboratorien des VEB Jenapharm, Jena, DDR

Zusammenfassung

Es werden Möglichkeiten für die Trennung racemischer Steroidverbindungen mit Hilfe mikrobiologischer Methoden und fermentative Umwandlungen enantiomerer Steroide beschrieben.

Die synthetischen Verfahren zur Herstellung von Steroidwirkstoffen haben in den vergangenen Jahren erheblich an Interesse gewonnen. Während die Arbeiten der Arbeitskreise um Smith, Windholz und Torgov hauptsächlich die C_{18}-Steroide zugänglich machen, werden bei den Synthesemethoden der französischen Schule um Velluz auch die C_{19}-Steroide erhalten.

Das Kernproblem, die moderne Synthesemethodik effektiv, d. h. mit hoher Ausbeute zu betreiben, besteht darin, die Asymmetriezentren möglichst direkt durch asymmetrische Synthesen aufzubauen, wie das z. B. von Gibian et al. publiziert wurde (Abb. 1).

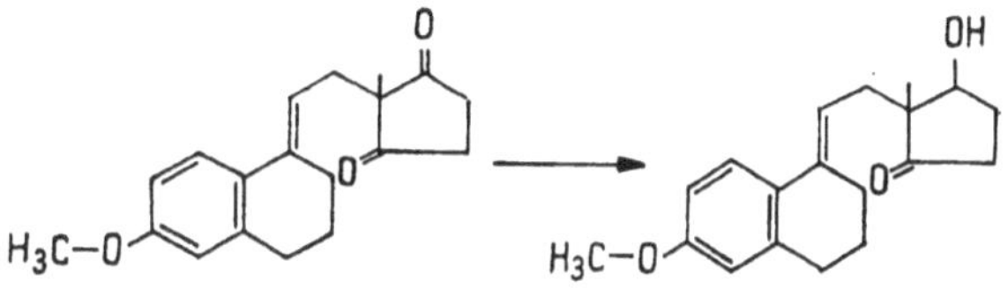

Abb. 1

Bei chemischer Cyclisierung erhält man Racemate (Abb. 2). Wir haben versucht, Racemate u. a. auf biologischem Wege zu trennen und Anwendungsmöglichkeiten auch für die enantiomere Form zu finden.

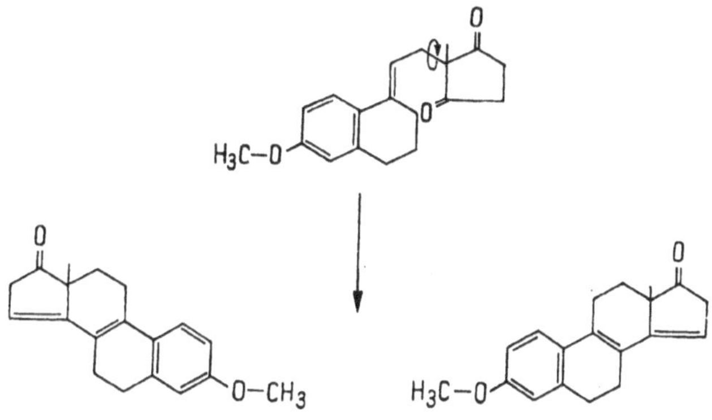

Abb. 2

Wir haben uns das Ziel gestellt,

1. Möglichkeiten der biologischen Racemattrennung bei einer Reihe von synthetisch gewonnenen Steroidkörpern zu untersuchen;

2. Mikroorganismen zu finden, deren Enzymsysteme außer der natürlichen Form auch die enantiomere Form umwandeln.

Biologische Racemattrennungen sind in der Literatur bisher wenig beschrieben worden. Zur Klärung der Fragestellung haben wir folgende Racemate eingesetzt (Abb. 3):

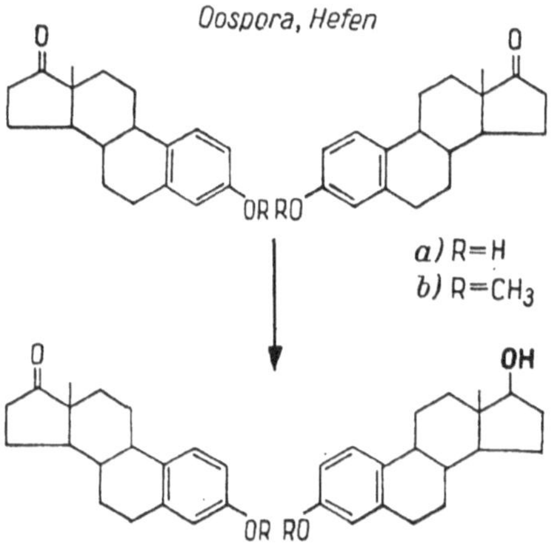

Abb. 3

Die Racematform wird aus Gründen der Übersichtlichkeit mit ihren beiden optischen Antipoden dargestellt.

Aus der Abbildung geht hervor, daß eine Trennung des Racemats durch selektive Reduktion der 17-Ketogruppe durch Hefen oder *Oospora*-Arten erreicht wird. Dabei wird ausschließlich die natürliche Komponente, die d-Form des Racemats angegriffen, während die l-Form unverändert zurückgewonnen wird.

Oospora, insbesondere *Oospora lactis*, wurde für Umwandlungen in der Steroidreihe noch nicht eingesetzt. Wir haben den aktivsten der für diese Reaktionen verwendeten Stämme aus Speisequark isoliert. Oospora ist wesentlich aktiver als die von uns geprüften Hefen. Die Umsetzungszeit ist kürzer, und die Ausbeuten sind im Vergleich zu den Hefen wesentlich höher.

Uns interessierte in diesem Zusammenhang noch die Frage, ob die Zahl der Doppelbindungen im Molekül einen wesentlichen Einfluß auf die Reaktionsgeschwindigkeit bei der Umsetzung und auf die Ausbeute selbst ausübt.

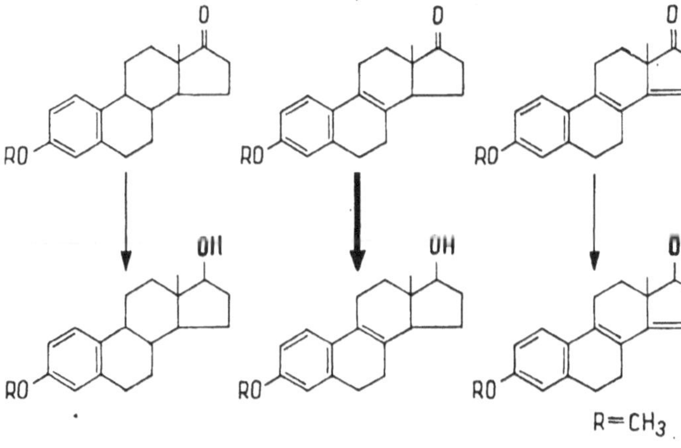

Abb. 4

Wir haben deshalb Östronmethyläther, rac. Dehydroöstronmethyläther und rac. Bisdehydroöstronmethyläther bei der Inkubation mit *Oospora* als Substrate verwendet (Abb. 4).

Wir haben gefunden, daß die Umsetzung des rac. 8-Dehydroöstron-3-methyläthers mit *Oospora* die höchsten Ausbeuten liefert. Sie liegen bei dieser Verbindung nahezu bei 100%, während der Östron-3-methyläther und der rac. 8.14-Bisdehydroöstron-3-methyläther nur zu einem geringen Teil in 17-Stellung reduziert werden. Daraus folgt, daß man eine biologische Racemattrennung mit *Oospora* am zweckmäßigsten mit dem rac. 8-Dehydroöstron-3-methyläther vornehmen kann. Hefen zeigen die gleiche Tendenz.

Interessant ist die Umsetzung von Racematen mit dehydrierend wirkenden Mikroorganismen, wie beispielsweise mit Mycobakterien. Es ist bekannt, daß Mycobakterien eingesetzt werden, um Hydrocortison zu Prednisolon zu dehydrieren. Wir haben für unsere Versuche racemisches 19-nor-Testosteron eingesetzt (Abb. 5).

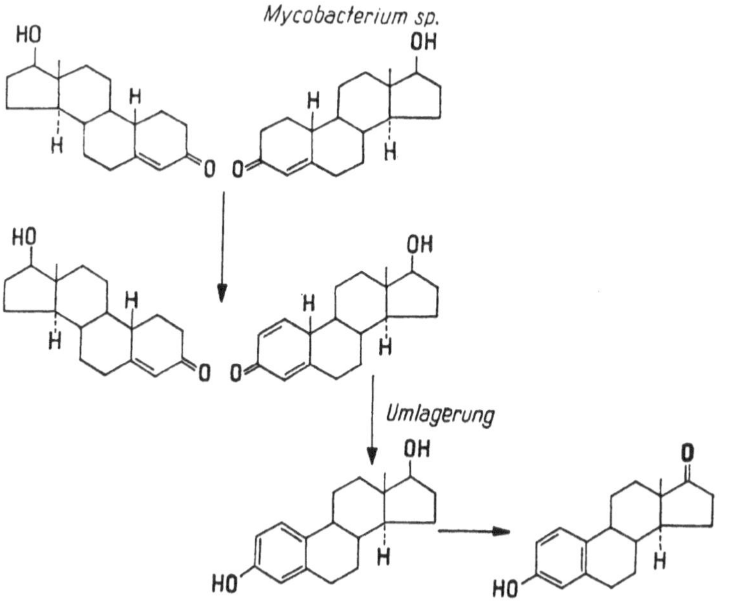

Abb. 5

Bei dieser Umsetzung erhielten wir ent-19-nor-Testosteron unverändert zurück, während die natürliche Komponente von *Mycobact. sp.* zunächst zwischen den C-Atomen 1 und 2 dehydriert wurde. Anschließend erfolgte eine nichtenzymatische Umlagerung zum Östradiol. Im weiteren Verlauf der Fermentation wird das Östradiol wiederum durch Einwirkung der Enzymsysteme zu Östron dehydriert.

Ein ähnliches Ergebnis wurde von Greenspan et al. bei Verwendung von *Corynebacterium simplex* erhalten.

Im Verlauf unserer Untersuchungen haben wir einen weiteren Mikroorganismus, und zwar einen speziellen *Circinella*-Stamm eingesetzt. Überraschenderweise erhielten wir mit diesem *Circinella*-Stamm und auch mit zwei *Rhizopus*-Stämmen sowohl einen Umsatz der natürlichen Form als auch des optischen Antipoden des eingesetzten Racemats (Abb. 6).

Wir haben im speziellen Fall rac. 19-nor-Testosteron verwendet und erhielten eine 100%ige Umsetzung, d. h., daß sowohl die natürliche Form als auch die enantiomere Form vollkommen umgewandelt wurde.

Wir standen diesem Ergebnis zunächst skeptisch gegenüber. Die Ergebnisse konnten jedoch bestätigt werden, nachdem jeweils die reine ent-Verbindung und die reine d-Verbindung mit *Circinella sp.* quantitativ in die entsprechende 9α-Hydroxy-Verbindung übergeführt wurde.

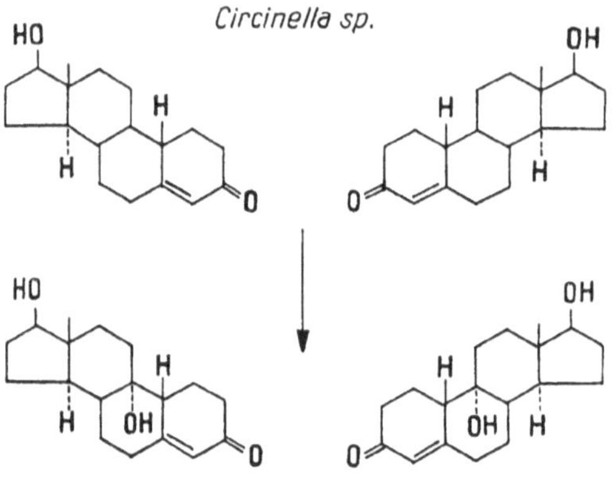

Abb. 6

Die Einführung der 9α-Hydroxygruppe in ein Steroidgerüst durch die Fermentsysteme des Pilzes *Circinella sp.* wurde von uns bereits in früheren Jahren beschrieben. Es wurden jeweils die natürliche Form des Progesterons, des Testosterons und des 19-nor-Testosterons an dem C-Atom 9 hydroxyliert.

Das aus dem enantiomeren 19-nor-Testosteron isolierte Produkt ist nach Schmelzpunkt, Infrarotspektrum und chromatographischem Verhalten identisch mit dem Produkt, das aus der natürlichen Form erhalten wird. Der Drehwert entspricht dem Wert der natürlichen Verbindung, jedoch mit umgekehrtem Vorzeichen.

d-9α-Hydroxy-19-nortestosteron Fp.: 205—215 °C $(\alpha)_D = +61°$ (Dioxan; c = 0.53)

l-9α-Hydroxy-19-nortestosteron Fp.: 206—216 °C $(\alpha)_D = -57°$ (Dioxan; c = 0.5)

In diesem Zusammenhang soll erwähnt werden, daß rac. Östron ebenfalls von *Circinella* umgesetzt wird. Allerdings findet beim Einsatz dieses Substrats keine Hydroxylierung, sondern wahrscheinlich eine Dehydrierung statt. Die Umsetzung erfolgt sehr langsam, und die Metaboliten konnten bisher wegen erheblicher Isolierungsschwierigkeiten noch nicht einwandfrei identifiziert werden.

Parallel zu unseren Untersuchungen wurden auch von L. L. Smith Umsetzungsprodukte von enantiomeren Steroiden erhalten, und zwar wurde die Einführung von Hydroxylgruppen in der 1β-Stellung, in 10β- und in der 11α-Stellung sowie die Dehydrierung zwischen den Kohlenstoffatomen 1 und 2 beschrieben.

Die enzymatische Umwandlung von enantiomeren Steroiden eröffnet einige neue Aspekte. Man kann erwarten, daß entgegen der bisherigen Ansicht biologisch aktive Verbindungen auch bei enantiomeren Formen entdeckt werden.

Bisher kann man feststellen, daß solche Veränderungen an enantiomeren Verbindungen, die durch Fermentsysteme ausgelöst werden, ausschließlich die Ringe A, B, C des Steroidgrundgerüstes betreffen. Der Bereich um das optisch aktive Zentrum am C-Atom 13 bleibt dabei unberührt. Die Fermentsysteme der Mikroorganismen verhalten sich demnach so, als ob der Ring D des Steroidgerüstes einschließlich der angulären Methylgruppe am C-Atom 13 nicht vorhanden wäre.

Literatur

[1] Gibian, H., et al., Tetrahedron Letters 1966, 2321.
[2] Greenspan, G., et al., J. Org. Chem. **31**, 2512 (1966).
[3] Padakandla, G. R., Diss. Mat.-Nat. Fak. Jena, 1962.
[4] Schubert, A., Proc. 4th Intern. Congr. Biochem., Vienna 1958, S. 120.
[5] Schubert, A., et al., Z. f. Chemie **2**, 289 (1962).
[6] Smith, H., et al., Experientia **19**, 394 (1963).
[7] Smith, L. L. et al., J. Am. Chem. Soc. **88**, 3120 (1966).
[8] Torgov, J. V., Pure appl. Chem. **6**, 525 (1963).
[9] Torgov, J. V., Recent Developments Chem. natur. Carbon Components (Budapest) **1**, 233–319.
[10] Velluz, L. et al., Angew. Chem. **77**, 185 (1965).
[11] Velluz, L. et al., Tetrahedron 1966, 495.
[12] Windholz, T. B., and M. Windholz, Angew. Chem. **76**, 249 (1964).

Les "Récepteurs Hormonaux". Mise en Evidence de la Liaison Spécifique de l'Oestradiol à des Protéines Uterines

E. E. Baulieu

Unité de Recherches sur le Métabolisme Moléculaire et la Physio-Pathologie des Stéroides — Service de Chimie Biologique — Hôpital de Bicêtre, France

Résumé

La liaison spécifique de l'Oestradiol à des protéines du ,,Surnageant" des cellules de l'endomètre de truie impubère est décrite. La spécificité est le fait:

1°/ de l'absence de compétition pour la liaison entre l'Oestradiol et les autres hormones stéroides.

2°/ de caractéristiques quantitatives indiquant une limitation de la capacité de liaison aux concentrations ,,physiologiques" d'Oestradiol.

3°/ de l'absence d'un système analogue dans une préparation correspondante de tissu pulmonaire.

Dans ce système, le Diéthylstilboestrol est équivalent à l'Oestradiol. L'étude de différents stéroides permet de montrer l'importance du groupe 17β-hydroxyl et du noyau phénol.

Le fractionnement des protéines en question est sommairement rapporté. Une méthode radio-chimique de dosage de l'Oestradiol, sensible au millimicrogramme est incidemment proposée.

L'étude du mécanisme d'action des hormones est à l'ordre du jour. Elle intéresse non seulement les biochimistes et les endocrinologistes, mais aussi les biologistes et les médecins qui observent et utilisent l'activité de ces messagers chimiques. De plus, les recherches qu'elle implique peuvent servir à mieux comprendre les effets des médicaments en général.

L'activité des hormones stéroides dépend strictement de leur structure. La moindre altération du noyau carboné ou des groupes chimiques qui y sont fixés modifie quantitativement et même parfois qualitativement les effets au niveau des organes cibles. En première approximation, mais avec quelque vraisemblance, on peut admettre que les hormones stéroides (de ,,petites" molécules) vont, pour produire leurs effets, entrer d'une façon ou d'une autre en contact direct avec des macromolécules[1]. On peut donc postuler qu'il existe au niveau des organes cibles, ceux qui répondent à l'incitation hormonale, des structures hautement différenciées capables de reconnaître le signal ,,hormone", de le recevoir, et de provoquer la (les) réaction(s) cellulaire(s). Ainsi le problème des récepteurs consiste particulièrement à rechercher ces structures spécialisées et à définir leur insertion dans la machine cellulaire. Il n'y a actuellement pas de preuve que les hormones stéroides s'insèrent dans une formation membranaire en provoquant ainsi une modification de l'afflux des substrats nutritifs (acides aminés, sucres, etc. . .). Il n'y a pas non plus de preuve qu'elles modifient initialement une ou des enzymes, que ce soit isostériquement ou allostériquement. Comme l'accroissement de synthèse des acides ribonucléiques et des protéines dans les organes cibles est un effet précoce des hormones stéroides, en particulier des hormones sexuelles qui sont avant tout morphogénétiques, on peut penser à leur intervention à un niveau quelconque de la séquence des

[1] Une alternative serait une modification de l'état de l'eau qui entraînerait secondairement une modification de certaines macromolécules.

évènements impliqués dans la synthèse protéique. Les travaux déjá publiés sont nombreux et d'ailleurs contradictoires. Certains ont permis de décrire les effets des hormones sur les acides déoxyribonucléiques, les histones, ou leur liaison. D'autres ont mis en évidence une augmentation de l'activité de la polymérase catalysant la synthèse des acides ribonucléiques, mais le système n'était pas purifié et on ne peut en déduire un effet sur l'enzyme elle-même. Enfin certaines expériences, d'ailleurs controversées, ont cherché à démontrer que l'hormone déréprime la synthèse des acides ribonucléiques en se liant à un répresseur (évidemment non isolé). On peut, dans l'ensemble, estimer que ces expériences n'ont pas encore mis en évidence un phénomène biochimiquement précisé et spécifique, c'est-à-dire strictement particulier pour une concentration physiologique d'une hormone donnée dans un tissu défini.

Jensen et ses collaborateurs ont eu le mérite de se servir les premiers d'hormones de haute activité spécifique (1, 2, 3), ce qui, entre autres, permet de s'en tenir à des doses ,,physiologiques". Etudiant le destin de l'Oestradiol tritié chez la Ratte impubère ou castrée, ils ont mis en évidence deux importants phénomènes. D'une part l'Oestradiol n'est pas chimiquement transformé dans l'utérus dont il provoque le développement, comme si l'hormone n'agissait que par sa seule présence avant de quitter l'organe. D'autre part, l'utérus capte, concentre et retient l'Oestradiol plus, et plus longtemps, que les autres organes ou tissus de l'économie. Il en est de même pour l'hypophyse, le vagin et les glandes mammaires (tumorales après administration de DMBA), ce qui fait coincider les notions d'organe cible et de récepteur, invitant à étudier le mécanisme, la spécificité et la signification de la capture, de la concentration et de la rétention de l'hormone.

Les équipes de Jensen (1, 2, 3), Gorski (4, 5), Ségal et Talwar (6), King (7), Bagget et Stone (8), Martin et Stone (9) et de ce laboratoire (10, 11), ont montré, essentiellement chez le Rat et la Souris, *in vivo* et *in vitro* en utilisant des *tissus* en survie, que l'endomètre est particulièrement actif pour capturer l'Oestradiol et synthétiser des acides ribonucléiques et des protéines, que différents oestrogènes de synthèse se comportent comme l'Oestradiol et entrent éventuellement en compétition avec lui pour occuper les ,,récepteurs", que l'Oestrone (analogue 17-cétonique de l'Oestradiol) n'était ni concentré ni retenu par l'utérus et que le 17α-Oestradiol (analogue 17α-hydroxylé de l'Oestradiol 17β-hydroxylé) l'est faiblement, enfin que les hormones autres qu'oestrogènes ne semblent pas entrer en compétition pour occuper les sites auxquels se fixe l'Oestradiol. La plupart des expériences ont été faites *in vivo*, et les résultats ont donc une grande signification physiologique. Cependant elles ne précisent pas les mécanismes en jeu et donc ne permettent pas de répondre aux questions fondamentales, car on ne peut distinguer ce qui revient au métabolisme (hépatique, rénal, etc. . .) des substances hormonales administrées, ce qui dépend de leur mode de transport dans le sang, ce qui tient à leur passage à travers les parois pour entrer dans les cellules cibles, et ce qui conditionne le transport, la concentration et la rétention de l'Oestradiol dans les cellules elles-mêmes. Les expériences *in vitro* avec des tissus en survie ont d'une part confirmé que l'Oestradiol était actif par lui-même (stimulation en culture organo-typique (12, 13)), et d'autre part permis d'étudier la spécificité et les caractéristiques quantitatives et cinétiques de la liaison de l'Oestradiol dans les tissus (3, 11, 14). Après fractionnement cellulaire suivant l'incorporation tissulaire d'Oestradiol radioactif injecté *in vivo* ou incubé avec les tissus, on trouve environ 30 à 50% de la radioactivité dans le surnageant obtenu après centrifugation des particules à $6 \cdot 10^6$ g $\times$ mn ou plus, et 40—70% dans les noyaux. Dans le ,,Surnageant" utérin, après injection d'Oestradiol à des Rattes impubères, Gorski a repéré par ultracentrifugation une fraction protéique de coefficient de sédimentation S = 9,5, saturée par une faible quantité (,,physiologique") d'Oestradiol alors que l'excédent va, le cas échéant, se fixer sur des composés de S = 4—6 (albumine en particulier).

Jensen et Jungblut (3) ont isolé à partir d'extraits cytoplasmiques et nucléaires d'endomètre de veau un ou des composé(s) à l'aide d'une colonne de chromotagraphie de p.aminobenzyl-cellulose à laquelle de l'Oestradiol était fixé par diazotation; l'élution est obtenue à l'aide de déoxycholate et on peut obtenir un anticorps qui diminue l'incorporation d'Oestradiol radioactif dans les noyaux d'endomètre incubés. TALWAR et SEGAL (6) ont utilisé une fraction de surnageant d'utérus de ratte castrée à laquelle se lie l'Oestradiol pour inhiber la synthèse d'acide ribonucléique dans une préparation de polymérase bactérienne. KING (7), enfin, a observé la fixation de radioactivité par certaines fractions de chromatine utérine après injection d'Oestradiol tritié. Dans notre laboratoire, la très forte liaison de l'Oestradiol à des protéines nucléaires étroitement jointives au DNA a été mise en évidence (14). Cependant aucune de ces observations ne démontre la spécificité de liaison d'une macromolécule en tant que telle avec l'Oestradiol, puisque les résultats les plus précis ont été obtenus chez l'animal entier ou avec un organe ou un tissu en survie, et qu'on ne peut exclure que la spécificité de liaison, quand elle est démontrée, ne soit due à l'organisation cellulaire et/ou tissulaire. De même les calculs cherchant à mesurer la capacité et l'affinité des récepteurs (4 et 11) sont-ils soumis à toute une série *d'a priori* très restrictifs.

Les faits démontrant la liaison spécifique de l'Oestradiol de protéines utérines en solution sont rapportés ici, qui prouvent pour la première fois l'existence de macromolécules tissulaires se liant spécifiquement à une hormone et permettent de rapporter au moins partiellement la *particularité* d'un organe cible à celle de certaines de ses protéines.

L'endomètre de truie impubère est homogénéisé au Potter. Les particules sont éliminées par centrifugation et ultracentrifugation successives. Le surnageant, sans autre traitement, est incubé à 37° avec des stéroides radioactifs et éventuellement non-radioactifs dont la pureté est systématiquement vérifiée par les techniques appropriées. On incube à la température du laboratoire jusqu'à équilibre, et on mesure alors la quantité de stéroides liés aux protéines après avoir absorbé ceux qui ne le sont pas à l'aide d'une suspension de charbon et dextran (16, 17) que l'on mélange à la préparation incubée et qu'après 15'' d'agitation on sépare par centrifugation. Dans la fraction surnageante, il ne reste que des stéroides liés aux protéines que l'on mesure par leur radioactivité à l'aide d'un compteur à scintillation. Théoriquement (et pratiquement) cette technique de compétition avec un charbon actif ne permet pas autre chose qu'une mesure relative (et par défaut) des stéroides radioactifs liés mais on peut l'utiliser avec profit dans des déterminations parallèles (14). Des mesures en utilisant la méthode de dialyse à l'équilibre, à 4 °C, ont servi de référence.

Les expériences de base sont faites en utilisant des concentrations „physiologiques" d'Oestradiol. La fig. 1 indique que la quantité d'Oestradiol dépend de celle du „Surnageant" présent dans le milieu d'incubation. Si on traite le „Surnageant" par des protéases (Pronase, Trypsine), sa capacité de liaison est très diminuée alors qu'il n'en est rien après ribonucléase ou déoxyribonucléase. Un pré-chauffage à 60° pendant une heure diminue la liaison de 50 à 70%. Il semble donc évident que dans le „Surnageant", ce sont des protéines qui sont responsables de la liaison de l'Oestradiol (17). Il s'agit d'un mélange complexe dans lequel" le récepteur" est très minoritaire, et il est remarquable qu'on puisse obtenir des résultats quantitativement et qualitativement aussi singuliers. Par exemple, à poids égal de protéine, l'albumine humaine, connue pour lier efficacement l'Oestradiol n'en fixe cependant que moins de 10 fois moins dans les conditions expérimentales précitées ($6{,}7 \times 10^{-10}$ M) d'Oestradiol et le préchauffage à 60° ne modifie pas sa liaison avec l'hormone. Par chromatographie sur DEAE cellulose et sur Sephadex G 25 et G 200, plusieurs fractions protéiques utérines ont été séparées; de l'albumine et des protéines de faible affinité pour l'Oestradiol ont été repérées et parmi d'autres, une fraction ayant une grande affinité et une étroite spécificité pour l'Oestradiol a été isolée. Son poids moléculaire d'après la chromatographie sur Sephadex ou la sédimentation en gradient de saccharose, est inférieur à celui de l'albumine; les détails concernent sa purification seront publiés ulterieurement.

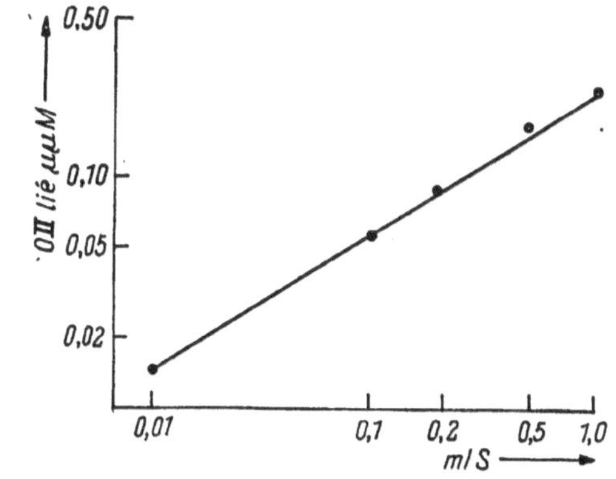

Fig. 1. Liaison de quantités variables de ,,Surnageant" avec l'Oestradiol.

1 ml de ,,Surnageant" (contenant environ 4 mg de protéines par ml) ou des fractions diluées à 1 ml avec du tampon phosphate 0,02 M, pH 7,4, sont incubés à la température du laboratoire avec 0,67 μμM d'Oestradiol tritié (A. S. environ 45 mC/μM) pendant 30 minutes. $6{,}7 \times 10^{-10}$ M est l'ordre de grandeur de la concentration physiologique de l'Oestradiol plasmatique. Prolonger l'incubation jusqu'à trois heures ne change pas les résultats.

O II: Oestradiol. S: Surnageant

Utilisant le ,,Surnageant" total, on observe, quand on augmente la concentration d'Oestradiol dans le milieu d'incubation un accroissement de sa fixation (fig. 2). On peut interpréter

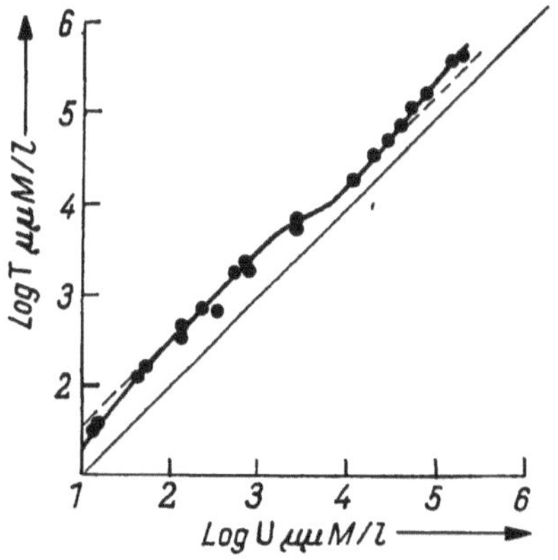

Fig. 2. Liaison de quantités variables d'Oestradiol aux proteines uterines.

Dialyse à l'équilibre à 4 °C utilisant 1,5 ml de ,,Surnageant".

U: Oestradiol libre. T: Oestradiol total (libre + lié).

Cette représentation sur laquelle nous ne pouvons insister ici (14) indique l'existence d'au moins 2 systèmes de liaison, dont un opératif aux concentrations faibles d'Oestradiol, de capacité limitée et de grande affinité, et l'autre de capacité pratiquement non limitée (et même croissante), aux concentrations les plus élevées de cette étude

les faits en postulant dans le ,,Surnageant" la présence des protéine(s) ,,spécifique(s)" de capacité limitée se saturant aux faibles concentrations hormonales, cependant que d'autres protéines ,,non spécifiques" (y compris l'albumine) peuvent lier de grandes quantités d'Oestradiol sans qu'apparaissent les phénomènes limitants. (14—17)

D'autres expériences confirment la spécificité de la liaison de l'Oestradiol dans le ,,Surnageant" utérin (18)) Le ,,Surnageant" utérin fixe la Testostérone sans montrer de système limitant, contrairement à ce qui se passe avec l'Oestradiol. La préparation équivalente de tissu pulmonaire des mêmes animaux fixe 60% de l'Oestradiol que lient les protéines utérines (à poids

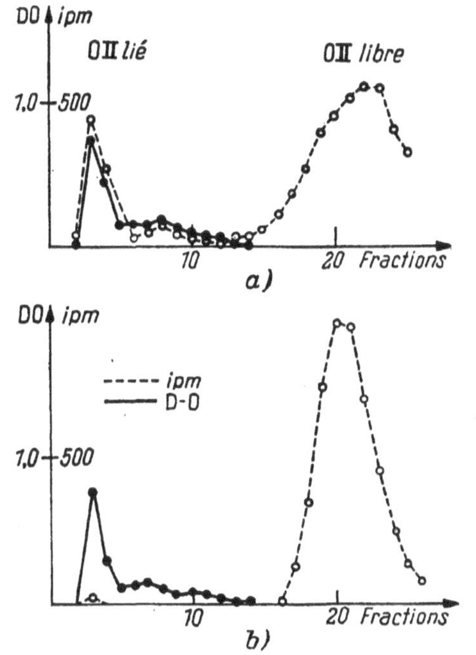

Fig. 3. Liaison de l'Oestradiol à des macromolécules du „Surnageant" uterin. Chromatographie sur Sephadex

0,3 ml de „Surnageant" sont incubés avec 61 µµg d'Oestradiol-^{3}H (15 mµC) (a), ou avec 3×10^5 µµg d'Oestradiol-^{3}H (15 mµC) (b). On a donc utilisé un traceur seul (a), et le même traceur auquel on a ajouté 0,3 µg d'Oestradiol (b). La chromatographie sur Sephadex G-25 indique la liaison de l'Oestradiol aux macromolécules (a). L'expérience (b) par une dilution isotopique d'environ 5.000 fois, montre que la capacité de liaison de l'Oestradiol et des macromolécules est limitée. Dans la technique au charbon, on élimine l'Oestradiol libre par adsorption, pour ne mesurer que l'Oestradiol lié.

DO: densité optique. ipm: impulsions par minute

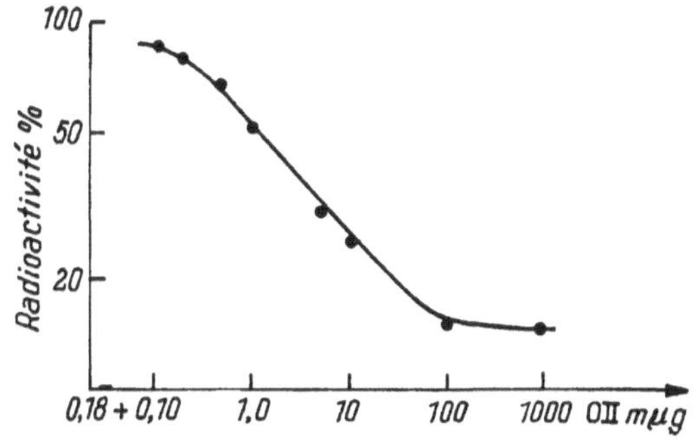

Fig. 4. Liaison d'Oestradiol radioactif à des protéines du „Surnageant". Un dosage de l'Oestradiol.

1 ml de „Surnageant" est incubé avec 0,18 mµg d'Oestradiol-^{3}H (A.S. 45 mC/µµM) seul (100% de radioactivité fixée par convention) et avec des quantités variées d'Oestradiol, de 0,1 à 1.000 mµg. Le pourcentage de la décroissance de la radioactivité liée est calculé

Sans modification, cette méthode permet une courbe d'étalonnage pour un dosage très sensible de l'Oestradiol, entre 0,5 et 50 mµg, ce qui n'est atteint — au prix d'efforts considérables — que par les meilleures et plus récentes méthodes utilisant les dérivés radioactifs et/ou la chromatographie en phase vapeur. La simplicité et la rapidité de la mesure de la liaison de l'Oestradiol radioactif aux protéines du „Surnageant", la spécificité implicite, la prévisible augmentation de la sensibilité en utilisant une préparation purifiée et concentrée, laissent prévoir le développement de cette méthode radiobiochimique

égal), et ne présente aucun signe de phénomène limitant (entre $6{,}7 \times 10^{-10}$ M et $3{,}7 \times 10^{-6}$M). Le „Surnageant“ d'utérus fixe trois fois moins de Testostérone que d'Oestradiol, alors que celui de poumon se lie plus à l'hormone mâle qu'à l'oestrogène.

Quand on incube une quantité fixe d'Oestradiol radioactif et des quantités croissantes d'Oestradiol non radioactif, la radioactivité fixée aux protéines diminue (fig. 3 et 4). Deux facteurs entrent en jeu : la dilution isotopique abaisse la radioactivité spécifique selon la quantité d'Oestradiol ajoutée et l'augmentation de la concentration entraîne une fixation accrue d'Oestradiol (fig. 2). Dans les conditions indiquées sur la figure 4 la radioactivité fixée aux protéines baisse avec l'addition de 100 μμg d'Oestradiol. Quand on atteint ou dépasse 100 mμg, on semble atteindre un plateau comme si les protéines avaient une capacité „non-physiologique“, non spécifique, de fixer de l'Oestradiol, approximativement proportionnelle à la concentration du stéroide, et sans limite dans les conditions décrites.

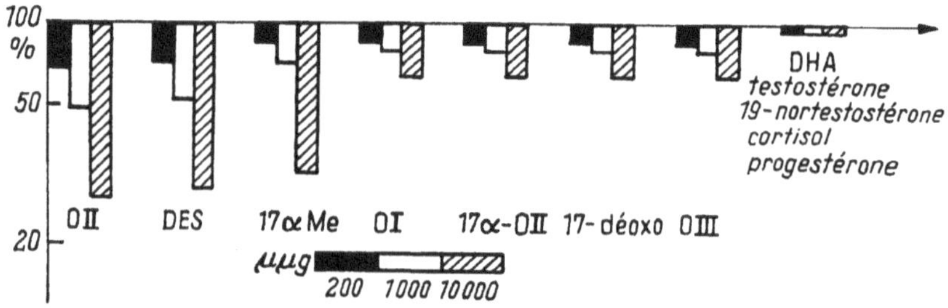

Fig. 5. Inhibition de la fixation d'Oestradiol à des protéines utérines par differents composes hormonaux.

On incube 1 ml de „Surnageant“ avec 183 μμg (0,67 μμM) d'Oestradiol-^{3}H seuls, ou avec 200, 1.000 et 10.000 μμg des stéroides indiqués sur la figure. 100% représente la quantité d'Oestradiol-^{3}H lié aux protéines après incubation des 183 μμg d'Oestradiol-^{3}H seul.

En présence de stéroide non radioactif, plus le pourcentage d'Oestradiol-^{3}H lié est abaissé, plus l'effet inhibiteur (par compétition probablement) du stéroide considéré est important.

O I, Oestrone; O III, Oestriol; 17α-Me, 17α-Méthyl-Oestradiol; 17-α-O II, 17α-Oestradiol; 17-Déoxo, 17-Déoxo-oestrone; DES, Diéthylstilboestrol; DHA, Déhydroépiandrostérone

Les expériences rapportées sur la fig. 5 résument les résultats obtenus en utilisant la même dose d'Oestradiol radioactif et trois concentrations différentes et non extrêmes de composés hormonaux. On peut relever que l'effet de Diéthylstilboestrol est pratiquement identique à celui de l'Oestradiol lui-même que le 17α-méthyl-oestradiol est très actif mais moins que le Diéthylstilboestrol, et que la Testostérone, la 19-nortestostérone, la Déhydroépiandrostérone, la Progestérone, et le Cortisol sont strictement sans effet. Quatre phénolstéroides faiblement oestrogènes sont actifs, mais beaucoup moins que l'Oestradiol. Les conclusions de ces résultats préliminaires sont très remarquables en ce qui concerne la spécificité de liaison des protéines utérines avec l'Oestradiol. Le seul phénolstéroide dont l'activité approche celle de l'Oestradiol est le dérivé 17α-méthyl et l'importance de l'intégrité du groupe 17β-hydroxyl est prouvé à contrario par la baisse d'activité que provoque son oxydation (Oestrone), son épimérisation (17α-Oestradiol), sa suppression (17-déoxo-oestrone), ou le voisinage d'un groupe 16α-hydroxyl (Oestriol).

Il est vraisemblable que la diminution de la liaison entre les protéines et l'Oestradiol radioactif au cours de ces essais relève d'une compétition au niveau des sites de liaison. C'est ce qui ressort des expériences au cours desquelles des quantités définies d'Oestradiol, d'Oestrone, et de Diéthylstilboestrol ont été incubées avec des quantités variables d'Oestradiol tritié dans la zone de concentration où c'est essentiellement le système de liaison spécifique et limité qui est en jeu (fig. 6). Tout se passe comme si l'Oestrone et le Diéthylstilboestrol se comportaient comme des inhibiteurs compétitifs de la fixation de l'Oestradiol (dans les conditions quantitatives indiquées). Un mμg d'Oestradiol, un mμg de Diéthylstilboestrol (bien que ce ne soit pas un stéroide) et 10 mμg d'Oestrone sont pratiquement équivalents (18). Ce résultat remarquable permet, entre autres, de reconsidérer les résultats obtenus *in vivo*, où le Diéthylstilboestrol est moins actif que l'Oestradiol, ce qui peut donc être dû aux évènements précédant son arrivée au niveau des récepteurs.

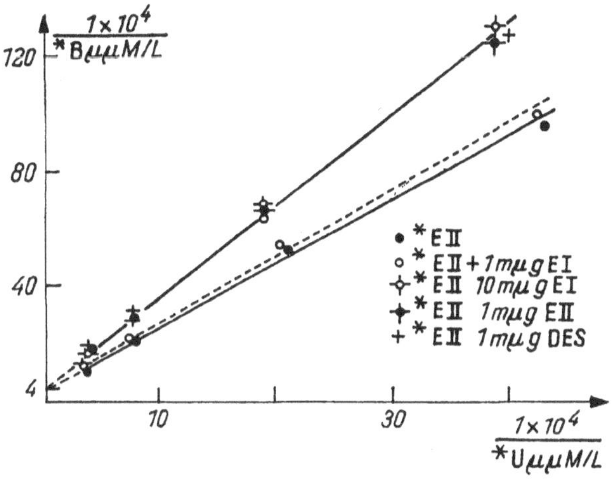

Fig. 6. Liaisons aux protéines du ,,Surnageant". Competition de l'Oestradiol, de l'Oestrone et du Diéthylstilboestrol avec l'Oestradiol-^{3}H.

Incubation avec 1 ml de ,,Surnageant". *U = Oestradiol-^{3}H libre. *B = Oestradiol-^{3}H lié. Plusieurs erreurs de chiffres et de graphique se sont glissées dans les notes préliminaires (17) et (18). Les valeurs que l'on peut déduire de la figure: capacité maximale de la liaison spécifique: 2.500 μμM/l (c'est à dire ∔ 0.6 μμM/mg de protéine) et constante d'affinité $\sim 2 \times 10^{8}$ l/M ont été vérifiées par d'autres techniques

Enfin des expériences ont indiqué la réversibilité de la liaison; en fixant dans un premier temps de l'Oestradiol sur les protéines utérines et en provoquant secondairement le relargage par dilution ou l'échange par incubation avec de l'Oestradiol non-radioactif (17). Ce résultat est une condition nécessaire pour pouvoir calculer une constante d'affinité entre une macromolécule et un ligand (19). Malheureusement elle n'est pas suffisante (19, 21), et sans entrer dans les détails, on peut citer la valeur de 5.10^{-9} comme constante de dissociation la plus probable à l'heure actuelle.

Ces résultats démontrent l'existence de protéine(s) tissulaire(s) liant spécifiquement dans un organe cible l'hormone qui en provoque le développement.

Les problèmes posés sont nombreux. La purification et l'étude physico-chimique de ces proteines est à faire. Leur localisation cellulaire est à préciser car les recoupements autoradiographiques (21) ne permettent pas d'exclure que leur présence dans le ,,Surnageant" ne soit un artefact du fractionnement cellulaire. Au reste la comparaison avec les macromolécules nucléaires détectées par Jensen (3) et dans ce laboratoire (14) est également nécessaire.

La signification des ,,récepteurs" dans la réponse tissulaire à l'hormone reste tout à fait conjecturale. Il n'est que trop facile d'imaginer une intervention de déclanchement, de répréssion, de protection, etc. . . aux différents niveaux de la synthèse des protéines ou des phospho-lipides ou encore dans les phénomènes réglant le fonctionnement énergétique de la cellule. Plusieurs niveaux de contrôle sont possibles et il n'est pas impossible que l'Oestradiol se lie aussi à d'autres molécules que des protéines, ou contracte des rapports fonctionnels définis avec de petites molécules. A la vérité, la seule démonstration d'un ou de plusieurs récepteurs ne résout pas le problème du mécanisme d'action des hormones, et elle peut éventuellement ne concerner qu'un évènement mineur, les protéines en question servant de réservoir ou de voie de dégagement. L'origine même de ces protéines est à découvrir et leur distinction des protéines plasmatiques de transport est à faire.

Il n'en reste pas moins que l'on peut espérer beaucoup de l'étude biochimique des „récepteurs" hormonaux, tant sur le plan théorique que pour comprendre mieux là „réceptivité" aux hormones des tissus normaux et pathologiques, ce qui ne peut que mener à de nouvelles conceptions thérapeutiques.

[1] Jensen, E. V., et H. I. Jacobson, In Biological Activities of Steroids in Relation to Cancer (G. Pincus and E. P. Vollmer) Academic Press N.Y. p. 191 (1960).
[2] Jensen, E. V., H. I. Jacobson, J. W. Flesher, N. M. Saha, G. N. S. Gupta, V. Colucci, D. Shiplacoff, H. G. Neumann, E. R. de Sombre, et P. W. Jungblut, Japanese—American Conferenco on steroids Dynamics. Academic Press N.Y. p. 133 (1966).
[3] Jensen, E. V., E. R. de Sombre, et P. W. Jungblut, Proceedings of the 2nd Intern Congr. Hormonal Steroids. Sous presse (1966).
[4] Noteboom, W., et J. Gorski, Arch. Biochem. Biophys. **111**, 559, 1965).
[5] Toft, D., et J. Gorski, Proc. Nat. Acad. Sci. **55**, 1574 (1966).
[6] Talwar, G. P., S. J. Segal, A. Evans, et O. W. Davidson, Proc. Nat. Acad. Sci. **52**, 1059 (1964).
[7] King, R. J. B., et B. R. Inman, J. Endocrin. **35**, 26 (1966).
[8] Stone, G. M., et B. Baggett, Steroids **6**, 277 (1965).
[9] Martini, L., et G. M. Stone, Steroids **6**, 475 (1965).
[10] Alberga, A., et E. E. Baulieu, C.R. Acad. Sc. **261**, 5226 (1965).
[11] Alberga, A., et E. E. Baulieu, J. Molec. Pharmacol sous presse (1968).
[12] Haffen, K., et C. Raynaud, Travail du laboratoire du Prof. E. Wolff et de notre groupe, à paraître (1966—1967).
[13] Maurer, H. R., D. E. Rounds, et C. W. Raiborn, Nature **204**, 182 (1967).
[14] Baulieu, E. E., A. Alberga, H. Rochefort, H. Richard-Foy, et J. P. Raynaud, Recherchos à paraître (1965—1967).
[15] Murphy, C. E. P., et E. Beverly, Nature **201**, 679 (1964).
[16] Nugent, C. A., et D. B. Mayes, J. Clin. Endocrin. **26**, 1116 (1966).
[17] Baulieu, E. E., A. Alberga, et I. Jung, C.R. Acad. Sc. Paris **265**, 354 (1967).
[18] Baulieu, E. E., A. Alberga, et I. Jung, C.R. Acad. Sc. Paris **265**, 501 (1967).
[19] Rosenberg, R. M., et I. M. Klotz, Analytical methods of protein chemistry 1: 131, (1961).
[20] Weber, G., et S. R. Anderson, Biochem. **4**, 1942 (1965).
[21] Stumph, W., et L. J. Roth, J. of Histochem. and Cytochem. **14**, 274 (1966).

Studies on the Mechanism of Action of Estrogen and Progesterone[1]

K. R. LAUMAS[2]

Reproductive Biology Research Unit All India Institute of Medical Sciences, New Delhi, India

Summary

The above evidence shows that circulating endogenous or exogenously injected estradiol has selective affinity to concentrate in certain target tissues of the body. The presence of high amounts of circulating estrogen increases the binding sites of estradiol in the uterus. In target tissues estradiol shows a heavy localization in the nuclear and $105{,}000 \times g$ supernatant fractions. Experimental evidence presented shows that the receptor present in the supernatant fraction is also able to bind DES. Investigations on the radioactivity present in the nuclear fraction show that estradiol-17β and DES show binding to a receptor present in the soluble fraction of the nucleus. Progesterone on the other hand behaves differently from estradiol in its uptake and disappearance from the target tissues. While estradiol is incorporated and retained in the uterus, vagina and pituitary, progesterone is rapidly taken up and disappears from these tissues.

Introduction

The first pure crystalline estrogen viz., estrone was isolated from pregnancy urine in 1929 by DOISY et al. [1] and BUTENANDT [2]. A short while after that the three classical estrogens, progesterone and testosterone were extracted and isolated from biological fluids. In spite of the fact that knowledge on the chemistry and pharmacology of these hormones has been understood for a considerable time yet the exact mechanisms by which these hormones bring about their profound biological action has evaded the scientists.

It is now known that minute amounts of hormones are effective in exerting marked anatomical and physiological changes in their target tissues. Such effects of the hormones are greatly altered by changes in the molecular structure and the stereochemistry of the hormone. An example of such a phenomena is the high potency of estradiol-17β in most of the biological assays and the lack of activity of the corresponding 17-alpha isomer namely estradiol-17α.

The work on elucidating the action of estrogens becomes difficult by the fact that there is a class of non-steroidal compounds which possess biological activity that is indistinguishable from natural estrogens. Examples of such compounds are stilbestrol, hexestrol, diethylstilbestrol etc. etc. Another characteristic of estrogenic response in the uterus depends on protein synthesis. Extensive work with inhibitors of protein synthesis [3, 4] has shown that puromycin and cyclo-hexamide sensitive changes in the metabolism of lipids, RNA and glucose are seen in the first hour after treatment with estrogen and before estrogen stimulates overall synthesis of proteins. Studies on the action of estrogen are still more complicated by the demonstration of protein synthesis in uterus by histamine similar to that induced by estrogen [5]. Further more estrogen induced uterine hyperemia and water inhibition can be simulated by intrauterine administration of histamine or histamine releasing agents.

Great strides in our understanding of the mechanism of estrogen action was achieved by JENSEN and JACOBSON [6] who showed that 100 pg. of estradiol is taken up in the uterus when

[1] Experimental work from this Laboratory was supported by a grant from the Ford Foundation.

[2] Lecture did not take place.

0.1 μg of 6:7-H^3-estradiol is given to immature rats. Thus it is an extremely minute amount of hormone which is capable of bringing about the various actions of the estrogen. The work of JENSEN and JACOBSON [6] opened new vistas for work with estrogen and its action at the molecular level. It would not be possible to cover in detail all the work of different investigators, however, appropriate references to pertinent observations would be given.

Constant infusion of 6:7-H^3-estradiol

It has been beautifully demonstrated by JENSEN and JACOBSON (6) that after a single injection of 0.1 μg of high specific activity 6:7-H^3-estradiol, the hormone is incorporated in uterus and vagina. On the other hand it shows a rapid uptake and clearance from liver and kidney. The chemical nature of the hormone after the administration of a physiological dose and the patterns of incorporation and retention have been investigated, whereby it was found that the accumulation of radioactivity in the uterus was maximal between 1 and 2 hours and declined slowly over the succeeding hours. Upto six hours after the administration of the hormone the only significant radioactive substance present in uterus and vagina was estradiol. Using the technique of local application of the hormone, STONE and MARTIN [7] have obtained similar results.

The approach to the problem of hormone action was initiated in our laboratory by constantly infusing the hormone at a slow rate in intact or ovariectomised animals [8]. It was considered that this would simulate the natural endogenous secretion of the steroid. Such an approach has been used to study the uptake, negative feed-back, sub-cellular distribution and binding of 6:7-H^3-estradiol to uterine receptors. One group of animals in estrus and another grouap in diestrus were infused with 6:7-H^3-estradiol for a period of 6 hours. After the termination of the infusion a number of tissues were taken out. The radioactivity in the tissues was determined with the help of a liquid scintillation counter. The results for different tissues are given in Figure 1. Similar results for the various brain parts investigated are described in Figure 2. It would be seen from Figure 1 and Figure 2, that in almost every tissue there was a higher uptake of estradiol in animals in estrus compared to animals in diestrus. The concentration of estradiol in the uterus of animals in estrus was 5.5×10^{-9} M and it was 1.9×10^{-9} M in the uterus of animals in diestrus.

Studies [6] on the fate of physiological doses of radioactive estradiol in immature rats have shown that estradiol exerts its growth promoting action without undergoing any metabolic transformation. Furthermore in those experiments where mixtures of estradiol-6:7-H^3 and

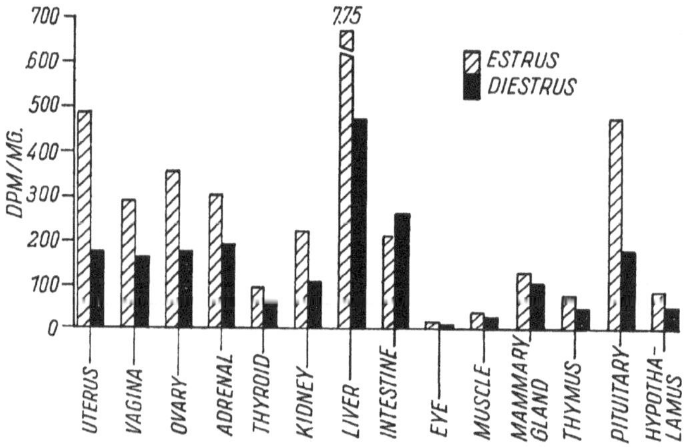

Fig. 1. The uptake of 6:7-H^3-estradiol in genital and other tissues of normal estrus and diestrus rats after infusion at a constant rate for 6 hours. The results are shown as disintegrations per minute per milligram wet tissue. Each point is the average of results in 5 or 6 animals

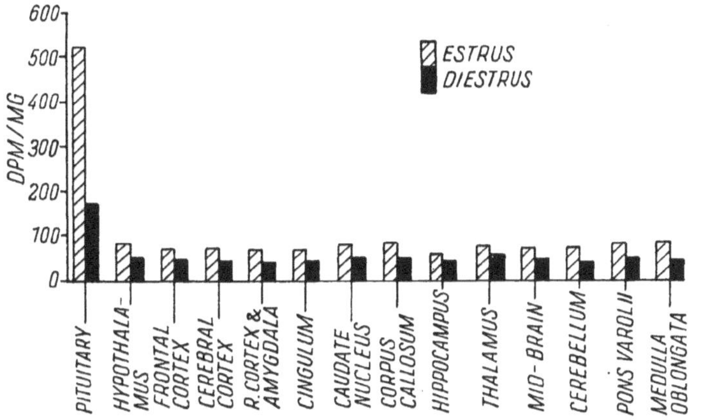

Fig. 2. The uptake of 6:7-H^3-estradiol in the various parts of the brain of normal estrus and diestrus rats after infusion at a constant rate for 6 hours. The results are shown as disintegrations per minute per milligram wet tissue. Each point is the average of results in 5 or 6 animals

estradiol-17-H^3 were employed it has been shown that neither the atrophic nor the actively growing rat uterus is capable of oxidising estradiol to estrone. On the other hand the human endometrium [9] has been shown to convert estradiol to estrone. Moreover, it has been found that human endometrium metabolises estradiol to estrone at a rate 40 times that of myometrium. The reverse reaction i.e. estrone → estradiol has been shown to be of considerably lower magnitude than that of the forward reaction. The significance of these observations on the biotransformation of estradiol in the human endometrium in terms of biologic action of estradiol have not yet been evaluated.

In the present studies, there was found a significantly high uptake of estradiol in pituitary, compared to total hypothalamus and other brain regions studied (Fig. 2). The uptake in pituitary was almost comparable to that in uterus. The comparatively high uptake in pituitary lends support to the hypothesis that this may be one of the sensitive sites for the negative feed-back of estrogen [10]. However in this investigation complete hypothalamus was used for estradiol uptake and it has been recently demonstrated by Kato and Villee [11], that estradiol is selectively localized in the anterior hypothalamus compared to posterior and middle regions of hypothalamus. The widely held concept of *indirect feed-back* of estrogen via hypothalamus is well documented [12] . However the work of Bogdanove [13] on pituitary cell reaction after castration and the studies of Palka et al. [14] on the implantation of tritiated estradiol in the hypothalamus and pituitary give some evidence for the direct feed-back of estradiol to the pituitary as shown by the present results.

An important point which is brought out by the results given in Fig. 1 is that animals in estrus which have more circulating estrogen show a higher uptake of estradiol in uterus. This indicates that pretreatment of the animals with estrogen or the presence of high amounts of endogenous estrogen is able to increase the binding sites for estradiol in the uterus.

Sub-cellular localisation and binding of 6:7-H^3-estradiol to receptors present in the supernatant and nuclear fractions of rat uterus

After constant infusion of 6:7-H^3 estradiol, the sub-cellular localisation and binding of estradiol to macromolecules present in rat uterus have been investigated. The results for sub-cellular localisation are presented in Table 1. It may be seen from the Table 1 that the target tissues for estrogen viz., uterus, vagina and pituitary show a heavy localisation of radioactivity in the nuclear and the supernatant fractions. The non-target tissues for estrogen like muscle show heavy localisation in the supernatant fraction, while comparatively little activity is present in the nuclear, mitochondrial and microsomal fractions. These results on sub-cellular localisation in adult intact animals in general, agree with the data obtained by other investigators [15, 16] who used immature or adult castrated animals.

Table 1

Localisation of radioactivity in the sub-cellular fractions of the normal adult female rat after constant infusion of 6:7-^{3}H-estradiol for 4 hours

Fractions	Uterus		Vagina		Pituitary		Muscle	
	DPM	% of total DPM	DPM	% of total DPM	DPM	% of total DPM	DPM	% of total DPM
Nuclear 1,500 × g/10 mts.	18,107	56.1	3,773	43.6	1,129	54.2	395	10.8
Mitochondrial 20,000 × g/30 mts.	1,403	4.3	206	2.3	49	2.3	144	4.2
Microsomes 105,000 × g/75 mts.	974	3.0	308	3.5	39	1.9	75	2.0
Supernatant	11,811	36.6	4,386	50.5	866	41.5	3,023	83.1
Total	32,295	100.0	8,673	99.9	2,083	99.9	3,637	100.1

TOFT and GORSKI [17], TALWAR et al. [16] and KING and GORDON [18] have demonstrated the presence of a macromolecular fraction in the 105,000 × g supernatant fraction of the rat uterus which showed binding to estradiol. The sensitivity to proteolytic enzymes of this receptor macromolecular component, its specificity in binding and its size make it probable that the receptor molecule is a large protein. Further work on the nature, binding characteristics and purification of this receptor is urgently needed. At present, it is not possible to say whether it is a single receptor or a mixture.

In our laboratory, studies have been carried out on the binding of 6:7-H^3-estradiol with the 105,000 × g supernatant and nuclear fraction of rat uterus after pretreatment of the animals with radioinert estradiol. As indicated above, pretreatment of animals with estradiol or the presence of high amounts of circulating estrogens were found to increase the number of binding sites for estradiol, therefore it was decided to use this preparation for investigations. The 105,000 × g supernatant fraction was chromatographed on Sephadex G-100 and the first protein peak was shown to bind with radioactive estradiol. The results were similar to those obtained by other workers [16, 17, 18]. The nuclear fraction obtained from rat uterus after the injection of 6:7-H^3-estradiol to estradiol pretreated rats was subjected to ultrasonication and fractionation. Various fractions sedimented at 100 × g, 10,000 × g and 100,000 × g. The soluble supernatant fraction after the sedimentation of the 100,000 × g fraction was chromatographed on Sephadex G-100 as before. The soluble fraction was found to contain a macromolecular component which showed binding to estradiol (Fig. 3). More than 60 per cent of bound estradiol, present in rat uterus is localised in the nuclear fraction. The demonstration of an estrogen binding macromolecular component in the nuclear fraction is therefore of great significance. KING et al. [19] have similarly shown the binding of estradiol to the nuclear fraction in the rat mammary adenocarcinoma tissues. With the presently available knowledge, it is not possible to say whether the nuclear receptor is the same or different from the receptor found in the supernatant fraction. It is possible that the two receptors are dissimilar and exist in a close physiologic relationship in the cytoplasm and the nucleus of the cell. However, the possibility of their common origin and possible separation during homogenisation and ultracentrifugation cannot be ruled out.

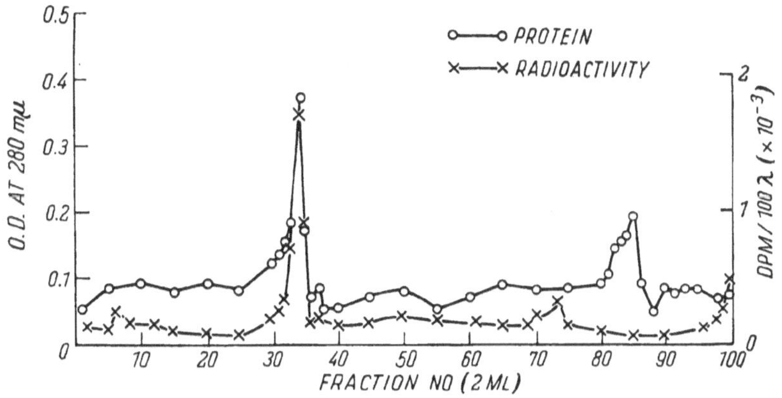

Fig. 3. Chromatography of the soluble fraction from rat uterine nucleus fraction on Sephadex G-100. Estradiol pretreated ovariectomised rats were injected (i.p.) with 6:7-H^3-estradiol, uterus removed and nucleus fraction isolated. The nuclear fraction was ultrasonicated and the soluble fraction obtained after centrifugation of the 100,000 × g sediment was used for chromatography

Uptake of diethylstilbestrol-H^3 after constant infusion

Diethylstilbestrol (DES) is a synthetic non-steroidal estrogen which is able to mimic almost all the effects of natural steroidal estrogens. It was of great interest to study the behaviour of this estrogen and to know whether diethylstilbestrol behaves the same way as estradiol in its uptake, sub-cellular localisation and binding to any specific receptors present in the uterus. The results (Fig. 4) show a great similarity to those presented for the estradiol uptake as given in Fig. 1. In the case of diethylstilbestrol there is a high uptake in uterus which is quite similar to that seen in pituitary. Vagina also showed a higher uptake compared to muscle. In respect of these observations, DES behaved like estradiol-17β. The high uptake of DES-H^3 observed in kidney and liver is due to the metabolic and clearance role of these organs.

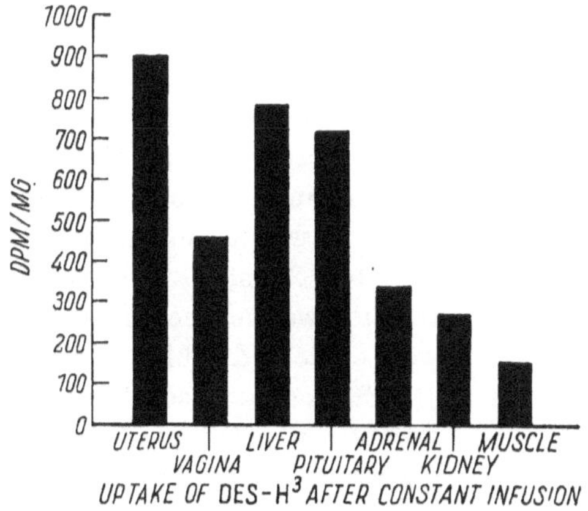

Fig. 4. Ovariectomised rats were infused with DES-H^3 at a constant rate for 5 hours. The results are expressed as disintegrations per minute per milligram wet tissue

Sub-cellular localization of DES-H^3

The sub-cellular localization of radioactivity in uterus, vagina, intestine, pituitary, liver and muscle of ovariectomized rats after an i.p. injection of DES-H^3 is given in Table 2. It may be pointed out that in the case of uterus, nearly 60 per cent of the radioactivity is

Table 2

Localisation of radioactivity in the sub-cellular fractions of ovariectomised rat uterus and other tissues after an i. p. injection of DES-H^3

Fractions	Uterus	Vagina	Pituitary	Intestine	Liver	Muscle
Nuclear 1,500 × g/10 mts.	61.0 ± 6.2 (5)[1]	43.5 ± 2.8 (4)	44.0 ± 8.0 (4)	13.5 ± 2.9 (4)	18.0 ± 3.0 (4)	13.0 ± 2.0 (3)
Mitochondrial 20,000 × g/30 mts.	4.5 ± 1.9 (5)	5.2 ± 2.1 (4)	13.0 ± 2.0 (4)	8.0 ± 2.5 (4)	7.0 ± 1.0 (4)	6.6 ± 0.8 (3)
Microsomal 105,000 × g/75 mts.	4.2 ± 0.3 (3)	3.5 ± 0.24 (4)	10.0 ± 4.0 (4)	4.5 ± 0.95 (4)	2.0 ± 0 (4)	2.4 ± 0.4 (3)
Supernatant	30.6 ± 5.3 (3)	48.0 ± 3.2 (4)	33.0 ± 3.0 (4)	74.0 ± 0.4 (4)	73.0 ± 0.3 (4)	78.0 ± 4.0 (3)
Total	100.3	100.0	100.0	100.0	100.0	100.0

[1] Figures in brackets represent the number of observations.

localized in the nuclear fraction and another nearly 30 per cent in the supernatant fraction. Only a small percentage of radioactivity was present in the mitochondrial and microsomal fractions. Almost similar results were seen in other target tissues for diethylstilbestrol viz. vagina and pituitary. However, there was a lesser percentage of radioactivity in the nuclear fraction of the later two tissues as compared to uterus. The supernatant fraction contained 48 per cent and 33 per cent of radioactivity in vagina and pituitary respectively. In general it may be mentioned that diethylstilbestrol showed an almost similar behaviour to estradiol in showing a selective localization in the nuclear and soluble fractions of target tissues. On the other hand non-target tissues for DES like intestine, liver and muscle showed very little localization in the nuclear fraction and nearly 75 per cent or more of radioactivity was present in the supernatant fraction. Therefore, in its sub-cellular localization in target and non-target tissues, DES behaved like estradiol-17β.

Binding of DES-H^3 to uterine receptors present in the supernatant and nuclear fractions of rat uterus

After an injection of radiochemically pure DES-H^3, the supernatant fraction obtained from rat uterus after the sedimentation of microsomal fraction at 105,000 × g, was chromatographed on a Sephadex G-100 column. The results indicated that a macromolecular fraction present in the supernatant fraction was binding to radioactivity.

The binding of DES-H^3 to the nuclear fraction was studied by using ovariectomised rats which were given i.p. injections of DES-H^3. Uteri were removed and the nuclear fraction separated by the usual techniques. It was homogenized and subjected to ultrasonication for 2 minutes at 20 kilocycles. The sonicated fraction was subjected to ultracentrifugation. Fractions were collected which separated at 100 × g, 10,000 × g, 100,000 × g and soluble fraction obtained after sedimentation of chromatin fraction at 100,000 × g. The distribution of proteins and radioactivity in the above fractions is shown in Table 3. Appreciable amounts of radioactivity and proteins were present in the fraction which sedimented at 100,000 × g and the soluble fraction. However the soluble fraction contained more than 50 per cent of the radioactivity and proteins. The soluble fraction as obtained above after the sedimentation of the chromatin fraction at 100,000 × g was chromatographed on a Sephadex G-100 column. The results indicated that the soluble fraction contained a protein fraction which showed binding to DES-H³.

Table 3

Percentage distribution of proteins and radioactivity in different fractions of ultrasonicated nuclear fraction from uterus after DES-H³ injections (i. p.)

	100 × g	10,000 × g	100,000 × g	Soluble fraction	Total
Proteins	5.0 ± 1.1 (6)[1]	12.0 ± 2.1 (6)	28.2 ± 2.8 (6)	54.8 ± 3.7 (6)	100.0
Radioactivity	4.1 ± 1.9 (5)	9.4 ± 2.2 (5)	24.3 ± 3.8 (5)	62.2 ± 4.5 (5)	100.0

[1] Figures in brackets refer to the number of observations.

The foregoing results show that DES-H³ is heavily localised in nuclear and supernatant fractions of rat uterus. The diethylstilbestrol has been shown to bind to macromolecules present in these fractions. DES shows binding to receptor macromolecules which have been shown to bind with estradiol in the uterus. In this respect DES behaves in the same fashion as estradiol. No stereo-chemical studies on the binding properties of the protein or the estrogen moiety have become available. It is possible DES may rearrange itself before it binds to the receptor protein.

Studies on uptake of 1:2-H³-progesterone

Work on the elucidation of the action of progesterone is complicated by the fact that the action of progesterone can only be elicited when the tissue has been previously acted upon by estrogen. Keeping this in view, ovariectomized and estradiol primed animals were used to study the uptake and disappearance of 1:2-H³-progesterone in the neural and genital tissues of rats [20]. The results are shown in Fig. 5. It is apparent that the maximum uptake of

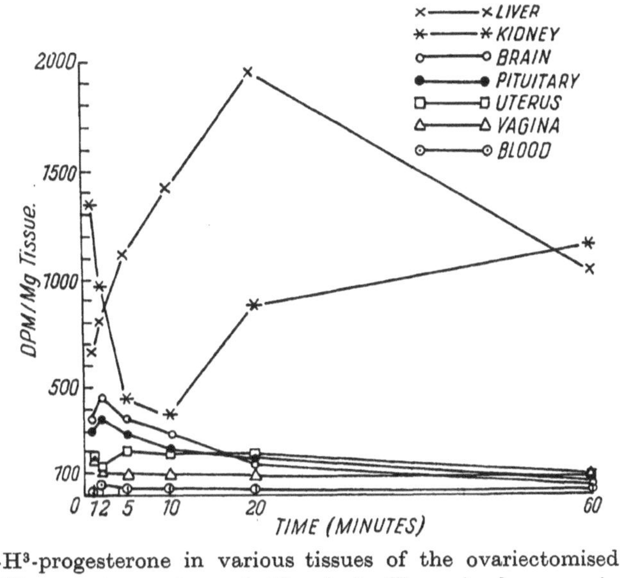

Fig. 5. Uptake of 1,2-H³-progesterone in various tissues of the ovariectomised rat after a single i.v. injection of 20 μc 1,2-H³-progesterone (sp. act. 90 μc/μg). The animals were primed with 0.1 μg. oestradiol/day in oil s.c. for 3 days. The results are expressed as mean disintegrations/min./mg. wet tissue and as mean disintegrations/min./μl/blood. Each point is the average of results in five or six animals. P-values for differences in uptake at 2 min. are <0.001 for brain v. vagina and pituitary v. vagina, and <0.005 for brain v. uterus and pituitary v. uterus; at 1 min. the values are <0.005 for brain v. vagina, <0.01 for pituitary v. vagina, <0.025 for brain v. uterus and <0.1 for pituitary v. uterus

progesterone and its metabolites was observed in uterus and vagina within 5 minutes of the i.v. injection. The uptake in uterus at the time of maximal uptake was of the order of 3.3×10^{-9} M after a single injection. Hypothalamus containing area of brain and pituitary showed maximum uptake within two minutes of the i.v. injection of 1:2-H^3-progesterone, and subsequently it disappeared rapidly. In these respects genital and neural tissues behaved differently in studies with progesterone compared to those reported by JENSEN and JACOBSON [6] for estradiol.

The binding of progesterone to macromolecules present in rat uterus after an i. p. injection of 1:2-H^3-progesterone, isolation of the supernatant fraction and Sephadex chromatography were carried out. The binding was greatly affected by the concentration of the steroid used and the pretreatment of the animal with estradiol. Further work is needed to elucidate this point.

Comments

The above evidence shows that circulating endogenous or exogenously injected estradiol has selective affinity to concentrate in certain target tissues of the body. The presence of high amounts of circulating estrogen increases the binding sites of estradiol in the uterus. In target tissues estradiol shows a heavy localization in the nuclear and $105{,}000 \times g$ supernatant fractions. On the other hand, in non-target tissues the major portion of radioactivity is localised in the supernatant fraction. Evidence from this and other laboratories show that the estradiol radioactivity present in the supernatant fraction is bound to receptor macromolecules. The receptor shows selective binding to estradiol-17β at the molecular level. Experimental evidence presented shows that the receptor present in the supernatant fraction is also able to bind DES. Investigations on the radioactivity present in the nuclear fraction show that estradiol-17β and DES show binding to a receptor present in the soluble fraction of the nucleus. It is not known whether the receptors present in the supernatant and nuclear fractions are the same or different. The physiologic role of these receptors in growth promoting processes within the cell in the presence of estrogens is not yet clear. More work is needed to characterise and extensively purify these receptor macromolecules. Progesterone on the other hand behaves differently from estradiol in its uptake and disappearance from the target tissues. While estradiol is incorporated and retained in the uterus, vagina and pituitary, progesterone is rapidly taken up and disappears from these tissues. Also progesterone is metabolised in the uterus, while estradiol remains unchanged. Furthermore estradiol induced changes in the uterus can be picked up within a few minutes of estradiol injection, progesterone response is delayed. Therefore in suggesting any possible action of these hormones, the above points must be taken into consideration.

References

[1] DOISY, E. A., C. D. VELER, and S. A. THAYER, Amer. J. Physiol. 90, 329 (1929).
[2] BUTENANDT, A., Naturwiss. 17, 879 (1929).
[3] MUELLER, G. C., In G. PINCUS and E. P. VOLLMER (eds.) Biological Activities of Steroids in relation to Cancer, p. 129, New York, Academic Press (1960).
[4] HAMILTON, T. H., Proc. Nat. Acad. Sci. USA, 49, 373 (1963).
[5] SZEGO, C. M., and D. A. LAWSON, Endocrinology, 74, 372 (1964).
[6] JENSEN, E. V., and H. I. JACOBSON, Recent Progr. Hormone Res., 18, 387 (1962).
[7] STONE, G. M., and L. MARTIN, Steroids 3, 699 (1964).
[8] LAUMAS, K. R., Proceedings of the Third Asia and Oceania Congress of Endocrinology, Jan. 2–6, 1967 Manila, Philipines, I, 124 (1967).
[9] SWEAT, M. L., M. J. BRYSON, and R. B. YOUNG, Endocrinology, 167, 81 (1967).
[10] LAUMAS, K. R., and A. FAROOQ, Abstract. Programme of the 46th Meeting of the Endocrine Society, Chicago (1966).

[11] KATO, J., and C. A. VILLEE, Endocrinology, 80, 561 (1967).
[12] FLERKO, B., and SZENTAGOTHAI, Acta Endocrinol. (Kbh), 26, 121 (1957).
[13] BOGDANOVE, E. M., Endocrinology, 72, 638 (1963).
[14] PALKA, Y. S., V. D. RAMIREZ, and C. H. SAWYER, Endocrinology, 78, 487 (1966).
[15] NATEBOOM, W., and J. GORSKI, Arch. Biochem. Biophy., 111, 1059 (1965).
[16] TALWAR, G. P., S. J. SEGAL, A. EVANS, and O. W. DAVIDSON, Proc. Nat. Acad. Sci. 52, 1059 (1964).
[17] TOFT, D., and J. GORSKI, Proc. Nat. Acad. Sci. 55, 1574 (1966).
[18] KING, R. J. B., and J. GORDON, J. Endocrin. 34, 431 (1966).
[19] KING, R. J. B., J. GORDON, D. M. COWAN, and D. R. INMAN, J. Endocrin. 36, 139 (1966).
[20] LAUMAS, K. R., and A. FAROOQ, J. Endocrin. 36, 95 (1966).

Action of Steroid Hormones on the Cell Nucleus

C. E. Sekeris, D. Gallwitz, I. Lukács, J. Homoki and G. Schütz
Physiologisch-Chemisches Institut der Universität Marburg/Lahn, DBR

This presentation deals with the early effects of some steroid hormones on isolated nuclear systems. We will present evidence that shows that the primary site of action of some steroid hormones is in the cell nucleus. This work is a direct consequence of the hormone-gene hypothesis put forward six years ago (Karlson, 1961). The central postulate of this concept is a primary interaction of the hormone with the genetic material as the first step of a sequence of reactions leading to the stimulation of RNA and protein synthesis and thus to the expression of hormone activity (see Fig. 1).

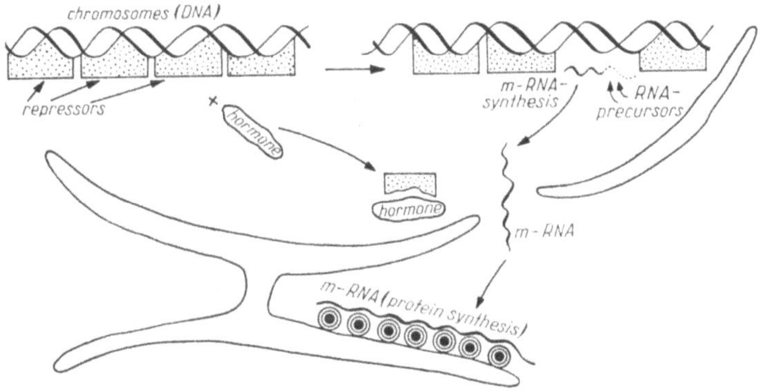

Fig. 1. Schematic representation of the hormone-gene hypothesis according to Karlson. The hormone interacts directly with the genetic material leading by derepression to stimulation of messenger RNA synthesis. The messenger RNA, in the cytoplasma, then codes for specific proteins

In this way our attention was focused on the nucleus. Our experiments were carried out with corticosteroids, mainly cortisol and with the insect steroid hormone ecdysone. In this paper the experiments dealing with the action of cortisol on isolated rat liver nuclei will be described. The experiments with ecdysone have been presented in a recent review (Karlson and Sekeris, 1966).

It is known that injection of cortisol to rats leads within 20—30 minutes to an increase in the activity of the liver RNA polymerase (Lang and Sekeris, 1964, Barnabei and Sereni, 1964), assayed as aggregate enzyme (Weiss, 1960) (Fig. 2).

This was followed by an increase of nuclear RNA synthesis as measured by the incorporation of ^{32}P into RNA (Fig. 3).

This in vivo effect is evident within 30—45 min after hormone injection. If the primay effect(s) leading to stimulation of RNA synthesis take place in the nucleus, it should be able to stimulate RNA synthesis by adding directly the hormone to the liver nuclei in vitro.

In order to test this possibility we faced the problem of isolation of rat liver nuclei in a pure and functionally intact form. The main steps of the method used (see Dukes et al., 1966 for

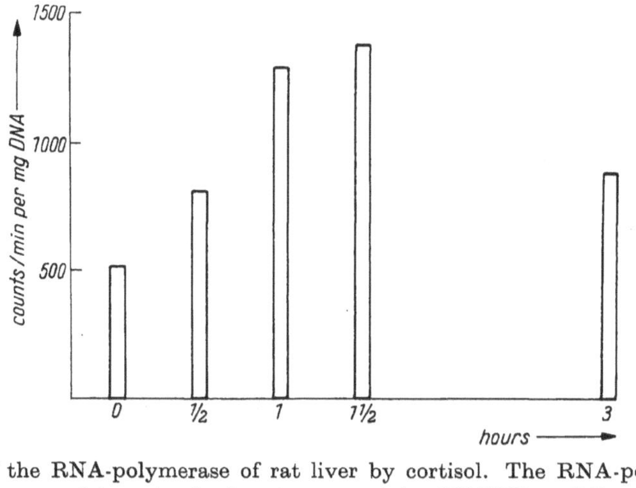

Fig. 2. Stimulation of the RNA-polymerase of rat liver by cortisol. The RNA-polymerase, isolated as aggregate enzyme, was assayed in a standard system containing ^{14}C-UTP at different time intervals after injection of 2.5 mg cortisol/100 g rat

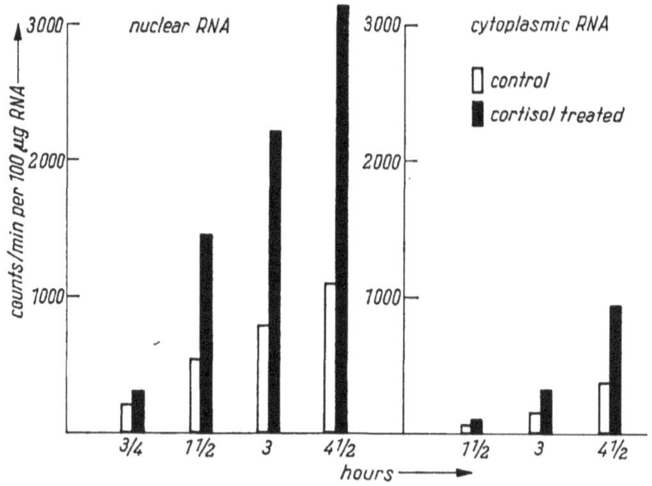

Fig. 3. Stimulation of incorporation of ^{32}P in liver RNA by cortisol. Cortisol (2.5 mg/160—180 g rat) and ^{32}P were injected simultaneously. At different time intervals the RNA was isolated by phenol extractions at 65° and its specific activity determined

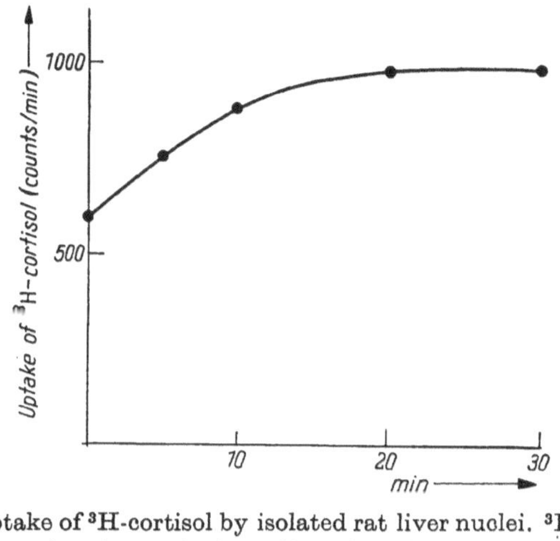

Fig. 4. Time course of the uptake of ^{3}H-cortisol by isolated rat liver nuclei. ^{3}H-cortisol was incubated with rat liver nuclei and at different time intervals the radioactivity incorporated in the nuclei was determined

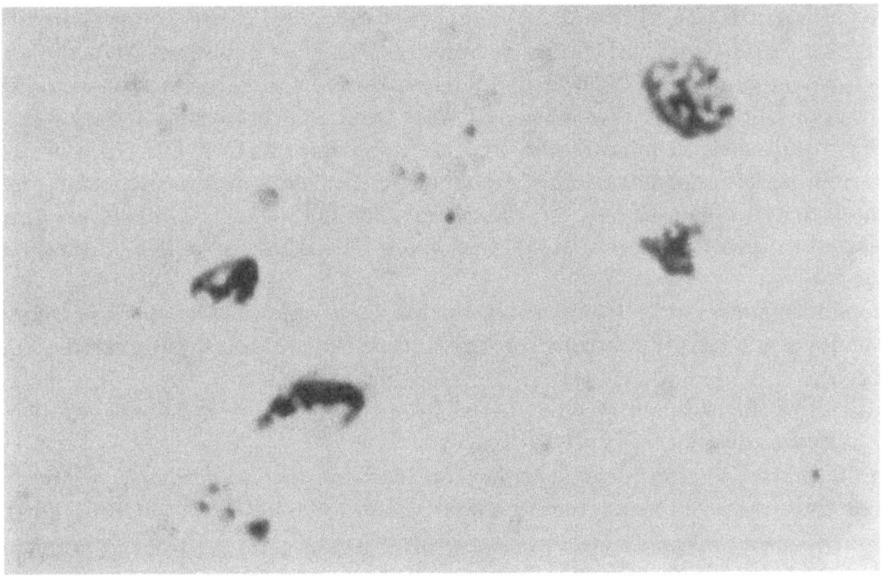

Fig. 5. Autoradiography of rat liver nuclei incubated for 10 min at 37° with ^{3}H-cortisol

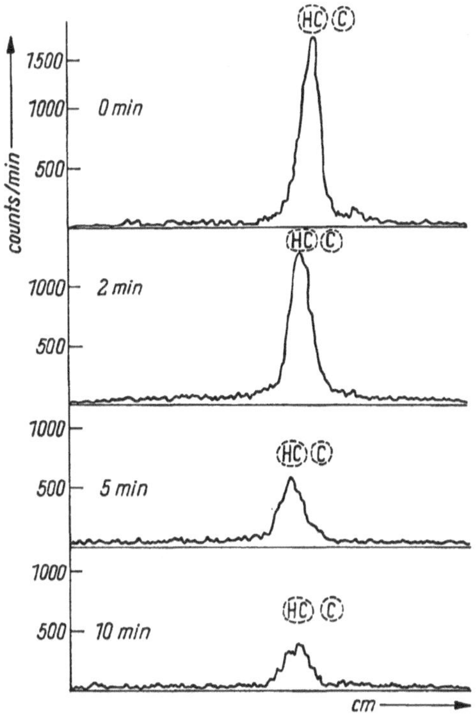

Fig. 6. Thin layer chromatography of methanol extracts of nuclei incubated with ^{14}C-cortisol. HC: localization of authentic cortisol run with the extract (from TIEMER and SEKERIS, 1967)

details) consists of homogenization of the liver in 0.25 M sucrose, 0.025 M KCl, 0.01 M $MgCl_2$, 0.065 M tris buffer, pH 7.55, filtering the homogenate through gauze and treating it with the detergent Triton X-100 at a final concentration of 0.5%. The detergent frees the nuclei from attached cytoplasmic impurities and leads to dissolution of the mitochondria and membranes. The nuclei prepared in this way are essentially free from contaminating material as seen from phase contrast and electron microscopy, by determination of RNA/DNA/protein ratios and also by determining the concentration of cytoplasmic enzymes (such as glucose-6-phosphatase and glutamate dehydrogenase) (see Sekeris, 1967, for details). The nuclei are functionally intact in respect to incorporation of RNA precursors into RNA as well as of amino acids into nuclear proteins.

Incubation of the nuclei with ^{3}H-cortisol leads to a very rapid uptake of the hormone (Fig. 4).

The hormone is not simply adsorbed on the nuclear membrane as suggested by autoradiography (Fig. 5).

Essentially all of the radioactivity extracted from the nuclei by methanol can be recovered as cortisol (Tiemer and Sekeris, 1967) (Fig. 6).

Addition of cortisol to the nuclear suspension leads within 30 seconds to increased RNA polymerase activity assayed as aggregate enzyme (Lukács and Sekeris, 1967) (Table 1).

Table 1

Stimulation of RNA-polymerase activity of rat liver nuclei after addition of 20 μg cortisol/ml in vitro

	Incubation time (min)	Activity (μμMol ^{14}C-UTP incorporated into RNA/mg DNA/min)
Experiment 1	0,5	
Control nuclei		89,7
hormone treated nuclei		112
Experiment 2	5	
Control nuclei		136
hormone treated nuclei		214
Experiment 3	10	
Control nuclei		75
hormone treated nuclei		173
Experiment 4	20	
Control nuclei		89,9
hormone treated nuclei		117,4

(from Lukács and Sekeris, 1967)

Maximal stimulation is seen after 10 minutes of incubation. The effect is dose dependent as shown in Table 2.

Table 2

Dose dependence of the action of cortisol on the RNA polymerase of isolated rat liver nuclei

	Cortisol (γ/ml)	Activity (μμMol ^{14}C-UTP incorporated in RNA/mg DNA/min)
Control nuclei	—	75
hormone treated nuclei	1	124
hormone treated nuclei	10	173
hormone treated nuclei	20	215

(after Lukács and Sekeris, 1967)

Stimulation of the polymerase is very specific. It cannot be evoked by structurally related steroids devoid of hormonal activity (Table 3).

As seen from Table 3, of the steroid hormones tested, only cortisol and testosterone show a stimulation of polymerase activity.

Table 3

Specificity of the action of different steroids on the RNA-polymerase activity of isolated rat liver nuclei

	Activity ($\mu\mu$Mol ^{14}C-UTP incorporated in RNA/mg DNA/min)
Experiment 1	
Control nuclei	111
Nuclei + cortisol	162
Nuclei + testosterone	163
Nuclei + androsterone	116
Experiment 2	
Control nuclei	117
Nuclei + cortisol	220
Nuclei + progesterone	146
Nuclei + pregnenolone	71

(after Lukács and Sekeris), 1967)

A direct consequence of the activation of the polymerase is an increased nuclear RNA synthesis as demonstrated by the increase of incorporation of ^{14}C-2-uracil into RNA of hormone treated nuclei in comparison to the controls (Dukes and Sekeris, 1965) (Fig. 7).

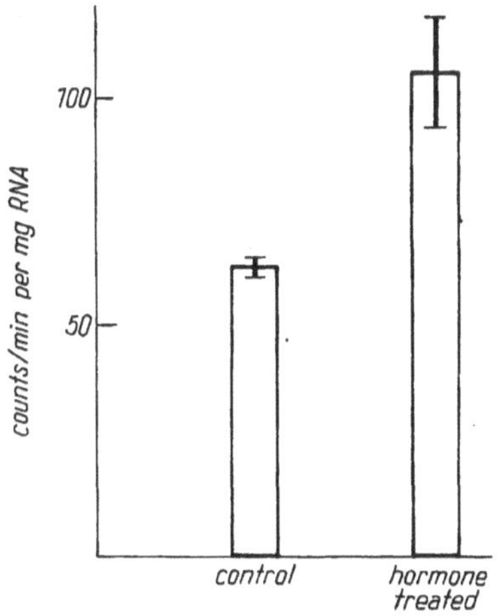

Fig. 7. Stimulation of incorporation of ^{14}C-2-uracil into nuclear RNA by cortisol.
Cortisol (10 μg/ml) and ^{14}C-uracil (25 μC/ml) were added at 0 time to the nuclei and the incorporation of uracil into nuclear RNA was measured after 2 hours of incubation at 37° (from Dukes and Sekeris, 1965)

A great part of the newly synthesized RNA in response to the hormone has messenger RNA characteristics as seen from its stimulatory action in a in vitro protein synthesizing system. The RNA from hormone treated nuclei shows a greater template activity than the control RNA (Dukes et al., 1966) (Table 4).

Stimulation of messenger RNA synthesis is observed 15 minutes after addition of cortisol to the nuclei (Schmid et al., 1967) i.e. a few minutes after maximal stimulation of the polymerase.

Table 4

Increase of template activity of nuclear RNA by cortisol in vitro

	^{14}C-leucine incorporated into protein cpm/mg ribosomal RNA
Ribosomal system alone	545
Ribosomal system + RNA from control nuclei	1070
Ribosomal system + RNA from cortisol treated nuclei (15 γ/ml)	1322

The results described above prove thar the effects of cortisol on RNA metabolism of the liver cell are due to a direct interaction of the hormone with the nuclei. The primary site of action of cortisol should be therefore sought within the confines of the nucleus. Elucidation of the nature of activation of the polymerase (i.e. increased template activity of chromatin, allosteric effects on the enzyme, de novo synthesis of the polymerase etc.) will bring us closer to this goal. The in vitro system described offers us the possibility of studying these different mechanisms.

Similar in vitro effects on nuclear RNA synthesis have been obtained with the system ecdysone-nuclei from epidermis tissue of the blowfly and it is very probable that the same may be true for other hormonal systems. Work in this direction is currently pursued in our laboratory.

Acknowledgements

We thank Professor P. KARLSON for his continued interest in our work and Mrs. U. VON HANSEN, Mrs. B. GALLWITZ and Miss I. HELLMANN for competent technical assistance. This work was supported by the Deutsche Forschungsgemeinschaft.

References

[1] BARNABEI, O., and F. SERENI, Biochim. Biophys. Acta, **91**, 239 (1964).
[2] DUKES, P. P., and C. E. SEKERIS, Z. Physiol. Chem., **341**, 149 (1965).
[3] DUKES, P. P., C. E. SEKERIS, and W. SCHMID, Biochim. Biophys. Acta, **123**, 126 (1966).
[4] KARLSON, P., Dtsch. Med. Wschr., **86**, 668 (1961).
[5] KARLSON, P., and C. E. SEKERIS, Rec. Progress. Horm. Res., **22**, 473 (1966).
[6] LANG, N., and C. E. SEKERIS, Life Sciences, **3**, 391 (1964).
[7] LUKÁCS, I. and C. E. SEKERIS, Biochim. Biophys. Acta, **134**, 85 (1967).
[8] SCHMID, W., D. GALLWITZ, and C. E. SEKERIS, Biochim. Biophys. Acta, **134**, 80 (1967).
[9] SEKERIS, C. E., 18. Mosbacher Kolloquium, Springer Verlag, Berlin, (1967) p. 126.
[10] TIEMER, K. W., and C. E. SEKERIS, in preparation.
[11] WEISS, S. B., Proc. nat. Acad. Sci., **46**, 1020 (1060).

The Mechanism of Action of Steroid Hormones: Some Aspects of the Contemporary Problem

A. E. Kellie

The Courtauld Institute of Biochemistry, the Middlesex Hospital Medical School,
London, England

Summary

With reference to the action of steroid hormones at the cellular level, the view is expressed that the action of steroids on the biochemical events in the cell is of more importance than the action of tissues on steroid hormones. Evidence is accumulating that hormones react with receptor molecules in the tissues and may act physiologically by releasing genetic information already stored in the cell. This idea is examined with particular reference to aldosterone and oestradiol.

A voluminous literature exists relating to the action of mammalian tissues on the metabolism of endogenous steroid hormones. On the anabolic side, knowledge of the biosynthetic steps leading to the formation of the hormones is reasonably complete and this knowledge has given considerable insight into clinical conditions which arise from hypo- and hyper-function of the steroidogenic endocrine organs. Information is also available concerning the metabolic fate of the secreted hormones, the inactivation by reduction and hydroxylation, the conjugation and ultimate excretion in urine and faeces. Although a considerable effort has been made to acquire this detailed knowledge there is no evidence that the catabolic changes observed have any connection with the mechanism of action of the hormones at the cellular level; on the contrary, circumstantial evidence suggests that the changes recorded represent the disposal of hormone which is in excess of the physiological requirement.

Perhaps of greater importance is the action of steroid hormones on tissue, for this is undoubtedly related to physiological function. The scope of the title is so wide that no attempt will be made to review all aspects of steroid action; comment will be limited to general considerations of steroid-tissue effects and to the particular consideration of the action of two steroid hormones, aldosterone and oestradiol, fields in which recent progress has been made.

The action of steroid hormones on tissues is very varied, ranging from that of cortisol which affects many tissues and induces complex changes, to that of aldosterone which, as far as is known, exercises only one physiological role in regulating the osmotic balance of the cell. The action of oestradiol and testosterone differs in that these hormones affect selected target organs initiating in these tissues the entire growth process. Evidence obtained *in vivo* suggests that the continued presence of these hormones is necessary to sustain growth in the target organ. Not only do some hormones affect many tissues but it is also true that some tissues are affected by a variety of hormones; thus protein synthesis in rat liver cells may be stimulated by growth hormone, thyroxine and testosterone. These growth effects are independent and additive and as they can be demonstrated consecutively in the same tissue and have different times of response, it is clear that separate mechanisms are involved. It is noteworthy that the demonstration of the steroid-tissue effects can only be made with viable tissue; it appears that the response is dependent on the integrity of the cellular structure.

Mechanism of Action of Aldosterone

The action of aldosterone on sodium transport is unique in that the effect can be studied under *in vitro* conditions with surviving frog skin (McAfee and Locke, 1961) or with the isolated toad bladder (Crabbe, 1961). Bladder tissue can be set up to form a boundary between two compartments separating different molar concentrations of sodium ions; if aldosterone (10^{-8} M) is added to one compartment, active transport of sodium ions against the osmotic and electrical gradient commences after a delay of 60—100 minutes and the effect may last for 5 hours. If [^{3}H]-aldosterone is used in this type of experiment it is possible to demonstrate that maximum concentration of the hormone occurs in the cells in less than 30 minutes, long before the effect of the hormone is observed. Moreover, after exposing the tissue to minimal amounts of labelled aldosterone, the bulk of the radioactivity ($>95\%$) can be washed from the tissue without influencing the period of delay or the magnitude of the response. Similar response is given by other steroid compounds which act as mineralocorticoids and the magnitude of the *in vitro* effect is related to that *in vivo*.

Pre-incubation of bladder tissue with a steroid-free medium (15 hours) enhances the sensitivity of the response and it has been shown that active transport requires the availability of some source of metabolic energy, e.g. ATP. Some insight into a possible mechanism for the transport of cations through the cell walls is given by the work of Skou (1962) and Judah and Ahmed (1964) which describes a membrane-bound Na^+—K^+—Mg^{++}-activated ATPase system.

The location of aldosterone at a cellular level in the isolated toad bladder has been studied by autoradiography (Edelman et al., 1963) and has shown concentration of radioactivity in the nuclei of the mucosal epithelial cells. To a lesser extent, radioactivity was found in the cytoplasm of these cells and in the submucosal and serosal layers. The presence of the hormone in the nuclei of the cells lends support to the hypothesis that steroid hormones may act at a nuclear level to modify the control of genetic information. Aldosterone stimulates the synthesis of DNA-dependent RNA; it increases the rate of incorporation of [^{3}H]-uridine into RNA and promotes the incorporation of amino acids into bladder protein. Actinomycin D which inhibits DNA-dependent RNA synthesis also inhibits the *in vitro* effect of aldosterone on sodium transport.

Possible mechanisms

Because of the delay in the effect produced by aldosterone, it is extremely unlikely that the hormone plays a direct role in sodium transport. Even if it was possible to demonstrate that the primary action of aldosterone was the *de novo* synthesis of protein it would still be necessary to postulate how the newly synthesized protein(s) or enzyme(s) could promote sodium transport. Suggestions that transport is influenced by a change in the permeability of the cell wall, merely replace one problem by another. *In vitro* the mineralocorticoid effect is dependent upon substrate supply, suggesting that the effect is linked to metabolic events. This view receives some support from the observation that there is a rise in the ATP content of aldosterone-treated toad bladders (Edelman, 1966). Aldosterone does not appear to increase the capacity of the ATPase system by protein synthesis but to control the activity by increasing the energy available.

The data accord with the hypothesis that aldosterone is bound to the tissue by a specific receptor but that it does not play a primary role in sodium transport. The latter effect is a remote consequence of the primary action which requires a source of energy and which may involve transcription and translation of genetic information.

Mechanism of Action of Oestradiol

It has been recognised for a long time that the growth and development of some tissues in the adult mammal is dependent on the continued secretion of the steroid sex hormones, oestradiol and testosterone. These tissues do not develop in the immature animal and their growth is not maintained in the adult after gonadectomy. Although the ultimate secondary sex changes are well recognised, the mechanism whereby the steroid hormones induce the changes at the cellular level is not understood, nor are we clear about the difference between a hormone-dependent and a hormone-independent tissue.

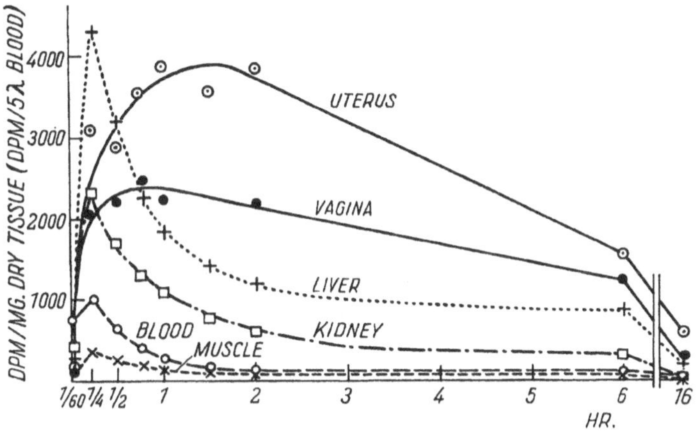

Fig. 1. Concentration of radioactivity in rat tissues after single subcutaneous injection of 0.098 μg. (11.5 μc.) of [6,7-^{3}H]-oestradiol in 0.5 ml. saline. Liver and kidney points are mean values of 3 aliquots of dried pooled tissue; other points are median values of individual samples from 6 animals. Muscle is *M. Quadriceps femoris* (JENSEN and JACOBSON, 1962)

In a classical approach to this problem, JENSEN and JACOBSON administered minute doses of [6,7-^{3}H]-oestradiol (0.1—0.01 μg) to immature rats and studied the tissue distribution of the radioactivity as a function of time (JENSEN and JACOBSON, 1962). They were able to show that the level of radioactivity rose more rapidly in liver and kidney tissue than in uterus and vagina but that the level was not maintained (Fig. 1). Although not more than 0.1% of the dose was present in the uterus, the concentration of oestradiol in this organ (10^{-8} to 10^{-11} M) was greater than that in plasma (5×10^{-11} M) and remained above normal for a prolonged period. Examination of an extract of the uterus showed that the radioactivity was entirely due to unchanged oestradiol. In similar experiments, carried out with [6,7-^{3}H]-oestrone, retention of radioactivity by the uterus was much lower and examination of the extract showed clearly that the oestrone had been partially converted into oestradiol. It is possible that oestrone becomes biologically active by conversion to oestradiol. Clear-cut evidence that oestradiol does not undergo reverse oxidation to oestrone has been provided by JENSEN et al. (1966) who administered a mixture of [6,7-^{3}H]- and [17α-^{3}H]-oestradiol to immature rats; the ratio of the hormones isolated from the uterus was identical with that of the injected hormones, implying that the reaction (b) (Fig. 2) did not occur.

Uterine absorption tests carried out with [6,7-^{3}H]-oestriol showed an affinity for this oestrogen but the radioactivity was cleared quickly from the blood and was retained by the target organ for a shorter time than with oestradiol. Other active oestrogens, 17α-methyl-oestradiol, 17α-ethynyloestradiol and hexoestrol, were bound in a manner similar to that of oestradiol and were recovered unchanged from the target organ. The injection of 1 mg of unlabelled oestradiol 2 hours before the administration of the tritiated oestrogens inhibited the uptake of radioactivity, presumably because all the binding sites were occupied. Stilboestrol, a non-steroidal oestrogen, could be substituted for oestradiol in this type of experiment, but 17α-oestradiol, testosterone and progesterone were not effectively absorbed. Puromycin and actinomycin D did not affect the binding of oestradiol by uterine tissue.

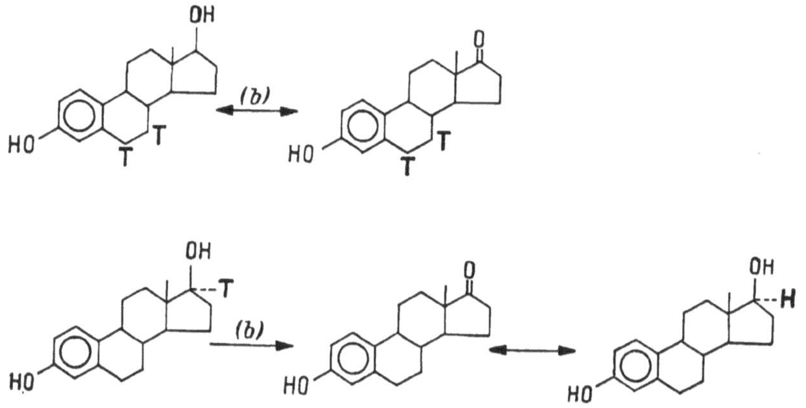

Fig. 2. The consequence of reversible oxidation of oestradiol on the tritium label in the two molecular species [6,7-^{3}H]-oestradiol and [17α-^{3}H]-oestradiol. In the former case the tritium would not be lost; in the latter all tritium would be lost. The fact that the ratio [6,7-^{3}H]:[17α-^{3}H] does not change on injection implies that reaction "b" does not occur

In the uterus, autoradiography of the tissue showed that activity is partly in the endometrium (lamina propria) and partly in the myometrium, but the concentration in the former was approximately twice that in the latter. Studies of the subcellular distribution indicate binding in the cytoplasm (Talwar et al., 1964; Dingman et al., 1965) and with the chromatin of the nucleus (Noteboom et al., 1965; Jensen, 1965).

The specific binding molecule

The first clear evidence of the presence of a specific binding site for oestradiol in the uterine endometrium was obtained by the injection of labelled oestradiol *in vivo* followed by removal and processing of the uteri. Homogenates prepared from the tissue were centrifuged at high speed in a sucrose gradient to separate subcellular particles. Radioactivity was associated principally with the nuclear pellet and with the cell supernatant. This procedure has been replaced by incubating the excised tissue (from immature rat uteri or calf uterine endometrium) *in vitro* with tritiated oestradiol (15 min. at 38°) and under these conditions binding of the hormone to the specific receptor occurs. In this way several groups (Toft et al., 1967; Maurer et al., 1967) have been able to prepare homogenous material which apparently binds specifically to biologically active oestrogens. Homogenates of rat uteri exposed to tritiated oestradiol and processed in this manner yielded two oestradiol peaks, one associated with a component 9.5S which has the properties of a hormone receptor and a component 3-6S which showed nonspecific binding.

The 9.5 component, with a molecular weight of approximately 200.000, is saturated with oestradiol at extremely low concentrations (14×10^{-7} μM) corresponding to a dissociation constant (k) of 7×10^{-10}. Kinetic studies show that the binding sites are homogenous, i.e. all of one kind, and on the assumption that they are uniformly distributed throughout the uterus, it has been calculated that this corresponds to approximately 2.500 sites per cell. Binding to the receptor molecule is destroyed by the proteolytic enzymes pronase and trypsin but not by deoxyribonuclease or by ribonuclease. Binding is optimal at pH 7 and is significantly lower at pH 6 and 8. Heat treatment (60°, 4 min.), detergents (0.5% sod. dodecyl sulphate or 1% taurocholate), — SH reacting agents all destroy binding capacity. The properties are compatible with the component 9.5S being a protein or at least partly a protein. Binding occurs in the cold and does not require energy or intact cellular organization.

Biochemical events associated with oestradiol stimulation

Within one hour of the injection of oestradiol *in vivo* it is possible to detect biochemical changes in the uterine tissue *in vitro* which appear to follow fixation of oestradiol. Mueller et al. (1958) noted that an early response of the uterus was generalised hyperaemia of the tissue followed by the imbibition of fluid throughout the structure. It was noted that within hours there was an increased rate of uptake of uridine and thymine which preceded increased incorporation of isotope-labelled amino acids. A modern form of this experiment illustrated in Fig. 3 (Tata, 1967), shows early events causally related to the advent of the hormone; these include increased rate of synthesis of rapidly-labelled nuclear RNA, increase in the incorporation of amino acids into protein. Excellent contemporary reviews of this work are available (Gorski et al., 1965, Korner, 1967).

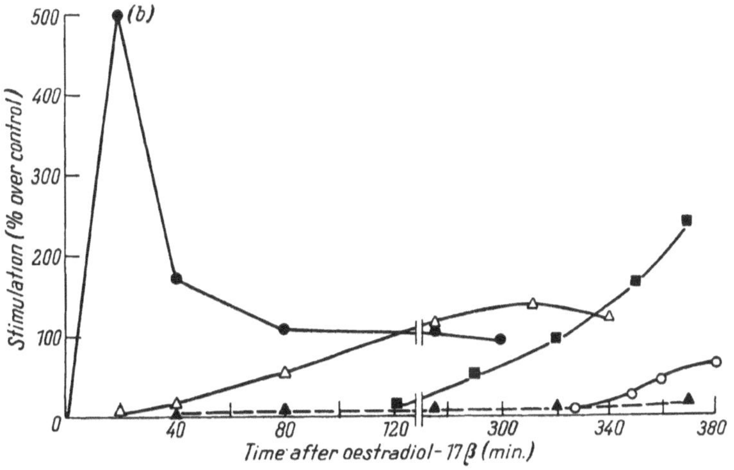

Fig. 3. Sequential events on RNA and protein synthesis in the uterus of ovariectomised rats after 2 – 10 μg oestradiol (Tata, 1967). ● — ●, rate of synthesis of rapidly labelled nuclear RNA *in vivo*; △ — △, Mg^{2+}-activated polymerase activity; ▲ — ▲, $Mn^{2+}/(NH_4)_2SO_4$-stimulated RNA-polymerase activity; ■ — ■, accumulation of newly formed (labelled) ribosomes in the cytoplasm; ○ — ○, incorporation of ^{14}C-labelled amino acids into protein/mg of RNA *in vitro* by ribosomes or microsomes

The essential steps whereby information recorded in the genes finds expression, by replication of DNA in the nucleus, transcription of RNA, of various types, in the nucleus and translation of the RNA as proteins at the ribosomes, is shown graphically in Figs. 4 and 5. Although increase in DNA during the initial phase of oestradiol stimulation is negligible, some of the earliest effects *in vivo* are on RNA synthesis; an increased rate of turn-over of RNA is a possible way in which the primary effects of the hormone could be multiplied in terms of metabolism and growth. It is not possible to say at this stage whether the primary effect is nuclear or cytoplasmic, as the transcription of RNA occurs in the nucleus whereas the majority of protein synthesis occurs in the cytoplasm at the polysomes formed by the organized assembly of ribosomes at intracellular membranes. A current shift of emphasis from transcription to the translational process is associated with a growing feeling of the importance of ribosomal protein and the mechanism of transport of m-RNA from the nucleus to the cytoplasm. Information is also required concerning transcription and translation with special reference to chain initiation and termination (Watson, 1966). This kind of information is likely to be obtained, in the first place, by a study of unicellular organisms and there is a constant need to relate such findings to the cells of metazoa and the mammalia.

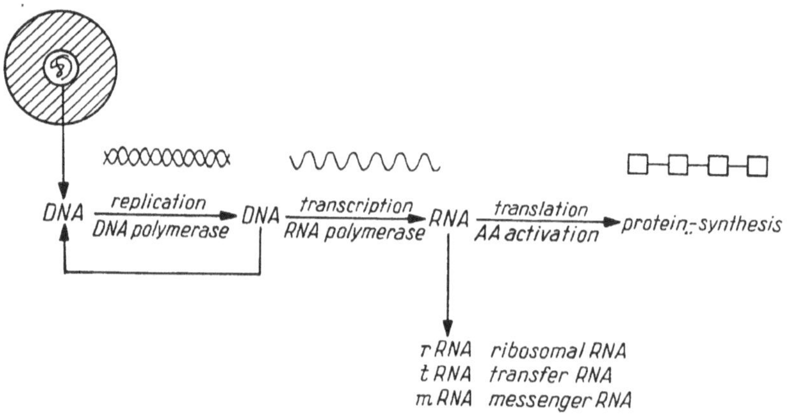

Fig. 4. The mechanism of release of genetic information culminating in protein synthesis

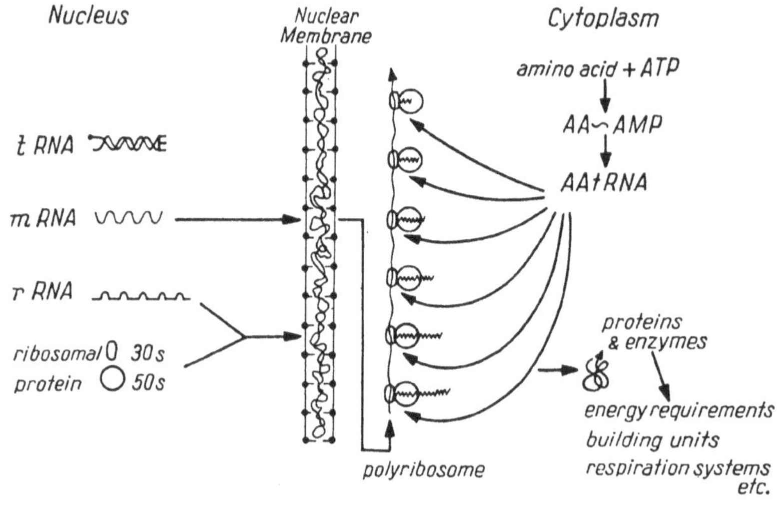

Fig. 5. Protein synthesis at the cellular level

The close association of both aldosterone and oestradiol with the nucleus of the cell together with the evidence of metabolic priming has led to speculation, along the lines of the Jacob and Monod hypothesis (1961), that the hormones may in some way unmask the repressed genetic information in the cell nucleus. Certainly the early changes in RNA and metabolism fit in with such a hypothesis, although they do not yet explain what change occurs in the receptor molecule after complex formation with the hormone.

References

[1] Crabbe, J., J. clin. Invest. **40**, 2103 (1961).
[2] Dingman, C. W., and M. B. Sporn, Science **149**, 1251 (1965).
[3] Edelman, I. S., Steroid. Dynamic., Edit. Pincus, G., Nakar, T., and Tait, F. J., Acad. Press. (1966).
[4] Edelman, I. S., R. Bogoroch, and G. A. Porter, Proc. Natl. Acad. Sci. U.S., **50**, 1169 (1963).
[5] Gorski, J., W. D. Noteboom, and J. A. Nicolette, Cell. & Comp. Physiol. **66**, 91 (1965).
[6] Jacob, F., and J. Monod, J. Mol. Biol. **3**, 318 (1961).
[7] Jensen, E. V., Proc. 2nd Int. Congr. Endocrin. 1964, Excerpta Med. Found., p. 420. Amsterdam (1965).

[8] JENSEN, E. V., and H. I. JACOBSON, Recent Prog. Hormone Res., **18**, 387 (1962).
[9] JENSEN, E. V., H. I. JACOBSON, J. W. FLASHER, N. N. SAHA, G. N. GUPTA, S. SMITH, V. COLUCCI, D. SHIPLACOFF, H. G. NEUMAN, E. R. DESOMBRE, and P. W. JUNGBLUT, Steroid Dynamics, Edit. Pincus, G., Nakao, T. and Tait, F. J., Acad. Press (1966).
[10] JUDAH, J. D., and K. AHMED, Biol. Rev. Cambridge Phil. Soc. **39**, 160 (1964).
[11] KORNER, A., Prog. Biophys. and Mol. Biology **17**, 61 (1967).
[12] MCAFEE, R. D., and W. LOCKE, Am. J. Physiol. **200**, 797 (1961).
[13] MAURER, H. R., and G. R. CHALKLEY, J. Mol. Biol. **27**, 431 (1967).
[14] MUELLER, G. C., A. HERRANEN, and K. F. JERVELL, Recent Progr. Hormone Res. **14**, 95 (1958).
[15] NOTEBOOM, W. D., and J. GORSKI, Arch. Biochem. Biophys. **111**, 559 (1965).
[16] SKOU, J. C., Biochim. biophys. Acta **58**, 314 (1962).
[17] TALWAR, G. P., S. J. SEGAL, A. EVANS, and O. W. DAVIDSON, Proc. Nat. Acad. Sci. Wash. **52**, 1059 (1964).
[18] TATA, J. R., Biochem. J., 2nd Colworth Lecture, **104**, 1 (1967).
[19] TOFT, D., G. SHYMALA, and J. GORSKI, Proc. Nat. Acad. Sci. U.S. **57**, 1740 (1967).
[20] WATSON, J. D., Molecular Biology of the Gene, Benjamin, New York (1965).

Effect of Steroids of Different Type on Transport ATPase

H. J. Portius and K. Repke

Institut für Biochemie, Berlin-Buch, der Deutschen Akademie
der Wissenschaften zu Berlin, DDR

Summary

The $Na^+ + K^+$-stimulated, Mg^{++}-dependent ATPase of cell membrane from guinea-pig cardiac muscle shown to be part of the active transport system for Na^+ and K^+ across cell membrane, was used as an *in vitro*-model to study the molecular nature of the known corticosteroid effects on cation transport. Over a broad range of concentrations, d-aldosterone, deoxycorticosterone or cortisol do not significantly and regularly stimulate the transport ATPase-system. Under various conditions studied, neither aldosterone nor cortisol are able to antagonize the inhibitory effect of the cardiac steroids ouabain or digitoxigenin on the enzymic system. The results do not favour the suggestion that the stimulatory corticosteroid effect on cation transport and the antagonism between corticosteroids and cardiac steroids occurs at the level of transport ATPase. High concentrations of corticosteroids may slightly inhibit the enzymic system, but this effect is also obtained with spirolactone and progesterone. The inhibitory activity of hormone type steroids is at least thousand times lower than that of cardiac steroids. The low affinity of common steroids to transport ATPase is due to the trans-junction of the rings C and D *and* the absence of the lactone ring at C-17.

There is increasing evidence in literature indicating that corticosteroids are of importance for cation transport not only in the kidney but in other organs and cells as well. For example, aldosterone is known to stimulate the active up-hill transport of sodium across biological membranes in several tissues including mammalian cardiac and skeletal muscle, in kidney tubule and in amphibian bladder and skin (Losert et al., 1964, Woodbury et al., 1957, Vander et al., 1961, Myers et al., 1961, Sharp et al., 1964, Fanestil et al., 1967, Crabbé, 1964, Crabbé et al., 1965). In contrast to corticosteroids, cardiac steroids such as ouabain are known to inhibit the active cation transport of tissues including cardiac muscle (for a review see Wilbrandt, 1963). Since corticosteroids and cardiac steroids appear to exert opposite effects on ionic transport, several investigators have postulated that corticosteroids and cardiac steroids may mutually antagonize their actions. As a matter of fact, for example an antagonism of the inhibitory effect of cardiac steroids by corticosteroids has been reported for potassium loss from the isolated guinea-pig heart (Kunz et al., 1963), for cation transport in rat kidney (Sulser et al., 1959) and for sodium transport across isolated frog skin (Wilbrandt et al., 1960).

There are several conflicting ideas in literature concerning the biochemical basis of the stimulating corticosteroid action on cation transport and the nature of the antagonistic effect of cardiac steroids. Chemical studies by Bernauer et al. (1957) indicating the possibility of chelation by corticosteroids led Wilbrandt (1959) to suggest that corticosteroids might act as cation carriers by cation chelation. On account of the apparent resemblance between the structures of the supposed corticosteroid chelate and digitoxigenin, a typical cardiac steroid, Wilbrandt (1959) further suggested that corticosteroids and cardiac steroids might compete for the same receptor in the cell membrane and that this could be the basis of their antagonistic effects.

We decided to check these possibilities by a more direct approach using the isolated cell membrane of guinea-pig cardiac muscle. As shown earlier (PORTIUS et al., 1963), the plasma membrane contains a $Na^+ + K^+$-activated, Mg^{++}-dependent ATPase which is part of the active transport system for Na^+ and K^+ across cell membrane. Thus, this enzymatic system will be called the transport ATPase. We have further shown that the transport ATPase is the receptor enzyme of cardiac steroids (REPKE et al., 1963). Therefore, it should be a useful model to analyse the nature of the stimulating corticosteroid effect on cation transport and the possibility of an antagonistic competition between corticosteroids and cardiac steroids for this receptor enzyme.

In the first step, the influence of d-aldosterone, deoxycorticosterone or cortisol on transport ATPase of cell membrane was studied (Fig. 1). With each compound, several enzyme prepara-

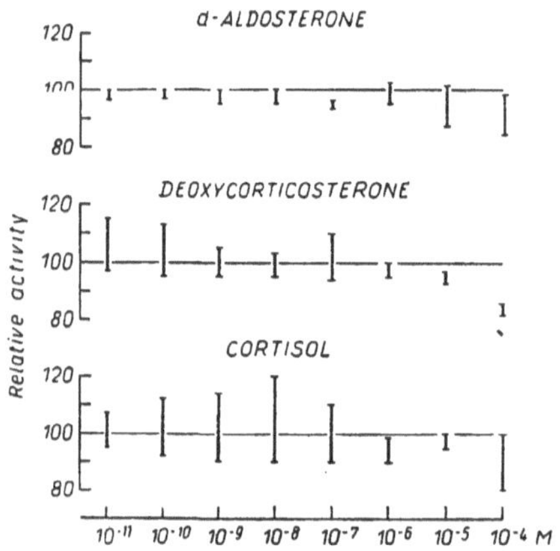

Fig. 1. Influence of various concentrations of corticosteroids on transport ATPase of cell membrane from guinea-pig cardiac muscle. Experimental conditions as described in Fig. 2, but incubation time was 15 min only. The enzyme activity in the absence of corticosteroids is arbitrarily set to be 100

tions were checked. All deviations from basic activity are included in this diagrammatic representation of corticosteroid effects. Within the range between 10^{-11} to 10^{-7} M, aldosterone does not stimulate transport ATPase. In the same range, deoxycorticosterone and cortisol may be found to promote slightly the activity of some enzyme preparations, but with others no stimulation or even an inhibition of the enzymatic system may be obtained. On the whole, there is no significant and regular stimulation of transport ATPase by corticosteroids. Therefore, the suggestion that corticosteroids might promote active cation transport by acting as cation carriers, although not disproved, is not supported by our results with the isolated system.

There are several hints in literature speaking in favour of the idea that the transport promoting effect of corticosteroids might be based on a reduction of passive membrane permeability (CRABBÉ, 1963, STREETEN et al., 1954). Since the leak-and-pump ratio determines the net size of up-hill cation transport, the supposed diminution of cation permeability by corticosteroids should increase the net efficiency of transport ATPase. Of course, this idea cannot be checked in our model system using fragments of cell membranes.

In the next step, we examined the question whether or not corticosteroids and cardiac steroids may compete for a common site in the transport ATPase system. Figure 2 shows the influence of various concentrations of d-aldosterone or cortisol on the inhibition of transport ATPase by ouabain or digitoxigenin. It is apparent that in the whole concentration range neither aldosterone nor cortisol are able to antagonize the inhibitory action of the two cardiac steroids. This negative result might be supposed to be arbitrary since we used only one concentration of ouabain or digitoxigenin effecting an inhibition of transport ATPase by approximately fifty and thirty-five percent respectively.

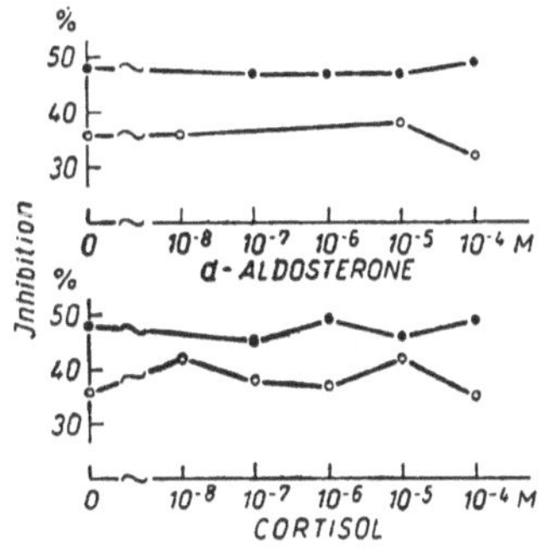

Fig. 2. Influence of various concentrations of d-aldosterone or cortisol on the inhibition of transport ATPase by 0.5 μM ouabain (filled signs) or 1.0 μM digitoxigenin (open signs). Experimental conditions: The final volume of 2.5 ml contained a mixture of the following components: 5 mM KCl, 135 mM NaCl, 2 mM $MgCl_2$, 9 mM Tris/HCl-buffer pH 7.4, 2 mM ATP, 0.2 to 0.6 mg of membrane protein. After preincubation in the absence of ATP (15 min at 20 °C) the incubation (60 min at 37 °C) was started by the addition of ATP. Enzyme activity was measured by determining the amount of the inorganic phosphate

Figure 3 shows the inhibitory activity of various ouabain or digitoxigenin concentrations in the absence or presence of 100 μM aldosterone. Apparently, neither with low nor with high concentrations of the two cardiac steroids, aldosterone is able to antagonize their inhibitory effect. The same is also true for cortisol (Fig. 4). For example, at 10^{-7} M ouabain or digitoxigenin, cortisol does not exhibit an antagonistic effect although its concentration is thousand times higher than that of the two cardiac steroids.

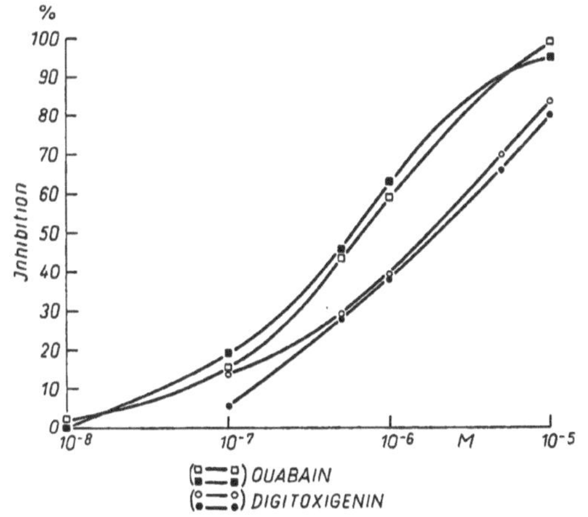

Fig. 3. Concentration/inhibition curves for ouabain or digitoxigenin in the absence (open signs) or presence (filled signs) of 100 μM d-aldosterone. Experimental conditions as described in Fig. 2

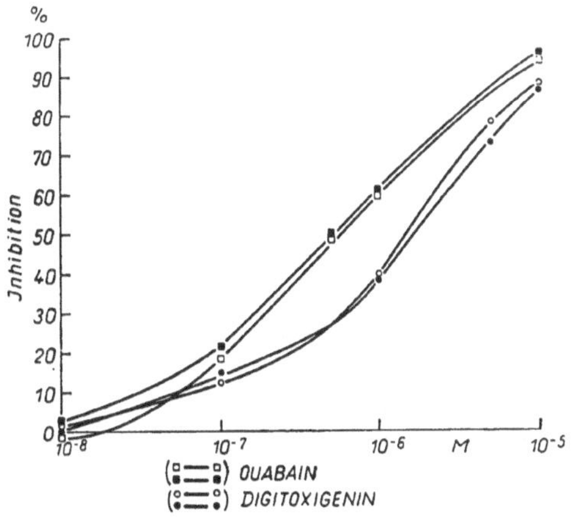

Fig. 4. Concentration/inhibition curves for ouabain or digitoxigenin in the absence (open signs) or presence (filled signs) of 100 μM cortisol. Experimental conditions as described in Fig. 2

Summing up this part of our paper, corticosteroids under the various conditions studied do not antagonize the inhibitory activity of cardiac steroids on transport ATPase. The results appear to exclude the possibility that the different steroid types compete for a common site in this enzymatic system. Apparently, corticosteroids and cardiac steroids have different points of attack. The corticosteroid antagonism against cardiac steroid inhibition of cation transport might possibly be effected by the supposed reduction of passive cation permeability resulting in an improvement of the net efficiency of transport ATPase.

High concentrations of d-aldosterone, deoxycorticosterone and cortisol may be able to inhibit transport ATPase (Fig. 1). The highest inhibitory effects of various steroids obtained with different enzyme preparations, are demonstrated in Table 1. The inhibitory action of corticosteroids is no specific effect since it is also obtained with the aldosterone antagonist spirolactone and the gestagen progesterone. On the other side, prednisolone is completely inactive. Since the inhibition of transport ATPase by hormone type steroids requires very high concentrations and shows no correlation to their biological functions, we may conclude that it does not participate in their physiological actions.

Table 1

Influence of high concentrations of various C/D trans steroids on transport ATPase. Experimental conditions as described in Fig. 2

Compounds	Concentration (μM)	Inhibition (%)
d-Aldosterone	100	16
Deoxycorticosterone	100	18
Cortisol	100	19
Spirolactone (3-[3-oxo-17β-hydroxy-Δ^4-androstene-17α-yl]-propanoic acid lactone)	100	25
Progesterone	100	25
Prednisolone	100	0
Androstane-3,17-dione	100	9

Table 2

Comparison between the inhibitory effects of steroids of different type on transport ATPase. Experimental conditions as described in Fig. 2

Compounds	Structural features			Concentration (μM)	Inhibition (%)
	Ring fusions		Butenolide ring		
	A/B	C/D			
d-Aldosterone	planar	trans	absent	100	16
Androstane-3,17-dione	trans	trans	absent	100	9
Digitoxigenin (14β-Testane-3β,14β-diol-17β-butenolide)	cis	cis	17β	0.1	15
14β-Testane-3β,14β-diol-17-one	cis	cis	absent	100	7
Androstane-3β-ol-17β-butenolide	trans	trans	17β	100	10
Testane-3β-ol-17β-butenolide	cis	trans	17β	100	20
Uzarigenin (14β-Androstane-3β,14β-diol-17β-butenolide)	trans	cis	17β	0.7	15

Table 2 gives a comparison between the inhibitory effects of steroids of different type on transport ATPase. It is evident that the inhibitory activity of d-aldosterone or androstane-3,17-dione is at least thousand times lower than that of digitoxigenin. This statement leads us to the question: why do the hormone type steroids display only such a weak affinity to transport ATPase? On the first view, it might be argued that the activity differences between corticosteroids and cardiac steroids are due to the differences in the junctions of the rings C and D. As well-known, their fusion is trans in hormone type steroids and cis in cardiac steroids. However, such an assumption is disproved by the fact that 14β-testane-3β,14β-diol-17-one is practically without any inhibitory activity although it shows the correct cis-junction of the rings C and D. In this synthetic derivative of digitoxigenin, the butenolide ring at C-17 so typical for cardiac steroids is replaced by a carbonyl. The tempting conclusion, however, that the butenolide ring is the *sole* prerequisite for a strong inhibitory effect does not prove right since androstane-3β-ol-17β-butenolide and testane-3β-ol-17β-butenolide are found to be almost inactive. These compounds have trans-trans or cis-trans fusion of the rings A/B and C/D. Their inactivity must be due to the trans-junctions of the rings C and D, as the cardiac steroid uzarigenin showing trans-junctions of the rings A and B is still highly active.

We emerge with the conclusion that corticosteroids and other hormone type steroids do not display a specific affinity to the receptor site of transport ATPase, because firstly their steroid framework is practically planar and extended whereas it is considerably bent at either end in cardiac steroids, and secondly their structure at C-17 is highly different from the butenolide or hexadienolide ring of cardiac steroids. These spatial and structural differences appear to explain that corticosteroids do not compete with cardiac steroids for their receptor site at transport ATPase as shown in this paper.

References

[1] Bernauer, K., and S. Fallab, Helv. Chim. Acta **40**, 1690 (1957).
[2] Crabbé, J., Nature **200**, 787 (1963).
[3] Crabbé, J., Endocrinology **75**, 809 (1964).
[4] Crabbé, J., and P. de Weer, J. Physiol. (London) **180**, 560 (1965).
[5] Fanestil, D. D., G. A. Porter, and I. S. Edelman, Biochim. Biophys. Acta **135**, 74 (1967).
[6] Kunz, H. A., and W. Wilbrandt, Helv. Physiol. Acta **21**, 83 (1963).
[7] Losert, W., C. Senft, and G. Senft, Arch. exp. Path. Pharmak. **248**, 450 (1964).
[8] Myers, R., W. Bishop, and B. Scheer, Amer. J. Physiol. **200**, 444 (1961).
[9] Portius, H. J., and K. Repke, Mber. Dt. Akad. Wiss. **5**, 193 (1963).
[10] Repke, K., and H. J. Portius, Experientia **19**, 452 (1963).
[11] Sharp, G. W. G., and A. Leaf, Nature **202**, 1185 (1964).
[12] Streeten, H. P., and A. K. Solomon, J. gen. Physiol. **37**, 643 (1954).
[13] Sulser, F., H. A. Kunz, R. Gantenbein and W. Wilbrandt,, Arch. exp. Path. Pharmak. **235**, 400 (1959).
[14] Vander, A., W. Wilde, and R. Malvin, J. theoret. Biol. **2**, 236 (1961).
[15] Wilbrandt, W., Schweiz. Med. Wschr. **89**, 363 (1959).
[16] Wilbrandt, W., and E. M. Weiss, Arzneim.-Forsch. **10**, 409 (1960).
[17] Wilbrandt, W., Arch. exp. Path. Pharmak. **245**, 28 (1963).
[18] Woodbury, D. M., and A. Koch, Proc. Soc. exp. Biol. (N.Y.) **94**, 720 (1957).

Charakterisierung eines Steroidrezeptors bei *Streptomyces hydrogenans*

L. Träger und A. Wacker
Institut für Therapeutische Biochemie der Universität Frankfurt/Main, DBR

Zusammenfassung

Der Mikroorganismus *Streptomyces hydrogenans* vermag in Gegenwart von Steroiden induktiv die Synthese eines steroidumwandelnden Enzymes, 20β-Hydroxysteroid-Dehydrogenase, anzuheben. Die induktiv wirksamen Steroide werden durch *Streptomyces hydrogenans* gebunden. Ein Anti-induktiv wirksames Steroid verdrängt die Induktoren von dem Rezeptor.

Der Rezeptor ist nach seinen chromatographischen, enzymatischen und elektrophoretischen Eigenschaften ein saures Protein. Das Protein vermag auch in vitro durch Gleichgewichtsdialyse Steroide zu binden. Es ist wahrscheinlich, daß die Bindung über eine allosterische Änderung erfolgt, da bereits gebundenes Steroid die Bindung weiterer Steroidmoleküle erleichtert.

Einleitung

Eine Klärung des Wirkungsmechanismus der Steroidhormone bei Säugetieren steht bis heute trotz intensiver Untersuchungen noch aus. Die Schwierigkeiten erwachsen hierbei in erster Linie aus den Eigenschaften des zur Untersuchung herangezogenen tierischen Gewebes, das erfahrungsgemäß zahlreiche funktionelle Wechselwirkungen zu bestimmten vor- und nachgeschalteten Geweben aufweist. Obwohl den einzelnen Steroiden in der Regel bestimmte Zielorgane zugeordnet werden, stellen diese Zielorgane nur diejenigen Gewebeabschnitte dar, an denen die größten und deutlichsten anatomischen und funktionellen Änderungen beobachtet werden können (Talalay, 1964). Darüberhinaus hat bereits Dirscherl darauf hingewiesen, daß unter Umständen die Spezifität der Hormonwirkung bei der Untersuchung am Tier oder am Gewebeschnitt durch geeignete experimentelle Bedingungen verändert werden kann (Dirscherl, 1965). Dirscherl diskutiert zwei praktisch brauchbare Wirkungsmechanismen für Steroidhormone, in denen sie entweder als Effektoren von Enzymen die biologische Aktivität der Enzyme verändern oder als Enzyminduktoren regulatorisch in die Biosynthese von Enzymen eingreifen. Für beide Möglichkeiten sind in der Vergangenheit Hinweise gegeben worden.

Wenn wir die Wirkung der Steroidhormone tatsächlich als eine direkte oder indirekte Aktivierung bestimmter Genorte auf der DNS verstehen (Hübener, 1962; Karlson, 1961; Kidsón, 1967), so sollte es auch möglich sein, die Angriffsorte zu beschreiben, an denen die Steroide ihre Funktionen ausüben, wenn ein einfaches biologisches System zu Verfügung steht.

Es lag daher nahe, mit Streptomyces hydrogenans den Wirkungsmechanismus von Steroiden näher zu untersuchen. Seit den Untersuchungen von Lindner (Lindner et al., 1958) und später von Hübener (Nesemann et al., 1960; Schmidt-Thome et al., 1962) war klar geworden, daß bei *Streptomyces hydrogenans* die Synthese des steroidumwandelnden Enzymes 20β-Hydroxysteroid-Dehydrogenase (HSDase) durch eine typische Induktion mit der auch bei Säugetieren beobachteten Kinetik und Hemmbarkeit stimuliert werden kann.

Streptomyces hydrogenans (ATCC 19631) ist nahe mit *Streptomyces albus* verwandt (Lindner et al., 1958) und vermag im Kulturmedium angebotene Steroide, die eine Dihydroxyaceton-Gruppierung in der Seitenkette an C-17 tragen, unter Umwandlung der C-20-Ketogruppe in die C-20-Alkoholgruppe umzuwandeln (HÜBENER et al., 1958). Bei Fermentierung in Gegenwart von Prednison erfolgt außerdem eine Hydrierung der Doppelbindung an C-1. Die als Induktoren für die HSDase aktiven Steroide stammen aus allen wichtigen Steroidhormonklassen (NESEMANN et. al., 1960). Der induzierenden Aktivität geht nicht die Affinität der Steroide zur HSDase oder deren Reduzierbarkeit parallel (SCHMIDT-THOME et al., 1962). Im Gegensatz zu den von CAPEK beobachteten Effekten beeinflussen die Steroide nicht das Wachstum von *Streptomyzes hydrogenans* (CAPEK et.al., 1966).

Die bei *Streptomyces hydrogenans* nachweisbare HSDase ist allerdings nicht mit der in der Säugetierleber wirksamen HSDase identisch, die Flavin als Cofaktor enthält und an Mikrosomen strukturgebunden ist (HÜBENER, 1957), während die HSDase aus *Streptomyces hydrogenans* NADH benötigt und sich im löslichen Überstand des Zellhomogenates befindet (HÜBENER, et al., 1959).

Ergebnisse

Der Actinomycet *Streptomyces hydrogenans* reagiert auf die Zufuhr bestimmter Steroidhormone zum Nährmedium mit der vermehrten Bildung der 20β-Hydroxysteroid-Dehydrogenase (HSDase), die in NADH-abhängiger Reaktion die Ketogruppe an C-20 bei C_{21}-Steroiden in eine β-Hydroxygruppe umwandelt. Drei Stunden nach Zugabe eines Induktors zum Nährmedium läßt sich das in geringen Mengen bereits konstitutiv vorhandene Enzym in zunehmendem Maße nachweisen (Tab. 1).

Tabelle 1

Induktion der HSDase nach Kultivierung von *Streptomyces hydrogenans* in Gegenwart von 100 μg Steroid pro ml Vollmedium

Steroid	HSDase[1]
Kontrolle	9
Testosteron	44
Progesteron	70
Cortisol	85
Diendiol[2]	310

Induktionsdauer: 3 Std. bei 30° in Fleischextraktmedium (Vollmedium) (RING et.al., 1966).

[1] HSDase-Konzentration wird nach HÜBENER et al. (1960) durch die NADH-abhängige Umwandlung von Cortison in Pregna-4-en-17α.20β.21-triol-3.11-dion (Reichstein U) bestimmt und als Enzymeinheiten (Umgesetzte nM Cortison/Min. × mg extrahiertes Protein) angegeben.

[2] Pregna-4.17(20)-dien-11β.21-diol-3-on.

Nach fünf Stunden werden maximale Werte für das induzierte Enzym gefunden (siehe auch NESEMANN et. al., 1960). Es besteht in der Regel ein linearer Zusammenhang zwischen der angebotenen Menge des Induktors im Nährmedium und der Menge des induzierten Enzyms.

Die durch Steroide bei Streptomyces hydrogenans ausgelöste Stimulation der HSDase-Synthese läßt sich durch eine Reihe von typischen Hemmstoffen der Protein- und Nucleinsäurebiosynthese hemmen, sofern diese Hemmstoffe durch die Zellen aufgenommen werden. In diesem Falle geht der Induktionshemmung in der Regel eine weitgehende Wachstumshemmung parallel.

Die durch Äthidiumbromid (als charakteristischem Hemmstoff der Nucleinsäuresynthese; WARING, 1965) ausgelöste Wachstums- und Induktionshemmung soll vertretend für ähnliche oder gleiche Wirkungen der Hemmstoffe Mitomycin C und Actinomycin D stehen. Dabei eilt die Induktionshemmung der Wachstumshemmung vor (Tab. 2).

Äthiumbromid hemmt die Enzyminduktion bei allen bisher untersuchten Steroiden. Hemmung durch Äthidiumbromid und Induktion durch die Steroide verlaufen wahrscheinlich nicht am selben Wirkungsort.

Wird die Induktion in Gegenwart der Alkaloide Ergocornin oder Reserpin durchgeführt, so läßt sie sich durch steigende Mengen der Alkaloide hemmen. Reserpin hemmt etwas stärker als Ergocornin (WACKER et al., 1967) (Tab. 3).

Die Hemmung durch beide Alkaloide erfolgt ebenfalls in Gegenwart aller bisher untersuchten Steroide. Interessanterweise läßt sich die Reserpin-Hemmung durch Serotonin aufheben.

Tabelle 2

Hemmung der Enzyminduktion und des Wachstums von *Streptomyces hydrogenans* durch Äthidiumbromid

Äthidiumbromid µg/ml	HSDase	Wachstum %	Bemerkungen
—	10	100	Kontrolle
—	192	100	Induktion mit 100 µg Diendiol pro ml
1	156	93	
2	57	81	
4	16	74	
6	17	55	
8	8	4	

Tabelle 3

Hemmung der Enzyminduktion durch Alkaloide in Gegenwart von 100 µg Testosteron/ml als Induktor und 10 µg Alkaloid/ml

Hemmstoff	HSDase
Kontrolle	8
ohne	40
Reserpin	24
Ergocornin	31

Während in der Regel die durch die Steroide ausgelösten Effekte sich annähernd addieren und induktions-inaktive Steroide (wie Östrogene, Cortison oder Cholesterin) keinen Einfluß auf die Induktion des gleichzeitig angebotenen Induktors haben, hat sich Cyproteron (1.2α-Methylen-pregna-4.6-dien-3.20-dion-6-chlor-17α-ol; NEUMANN et al., 1966) als ein sehr wirksames antiinduktives Steroid herausgestellt. Dieses auch beim Tier antimaskulin wirkende Gestagen (JUNKMANN et al., 1964) hemmt kompetitiv die durch Testosteron, Diendiol, Desoxycorticosteron und anderen ausgelösten Induktionen, ist dagegen — wie zu erwarten — unwirksam bei Progesteron (Tab. 4). Auch eine Reihe weiterer unnatürlicher Steroide, wie 17α-Methyltestosteron (RUZICKA et al., 1935) und 7α.17α-Dimethyltestosteron (STUCKI et al.,

Tabelle 4

Hemmung der Enzyminduktion durch Cyproteron

Testosteron µg/ml	Desoxycorticosteron µg/ml	Diendiol µg/ml	Progesteron µg/ml	Cyproteron µg/ml	HSDase
—	—	—	—	—	9
—	—	—	—	10	8
—	—	—	—	100	11
100	—	—	—	—	44
100	—	—	—	10	31
100	—	—	—	50	24
100	—	—	—	100	21
100	—	—	—	200	16
—	50	—	—	—	119
—	50	—	—	50	104
—	50	—	—	100	69
—	200	—	—	—	21
—	200	—	—	50	16
—	200	—	—	100	11
—	—	100	—	—	312
—	—	100	—	10	102
—	—	100	—	50	75
—	—	100	—	100	60
—	—	100	—	200	51
—	—	—	50	—	25
—	—	—	50	10	18
—	—	—	50	100	19
—	—	—	100	—	67
—	—	—	100	10	75
—	—	—	100	100	72

1965), die beide androgene Anabolika darstellen, sowie das ötsrogen wirkende Diäthylstilböstrol hemmen die Enzyminduktion (Tab. 5).

Tabelle 5
Hemmung der Enzyminduktion durch unnatürliche Steroide und Diäthylstilböstrol in Gegenwart von 100 μg Testosteron/ml Nährmedium

Hemmstoff	Konzentration des Hemmstoffs μg/ml	HSDase
Kontrolle (ohne Testosteron)	—	7
ohne	—	40
Cyproteron	100	21
17α-Methyltestosteron	100	29
7α.17α-Dimethyltestosteron	100	26
Diäthylstilböstrol	10	9

In all diesen Fällen (ausgenommen Diäthylstilböstrol) drängen steigende Mengen der antiinduktiv wirksamen Steroide die Enzyminduktion in Gegenwart verschiedener Steroide zurück. Gleichzeitig werden die induktiv wirksamen Steroide aus der Zelle verdrängt.

Auf der Grundlage der von Monod entwickelten Vorstellung über den Mechanismus der Enzyminduktion läßt sich die kompetitive Verdrängung des Induktors durch steigende Mengen eines Antiinduktors nicht erklären: Monod schließt eine Wechselwirkung der Induktoren mit den Repressoren unter Bildung von Übergangskomplexen aus (Monod, 1966). Eine kompetitive Verdrängung schließt aber mindestens die Beteiligung eines gemeinsamen Rezeptors für Induktor und Antiinduktor ein, von dem die induzierende Wirkung auf den Repressor ausgehen müßte. Oder man müßte doch eine chemische Wechselwirkung zwischen Repressor und Induktor annehmen, die zu meßbaren Komplexen führen müßte.

Wir haben daher zunächst untersucht, in welchem Maße es zur Bindung der Induktoren in der Zelle kommt. Dazu wurden die Zellen mit radioaktivem Induktor kultiviert und die Zellextrakte auf Sephadex G100 oder G200 filtriert. Man erhält im Ausschlußvolumen eine hochmolekulare Fraktion, die ca. 50% der von den Zellen aufgenommenen Radioaktivität

Tabelle 6
Einbau von H-3-Testosteron (spezif. Radioaktivität 1000 mC/mM) in die hochmolekulare Fraktion von *Streptomyces hydrogenans* in vivo

Versuchs-Nr.		1	2	3	4
Testosteron	μg/ml	0,1	100	100	100
Cyproteron	μg/ml	—	—	100	—
Diäthylstilböstrol	μg/ml	—	—	—	10
HSDase		10	25	14	6
Radioaktivität	Imp/Min./ml	6100	6600	4800	9000
im Extrakt	Imp/Min./mg Prot.	1020	1460	1300	1800
Radioaktivität an der hochmolekularen Fraktion bezogen auf die Gesamtradioaktivität	%	55	68	25	56

Streptomyces hydrogenans wurde in Vollmedium vorkultiviert. Die in der log. Wachstumsphase befindlichen Keime wurden abfiltriert, zu gleichen Teilen auf 4 Ruhemedien (30 ml Phosphatpuffer, 0,1 m, pH 6,8) verteilt und mit den Induktoren bzw. Hemmstoffen drei Stunden bei 30° weiter kultiviert.

enthält. Der Rest läuft als freies radioaktives Steroid nach. Insgesamt nehmen die Zellen durchschnittlich 0,01 bis 0,1% des angebotenen radioaktiven Steroides auf (Tab. 6).

Aus diesen Versuchen ergeben sich mehrere Folgerungen:

A) die Zugabe nichtradioaktiven Testosterons (Erhöhung der Testosteron-Konzentration von 0,1 auf 100 µg/ml) verstärkt die Bindung des Testosterons, während man erwarten sollte, daß die Radioaktivität durch die Zugabe des „kalten“ Testosterons verdünnt würde, und demnach die gebundene Radioaktivität abnehmen oder zumindest gleich bleiben sollte.

B) In Gegenwart des Hemmstoffs Cyproteron verdrängt dieses Steroid im gleichen Maße das radioaktive Steroid aus der hochmolekularen Fraktion, wie es die Enzyminduktion hemmt.

C) Diäthylstilböstrol erniedrigt die Testosteronbindung nicht. Demnach müssen wir annehmen, daß Diäthylstilböstrol an anderer Stelle in den Induktionsmechanismus eingreift als der Induktor.

Entsprechende Verdrängungsversuche mit radioaktivem Progesteron und Cyproteron bestätigen, daß zwischen diesen beiden Steroiden weder kompetitive Hemmung noch Verdrängung meßbar sind.

Werden die radioaktiven Extrakte nicht über Sephadex filtriert, sondern auf DEAE-Cellulose mit einem NaCl-Gradienten chromatographiert, so erhält man die gebundenes und freies Steroid enthaltenden Fraktionen ebenfalls getrennt. Während das freie Steroid am Beginn der Chromatographie mit 0,001 m Trispuffer eluiert wird, erscheint die Fraktion mit dem gebundenen Steroid bei 0,05—0,08 m NaCl. Die Chromatographie auf DEAE-Cellulose gestattet bereits eine engere Charakterisierung der Steroid-bindenden Fraktion. Zunächst zeigt sich, daß an der gleichen Stelle die Rezeptoren für Testosteron, Cortisol, Progesteron, und Cyproteron eluiert werden. Wird eine Kultur in Gegenwart von Cyproteron und radioaktivem Testosteron induziert, so bestätigt sich auch nach DEAE-Cellulose-Chromatographie die durch Gelfiltration nachgewiesene Verdrängung des (radioaktiven) Testosterons von der gebundenen Fraktion in die freie.

Die chromatographischen Befunde bezüglich der Verdrängung der Induktoren haben nur dann allgemeine Beweiskraft, wenn man die Möglichkeit ausschließen kann, daß die beobachteten Steroidbindungen Artefakte der Chromatographie selbst sind bzw. ihnen keine natürlichen Strukturen entsprächen. Obgleich diese Betrachtungsweise bereits durch die Analogie zwischen Radioaktivitätsaufnahme und Verdrängung aus der ganzen Zelle einerseits und in den chromatographischen Eluaten andererseits schon weitgehend auszuschließen ist, haben wir dennoch die Bindung der Steroide mit den gereinigten Fraktionen in vitro verfolgt.

Werden nichtinduzierte Zellen von Streptomyces hydrogenans homogenisiert und der zellfreie Extrakt auf DEAE-Cellulose aufgetrennt, so kann die steroidbindende Aktivität der einzelnen Fraktionen nach Chromatographie durch Dialyse oder Gleichgewichtsdialyse leicht gemessen werden. Dazu werden die Fraktionen (je 3 ml) entweder mit 0,5 µC H-3-Testosteron versetzt und 3 Stunden bei 18° gegen Tris-Puffer, pH 7,0, 10^{-3} m, dialysiert oder in einer Gleichgewichtsdialyse gegen ein radioaktives Außendialysat (mit ca. 6000 Imp/Min./0,5 ml) dialysiert. Die Radioaktivität in den Innendialysaten stellt ein Maß der Testosteronbindekapazität dar. Auch hier zeigt sich wiederum, daß die Fraktionen, die auch in vivo „gebundene“ Steroide zu tragen vermögen, in vitro ebenfalls das Steroid akzeptieren.

Die Anheftung der Steroide in vitro stellt eine Möglichkeit dar, den Rezeptor „spezifisch“ zu markieren. Da die Darstellung durch Bindung in vitro wesentlich unkomplizierter und kostensparender ist, wurde daher späterhin mit Zellextrakten gearbeitet, die entweder durch Gleichgewichtsdialyse, Dialyse oder Inkubation bei 0° oder 18° zur Akzeption der Steroide angeregt wurden.

Statt durch Dialyse läßt sich der Anteil der in vitro gebundenen Steroide auch durch Gelfiltration bestimmen (Tab. 7).

Tabelle 7

Filtration der zellfreien Extrakte von *Streptomyces hydrogenans* nach Inkubation mit verschiedenen Zusätzen auf Sephadex G 25

Testosteron µg	Cyproteron µg	Reserpin µg	Diäthylstilböstrol µg	Radioaktivität[1] %
0,5	—	—	—	5
100	—	—	—	8
100	100	—	—	3
100	—	100	—	3
100	—	—	10	4,5

4 ml Extrakt werden mit den kalten Steroiden bzw. Hemmstoffen versetzt und anschließend mit 0,5 µg H-3-Testosteron. Das Gemisch wird 10 Min. bei 0° belassen, dann auf Sephadex G 25 mit Trispuffer, pH 7,0, 10^{-3} m, filtriert.

[1] Radioaktivität im hochmolekularen Teil bezogen auf die gesamte eluierte Radioaktivität

Bei den Versuchen mit der Bindung der Steroide in vitro verringert sich zusätzlich die Chance, daß das akzeptierte Steroid metabolisiert wird. Wie Versuche mit in vivo angebotenem radioaktivem Testosteron ergaben, waren von dem als freies Steroid eluierten Induktor nur noch ca. 20% Testosteron. Mit ca. 65% hatte sich Androstendion und außerdem ein weiterer Testosteron-Metabolit (15%) gebildet. Das in vivo gebundene Steroid läßt sich nur zu 50 bis 65% mit organischen Lösungsmitteln wieder extrahieren. Die Radioaktivität konnte nicht mehr Testosteron zugeordnet werden. Es ist wahrscheinlich, daß es sich vielmehr um Steroidkonjugate (Phosphate oder Sulfate) handelt. Die Identifizierung steht noch aus. Das in vitro gebundene Steroid war dagegen völlig unverändertes Testosteron.

Werden Extrakte mit in vivo gebundenem Steroid vor Chromatographie mit RNase oder DNase behandelt, so wird das radioaktive Steroid an der gleichen Position wie die Kontrolle eluiert. Wird dagegen der zellfreie Extrakt mit Trypsin behandelt, so verschwindet die Radioaktivität aus der bindenden Fraktion. Dieser Befund weist darauf hin, daß die bindende Komponente ein Protein ist. In die gleiche Richtung weisen Gleichgewichtsdialyse-Teste mit Extrakten von Streptomyces hydrogenans, die durch CsCl-Dichtegradientenzentrifugation in RNS, DNS und Protein aufgetrennt wurden (Tab. 8).

Tabelle 8

Gleichgewichtsdialyse der durch CsCl-Dichtegradientenzentrifugation erhaltenen Nucleinsäure- bzw. Proteinfraktionen gegen H-3-Testosteron

Zeit Min.	RNS Imp/Min./ml	DNS	Protein	Außendialysat
0				5400
30	4000	4100	4400	5300
60	5400	5000	7100	5300

0,2 ml Homogenat wurden mit 2 ml CsCl-Lösung (Brechungsindex des Gemisches n = 1,4005) im SW 39 61 Std. bei 30000 Upm zentrifugiert, abgetropft, mit 1,5 ml Trispuffer versetzt und 15 Min. gegen Trispuffer vom anhaftenden CsCl befreit. Anschließend wird gegen H-3-Testosteron im Außendialysat äquilibriert (Visking-Schlauch, Type 20/32), Trispuffer, pH 7,0, 0,001 m.

Demnach bindet die Proteinfraktion eine signifikante Menge H-3-Testosteron, die beiden Nucleinsäurefraktionen reichern das Steroid nicht an. Aus den durch DEAE-Cellulose-Chromatographie erhaltenen bindenden Fraktionen läßt sich eine Abschätzung der gebundenen Steroid-Moleküle durch das Protein vornehmen. Aus dem Verhalten an Sephadex G 100 wird für das bindende Protein ein Molekulargewicht von mind. 150000 geschätzt. Bei einer spezif. Radioaktivität von 1000 mC/mM finden wir dann, daß in vivo pro 7,5 Proteinmolekülen (der wahrscheinlich heterogenen) Eiweißfraktion 1 Molekül Testosteron gebunden wird. Das gleiche Verhältnis erhält man nach Inkubation des zellfreien Extraktes in vitro. Inkubiert man jedoch in vivo oder in vitro mit Testosteron fallender spezif. Radioaktivität, so verschiebt sich dieses Verhältnis bis um den Faktor 60 und es werden ca. 10 Moleküle Testosteron durch 1 Molekül Protein gebunden.

Wird der mit H-3-Testosteron ,,markierte" Rezeptor auf Polyacrylamidgel elektrophoretisch aufgetrennt, so verhält er sich wie ein saures Protein und läuft relativ rasch zur Anode.

Diskussion

Die neuere Literatur hat in einer Reihe glänzender Arbeiten die Wirkungsweise der Steroidhormone als Aktivierung der genetisch gesteuerten Proteinbiosynthese interpretiert (Clever, 1961; Karlson, 1961). Bei Mikroorganismen fehlen jedoch noch weitgehend Anhaltspunkte dafür, daß Steroidhormone in unmittelbare räumliche Beziehung zu den Strukturen treten, die die genetische Information enthalten. Bei *Streptomyces hydrogenans* lassen sich nach Kultivierung in Gegenwart von Steroiden eine Reihe von Fakten erkennen, die für die Gültigkeit des Konzepts der Genaktivierung durch Steroide bei diesem Mikroorganismus sprechen. Diese Fakten lassen sich etwa wie folgt zusammenstellen:

Es findet eine Nettoneusynthese des induzierten Enzyms statt. Eine Aktivierung inaktiver HSDase-Vorstufen durch Steroide ist nicht nachweisbar.

Die induzierte Proteinsynthese ist mit einer vermehrten RNS-Synthese verbunden (Wakker et al., 1966).

Beide Beobachtungen lassen sich nach dem heutigen Stand unserer Kenntnisse auf eine durch Steroide ausgelöste selektive Aktivierung der Transkription der auf der DNS niedergelegten Information zurückführen. Die zur Induktion bei *Streptomyces hydrogenans* notwendigen Steroiddosen sind im Vergleich zu den beim Säugetier wirksamen Dosen relativ hoch. Jedoch muß man beachten, daß von den angebotenen Steroiden nur ein sehr geringer Anteil (meist nicht mehr als 0,1%) von den Zellen aufgenommen wird. Werden z. B. die Zellen in Gegenwart von 100 µg Testosteron/ml Nährmedium bebrütet, so findet man schließlich im Mittel 20 µg Steroid pro 100 mg Zellprotein im Homogenat wieder. Mit einer vergleichbaren Dosierung werden Corticoide im Tierversuch appliziert.

Folgt man der Hypothese der Genaktivierung als primäre Wirkung der Steroide bei Enzyminduktionen, so stellt die Hemmung der Enzyminduktion einen Eingriff dar, der theoretisch an einer Vielzahl unterschiedlicher Angriffspunkte möglich wäre. Die von uns untersuchten Hemmstoffe, wie das Anti-Androgen Cyproteron, die Alkaloide Ergocornin und Reserpin sowie das auch im Tierversuch antiandrogen wirkende Diäthylstilböstrol, führen zu einer Änderung der Ribosomenfraktionen unter Annäherung an die Eigenschaften der Ribosomen aus nicht induzierten Zellen (siehe auch Maturova et al., 1967).

Während bei Cyproteron und möglicherweise auch Reserpin eine direkte Wechselwirkung mit dem Steroidrezeptor unter Kompetition mit der Steroidbindung festgestellt wird, gilt dies nicht für Diäthylstilböstrol. Die in den Versuchen nachgewiesene Konkurrenz der Induktoren und des Antiinduktors Cyproteron um einen gemeinsamen Akzeptor bietet eine hinreichende Erklärung für dessen antiinduktiven Effekt. Danach würde der Antiinduktor zwar eine in der Größenordnung gleiche Affinität zum Akzeptor aufweisen, jedoch keinerlei induktive

Eigenschaften entwickeln. Die gleiche kompetitive Verdrängungsreaktion zwischen Testosteron und Cyproteron wird auch beim Säugetier beschrieben (Junkmann et al., 1964). Danach lagert sich Cyproteron an das Sexualzentrum von Säugetieren an, verdrängt dabei Testosteron und bewirkt eine vermehrte Ausscheidung von Gonadotropinen. Die durch Testosteron-Überschuß einsetzende Hodenatrophie (wegen mangelnder Gonadotropinausscheidung, insbesondere des ICSH) ist daher durch Cyproteron aufhebbar.

Bei *Streptomyces hydrogenans* beobachten wir parallel zu den Befunden beim Säugetier einen spezifischen Antagonismus des Cyproterons mit bestimmten Steroiden, wobei die physiologische Aktivität dieser Steroide ganz oder teilweise unterdrückt wird. Diese Eigenschaft des Antiinduktors läßt sich vorläufig nur schwer auf strukturelle Eigentümlichkeiten des Steroidmoleküls zurückführen. Bei Cyproteron wird die antiinduktive Wirkung durch Einführung der 1.2α-Methylenbrücke erzeugt, andere antiinduktive Steroide weisen auf die Bedeutung der an C-17 stehenden Substituenten hin.

Von besonderem Interesse dürfte die Zunahme der Testosteronbindung in vivo und in vitro bei steigendem Angebot des Induktors durch die Rezeptorfraktion sein. Sie erinnert in ihrem Ausmaß an das allosterische Verhalten bestimmter Proteine. Die Erhöhung der zu bindenden Komponente führt dabei zu einer Zunahme der Haftstellen am Protein. Ein vergleichbarer Effekt wurde kürzlich von Zimmering et al. (1967) mit steroidspezifischen Antikörpern beobachtet, deren Affinität zum Steroid durch bereits gebundenes Steroid erhöht wird. Dieser Effekt läßt sich auch bei der Bindung der Steroide in vitro beobachten.

Die fehlende Wechselwirkung zwischen Progesteron und Cyproteron in vivo und in vitro läßt sich durch die Annahme verschiedener Rezeptoren für die Steroide erklären. Allerdings wäre es ebenso denkbar, daß der Bindungsort für Testosteron (und Cyproteron) von dem für Progesteron auf dem gleichen Rezeptor verschieden ist. Die Möglichkeit, daß die Konkurrenz zwischen Cyproteron und den Induktoren auf der Stufe des Transportes stattfindet, läßt sich eindeutig durch die Bindungsversuche der Steroide in vitro ausschließen. Denn die Bindung von Steroiden an den chromatographischen Fraktionen in vitro, an denen in vivo ebenfalls die Steroide gebunden werden, schließt aus, daß die unterschiedliche Wirkung der Steroide auf ihre unterschiedliche Aufnahme(geschwindigkeit) zurückzuführen ist.

Aus einer Reihe von Experimenten ließ sich beweisen, daß die Rezeptorfraktion aus einem Protein besteht. Dies geht aus dem Einbau von radioaktiv markiertem Phenylalanin in diese Fraktion hervor, aus der Bindung der Steroide in vitro an diese Proteinfraktion, aus der Freisetzung der Induktoren aus der Rezeptorfraktion durch tryptische Verdauung und schließlich durch Gelelektrophorese und nachfolgende Anfärbung der Proteinbanden. Aus den elektrophoretischen Versuchen ist zu entnehmen, daß der Rezeptor ein saures Protein ist, das nach seinem Verhalten auf Sephadex ein Molekulargewicht von mindestens 150000 haben muß.

Für Steroidrezeptoren beim Säugetier sind ähnliche Eigenschaften beschrieben worden. So soll der nucleare Östradiol-Rezeptor ein saures Protein sein (King et al., 1965), dessen Molekulargewicht auf 200000 geschätzt wird (Toft et al., 1966). Auch Progesteron soll durch ein saures α_1-Glykoprotein gebunden werden (Westphal et al., 1961). Ein Rezeptor für Testosteron soll nach seiner Aminosäurenzusammensetzung ein basisches Protein sein (Lippert et al., 1967).

Zur biologischen Bedeutung des Rezeptorproteins für die Steroide lassen sich verschiedene Überlegungen anstellen. Es ist unwahrscheinlich, daß dessen Funktion bei *Streptomyces hydrogenans* auf eine bloße Kompartmentierung der Steroide hinausläuft, da die Bindung der Steroide deren Induktionsaktivität widerspiegelt. Gehen wir von der Vorstellung aus, daß der steroidbindende Rezeptor der Repressor selbst ist, so sollten Extrakte von nicht induzierten Zellen, deren Enzymsynthese also repremiert ist, eine höhere Repressoraktivität aufweisen als die von induzierten Zellen. Bei Versuchen zur Hemmung von RNS-Polymerase in vitro zeigte sich, daß tatsächlich diese Extrakte die höchste Hemmwirkung gegenüber der Polymerase aufweisen (Wacker et al., 1965a, 1965b). Die Hemmung kann in vitro durch Zugabe der Induktoren aufgehoben werden. Die Hemmwirkung der gereinigten Rezeptorfraktion auf die RNS-Polymerase bzw. deren Enthemmung durch Steroide in vitro wurde bisher noch nicht untersucht. Aus der Parallelität zwischen Induktionsauslösung und Bindung durch diese Fraktion läßt sich jedoch indirekt schließen, daß die steroidbindende Komponente in *Streptomyces hydrogenans* möglicherweise repressorische Funktionen bei der Induktion der HSDase ausübt.

Literatur

[1] Capek, A., O. Hanc, and M. Tadra, Microbial transformations of steroids, Academia Publishing House of the Czechoslovak Acad. of Sci., Prague 1966, S. 55.
[2] Clever, U., Chromosoma **12**, 607 (1961).
[3] Dirscherl, W., Mechanisms of Hormone Action (P. Karlson, Herausgeb.), Georg Thieme Verlag, Stuttgart 1965, S. 1.
[4] Hübener, H. J., Experientia **13**, 210 (1957).
[5] Hübener, H. J., Dtsch. Med. Wschrft. **87**, 438 (1962).
[6] Hübener, H. J., und C. O. Lehmann, Hoppe Seyler's Z. physiol. Chem. **313**, 124 (1958).
[7] Hübener, H. J., und F. G. Sahrholz, Biochem. Z. **333**, 95 (1960).
[8] Hübener, H. J., F. G. Sahrholz, J. Schmidt-Thome, G. Nesemann und R. Junk, Biochim. Biophys. Acta **35**, 270 (1959).
[9] Junkmann, K., und F. Neumann, Acta Endocrinologica Suppl. **90**, 139 (1964).
[10] Karlson, P., Dtsch. Med. Wochschr. **86**, 668 (1961).
[11] Kidson, C., Nature **213**, 779 (1967).
[12] King, R. J. B., J. Gordon und L. Martin, Biochem. J. **97**, 28P (1965).
[13] Lindner, F., R. Junk, G. Nesemann und J. Schmidt-Thome, Hoppe Seyler's Z. physiol. Chem. **313**, 117 (1958).
[14] Lippert, U., K. O. Mosebach und G. Krampitz, Nature **214**, 917 (1967).
[15] Maturova, M., H. Beckmann und A. Wacker, Z. Naturforschg. **22b**, 621 (1967).
[16] Monod, J., Angew. Chem. **78**, 694 (1966).
[17] Nesemann, G., H. J. Hübener, R. Junk und J. Schmidt-Thome, Biochem. Z. **333**, 88 (1960).
[18] Neumann, F., W. Elger, R. von Berswordt-Wallrabe und M. Kramer, Naunyn-Schmiedebergs Arch. Pharmak. u. exp. Path. **255**, 236 (1966).
[19] Ring, K., und E. Heinz, Biochem. Z. **344**, 446 (1966).
[20] Ruzicka, L., M. W. Goldberg und H. R. Rosenberg, Helv. chim. Acta **18**, 1487 (1935).
[21] Schmidt-Thome, J., G. Nesemann, H. J. Hübener und I. Alester, Biochem. Z. **336**, 322 (1962).
[22] Stucki, J. C., G. W. Duncan und S. C. Lyster, Hormonal steroids. Biochemistry, pharmacology and therapeutics, Proc. of the first international Congr. on hormonal steroids (L. Martini u. A. Pecile, Herausgeb.), Academic Press, New York, 1965, Vol. 2, S. 119.
[23] Talalay, P., The biochemical aspects of hormone action (A. B. Eisenstein, Herausgeb.), Little, Brown & Co., Boston 1964, S. 167.
[24] Toft, D., und J. Gorski, Proc. Natl. Acad. Sci. **55**, 1574 (1966).
[25] Wacker, A., und H. Feller, unveröffentlichte Ergebnisse (1966).
[26] Wacker, A., J. Drews, W. B. Pratt und P. Chandra, Angew. Chem. **77**, 172 (1965a).
[27] Wacker, A., L. Träger, P. Chandra und H. Feller, Biochem. Z. **342**, 108 (1965b).
[28] Wacker, A., L. Träger, M. Maturova und H. Beckmann, Naturwissenschaften **54**, 90 (1967).
[29] Waring, M. J., Mol. Pharmacol. **1**, 1 (1965).
[30] Westphal, U., B. D. Ashley und G. L. Selden, Arch. Biochem. Biophys. **92**, 441 (1961).
[31] Zimmering, P. E., S. Lieberman und B. F. Erlanger, Biochemistry **6**, 154 (1967).

Further Studies of Microbial Reduction of 20-Keto Steroids into 20α-Hydroxy Steroids

L. M. Kogan and E. A. Yelin

Institute for Chemistry of Natural Compounds, Academy of Sciences of USSR,
Moscow, USSR

Summary

A soluble 20α-hydroxysteroid: NADP oxidoreductase (20α-oxidoreductase) has been isolated from *Actinomyces roseochromogenus.* The ability of a number of steroids to induce the formation of the enzyme has been studied. After partial purification the 20α-oxidoreductase preparation has been separated from other steroid transforming activities. The enzyme catalyzes the reduction of 17-oxygenated 20-keto pregnanes and 17α-pregnanes into 20α-alcohols. It has been shown that the ability to reduce 17α-hydroxyprosterogesterone into pregn-4-en-17α,20α-diol-3-one is characteristic of and peculiar to those *Actinomyces* spp. which hydroxylate progesterone in the 16α-position.

Many microorganisms are known to reduce 20-keto steroids into corresponding 20β-hydroxy compounds. However, the formation of 20α-hydroxy steroids by microorganisms has been observed in a few instances only (Akhrem and Titov, 1965). In 1966 we found that Actinomyces roseochromogenus (according to the Krassilnikov's classification) reduced some 17α-hydroxy-20-keto steroids into 20α-alcohols (Kogan and Yelin, 1966).

The present report deals with further investigation of microbial 20α-reduction.

This reaction turned out to be peculiar not only to *Actinomyces roseochromogenus* but also to many other species of actinomycetes. At first we inspected quite a number of *Actinomyces* species which were known to hydroxylate progesterone (I) into 16α-position like *Actinomyces roseochromogenus.* All these strains reduced 17α-hydroxyprogesterone (IV) into 20α-alcohol (V) except two species which did not transform that substrate in any way. The latter two species are the only ones forming pregn-4-en-20β-ol-3-one (III) in addition to 16α-hydroxyprogesterone from progesterone (I). 8 species of this group of actinomycetes transformed 17α-hydroxyprogesterone (IV) into both 20α-alcohol (V) and 20β-alcohol (VI). The second group includes actinomycetes reducing progesterone (I) into its 20β-dihydro derivative (III). The fermentation of these microorganisms with 17α-hydroxyprogesterone (IV) gave rise to its 20β-dihydro

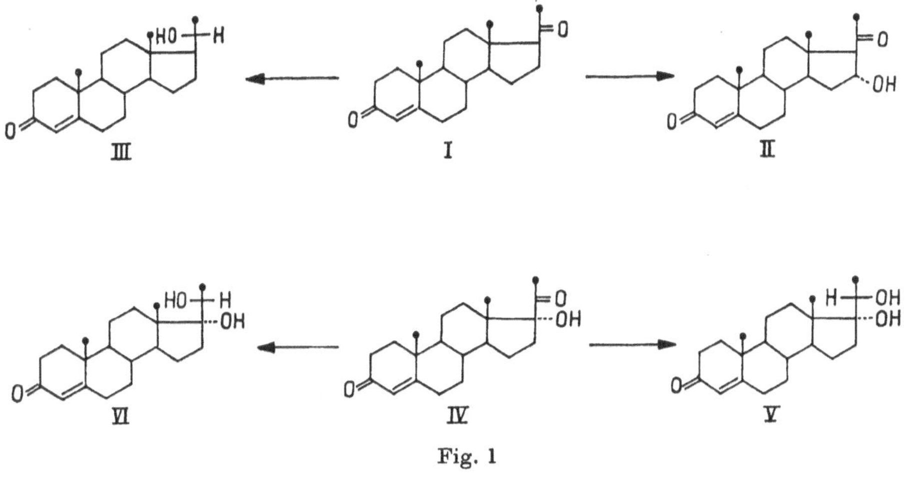

Fig. 1

Table 1

Actinomycetes transforming progesterone into		Reduction of 17α-hydroxy-progesterone into	
NN	16α-hydroxyprogesterone	20α-alcohol	20β-alcohol
	1	2	3
1	*A. bikiniensis* ATCC 11062	+	
2	*A. californicus* WKS	+	
3	*A. californicus* ATCC 3312	+	
4	*A. fimicarius* IPV-902	+	
5	*A. gelaticus* ATCC 3323	+	
6	*A. praecox* ATCC 3374	+	
7	*A. vinaceus* USSR RIA-591	+	
8	*A. vinaceus* NCJB-8852	+	
9	*A. vinaceus* NRRL B-2285	+	
10	*A. gibsonii* ATCC 6852	+	
11	*A. caprae* IPV-206	+	
12	*A. roseochromogenus* ATCC 3347	+	
13	*A. streptomycini*	+	
14	*A. parvus* IFO-3380	+	
15	*A. lipmanii* ATCC 3331	+	
16	*A. aureus* ATCC 3309	+	+
17	*A. flavus* ATCC 3369	+	+
18	*A. cinnamonensis* IAW	+	+
19	*A. venezuelae* ATCC 3534	+	+
20	*A. microflavus* ATCC 3332	+	+
21	*A. microflavus* IAW	+	+
22	*A. purpureochromogenus* ATCC 3343	+	+
23	*A. olivaceus* ATCC 11626[1]		
24	*A. viridis* IPV-894[1]		
	pregn-4-en-20β-ol-3-one		
25	*A. bobiliae* ATCC 3310	+	+
26	*A. tendae* ETH-14077	+	+
27	*A. flaveolus* IPV-869	+	+
28	*A. flaveolus* ATCC 3319	+	+
29	*A. flaveolus* WKS	+	+
30	*A. caeruleus* Bu-3,7	+	+
31	*A. mediocidicus* ETH-24350	+	+
32	*A. cellulosae* ATCC 3313	+	+
33	*A. citreus* WKS	+	+
34	*A. verne* WKS	+	+
35	*A. verne* ATCC 3353		+
36	*A. coelicolor* ATCC 3355		+
37	*A. chrysomallus* ATCC 3657		+
38	*A. griseolus* ATCC 3325		+
39	*A. rutgersensis* ATCC 3350		+
40	*A. annulatus* ATCC 3307		+
41	*A. diastaticus* ATCC 3315		+
42	*A. intermedius* ATCC 3329		+
43	*A. odorifer* ATCC 6246		+
44	*A. albus* ATCC 3006		+
45	*A. sulfureus* ATCC 3007		+
46	*A. gongerotii* ATCC 10975		+

[1] This organism produces both 16α-hydroxyprogesterone and pregn-4-en-20β-ol-3-one.

derivative (VI). 10 species of this group converted 17α-hydroxyprogesterone (IV) both into 20α- and 20β-alcohol.

Thus actinomycetes which 16α-hydroxylate progesterone (I) contain as a rule 20α-reducing enzyme. In no case we observed the formation of 20α-alcohol from progesterone (I). Some aspects of substrate specificity we have studied in more details with *Actinomyces roseochromogenus*. According to several communications the 16α-hydroxylation was recognized as a common reaction for Actinomyces roseochromogenus (Ref. in AKHREM and TITOV, 1965, Chapter VII). In our hands *Actinomyces roseochromogenus* did transform also progesterone (I), 11α-hydroxyprogesterone and deoxycorticosterone (XIV) into their 16α-hydroxy derivatives. The effect of *Actinomyces roseochromogenus* on certain 17-oxygenated pregnenes proved to be quite different. 17α-Hydroxyprogesterone (IV), 11α,17α-dihydroxyprogesterone (VII), 16α,17-epoxyprogesterone (XI) and cortexolone (VIII) have been reduced into 20α-hydroxy compounds. In these cases no 16α-hydroxylation took place. Incubation of 17α-acetoxyprogesterone in form of the free alcohol gave no transformation product.

Fermentation of 17α-methylprogesterone (XVI)[1] with *Actinomyces roseochromogenus* yielded 16α-hydroxy-17α-methylprogesterone (XVII). The location and configuration of the newly introduced hydroxy group were determined from mass spectrum and molecular optical rotation of the compound XVII and its acetate.

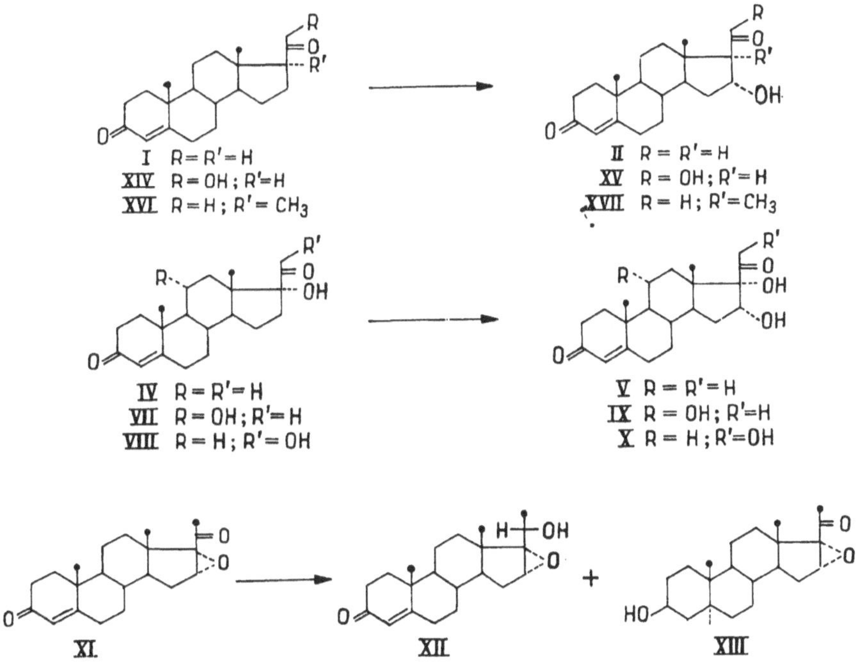

Fig. 2

To learn the influence of the side chain configuration on the course of reduction we carried out the 17β-hydroxy-17α-progesterone (XVIII)[2] fermentation. This compound was found to be reduced by *Actinomyces roseochromogenus* into the 20α-dihydro derivative (XIX) as well as by sodium borohydride reduction (NORYMBERSKI and WOODS, 1955). The 20-hydroxy group configuration in the compound was assigned on the basis of IR-spectra

[1] This compound was prepared from the pregna-5,16-dien-3β-ol-20-on-3β-acetate according to DEGHENGHI (DEGHENGHI et al., 1963).

[2] This compound was prepared from the 17-ethynyltestosterone by the method of AMIARD (AMIARD et al., 1961).

and Dreiding's models. Examination of molecular models indicates that the intramolecular hydrogen bonding in the case of the 20α-epimer (XIX) is possible. In contrast, the interaction of 21-methyl group with axial protons at C-12 and C-14 in the 20β-epimer (XX) prevents from hydrogen bonding. Indeed infra red data showed the intramolecular hydrogen bonding in the diol (XIX) rather than in diol (XX).

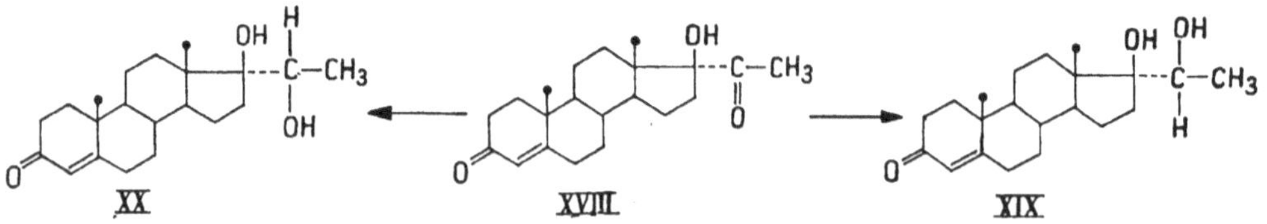

Fig. 3

Thus *Actinomyces roseochromogenus* is seen as reducing the 20-keto group only in steroids possessing a hydroxy or epoxy group in the 17-position. On the other hand 17-hydroxy groups make the possibility of 16α-hydroxylation very difficult. The substitution of 17-hydroxyl by a relatively bulk nonpolar methyl group prevents reduction of 20-carbonyl but not 16α-hydroxylation. This regularity has been also demonstrated with cell-free enzyme preparations, so it seems to be independent of the cell permeability.

In our opinion the importance of the 17-oxygen for 20α-reduction by *Actinomyces* species is of great interest. We must remember that in the previously reported microbial transformations yielding 20α-alcohols the determining influence of 17-oxygen was noticed or only the 17-oxygenated substrated were used (Carvajal et al., 1959; Chang et al., 1961; Takahashi and Uchibori, 1962; Suvorov et al., 1963). The presence of 17-oxygen appears to be of particular importance for putting into effect the microbial reduction of 20-keto steroids into 20α-alcohol in general.

It can be noticed an inhibitory effect of the 21-hydroxy group on both 20-ketone reduction and 16-hydroxylation and that of 11α-hydroxy group on 20-ketone reduction.

The side chain of pregnanes is considered by Engel and Rakhit (Rakhit and Engel, 1962) as being in one of the four prefered conformations: A_1, A_2, B_1, B_2. The type A conformations favour (according to the Cram's rule) the formation of 20α-alcohols in reductions of 20-ketones by metal hydride complexes. The B type conformations favour the formation of 20β-alcohols. In Engel's and Rakhit's opinion the side chain of the 17α-hydroxy pregnanes has the B type conformation. However, substituents near the 20-carbonyl group can change the side chain conformation because of intermediate complex formation. That is why the lithium aluminium hydride reduction of 17α-hydroxy steroids leads to 20α-alcohols.

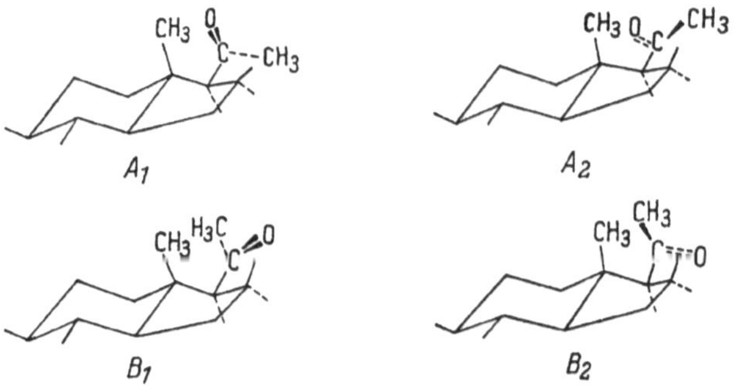

Fig. 4

17-Oxygen seems to involve in enzyme-substrate complex formation during microbial reduction of 20-keto steroids. It can be imagined enzymic reduction as occuring from less hindered side of steroid molecule. The complex involved enzyme and 17-oxygen moiety makes side chain change the conformation into that favouring α-epimer formation.

The 20α-oxidoreductase from *Actinomyces roseochromogenus* is constitutive in nature. We have investigated the ability of a number of steroids to induce "20-ketoreductase" formation, some data being listed in Table 2. The steroids tested can be divided into two groups. The steroids of the first one promote the reductase formation. This group involves progesterone, some monooxygenated progesterone derivatives and some C-19 steroids. The second group includes all 17α,21-dihydroxysteroids we tested and deoxycorticosterone and 16α-hydroxyprogesterone.

Table 2
Effect on levels of 20α-oxidoreductase of addition of some steroids

NN	Steroid	Relative activity
1	without inducer	0,49
2	11α-hydroxyprogesterone	1,00
3	17α-hydroxyprogesterone	1,30
4	11β-hydroxyprogesterone	0,81
5	11-ketoprogesterone	1,30
6	corticosterone	0,61
7	progesterone	0,55
8	adrenosterone	0,63
9	androstendione	0,81
10	16α-hydroxyprogesterone	0,00
11	deoxycorticosterone	0,10
12	cortexolone	0,00
13	hydrocortisone	0,00
14	prednisone	0,00
15	cortisone	0,09
16	prednisolone	0,00

While a cell-free preparation of non-induced culture of *Actinomyces roseochromogenus* possesses the relative activity 0,49, cell-free extracts prepared from mycelium grown with steroids of the second group have lack of reductase activity. Moreover, cortexolone being added together with a powerful inducer such as 17α-hydroxyprogesterone suppresses the enzyme formation. However, simultaneous incubation of both steroids with 20α-oxidoreductase cell-free preparation does not suppress the reduction of 17α-hydroxyprogesterone. These data seem to indicate that the steroids of the second group do inhibit 20α-oxidoreductase synthesis.

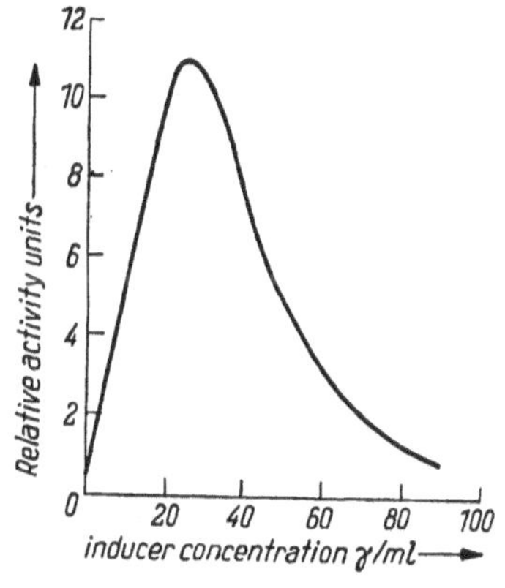

Fig. 5

Using 11α-hydroxyprogesterone as inducer we got a maximum of 20-reductase yield with 25 γ of inducer per mililiter of a cultural broth.

To obtain cell-free preparations the culture of *Actinomyces roseochromogenus* was grown in the presence of 11α-hydroxyprogesterone as an inducer. After 18 hours cells were harvested nad washed with saline, then phosphate buffer containing ethylenediaminetetraacetate and β-mercaptoethanol. The mycelium was then suspended in the same buffer and sonicated. The sonicate was centrifuged at 13500 g. Precipitation with ammonium sulfate showed no 20-ketoreductase activity being present in the 60% saturation fraction. The ketoreductase activity was precipitated between 60 and 90% of saturation. The precipitate was dissolved and subjected to dialysis through a column with Sephadex G-25.

The protein concentrations were determined by the KALCKAR's method (KALCKAR, 1947). Spectrophotometric measurements of the reduction of 17α-hydroxyprogesterone to the 20α-hydroxy compound by reduced nicotinamide adenin dinucleotide phosphate has been employed to assay the enzyme.

Thus 20α-oxidoreductase has been partially purified and separated from hydroxylase, 4,5-reductase and 17β-oxidoreductase activities. By above purification the 20-ketoreductase activity has been increased 6 fold.

References

[1] AKHREM, A. A., and YU. A. TITOV, "Microbiological Transformations of Steroids". Moscow, (1965), (in Russian).

[2] AMIARD, G., M. LEGRAND, J. MATHIEU, R. HEYMES, Truong van Thuong, Bull. Soc. Chim. France, 2417 (1961).

[3] CARVAJAL, F., O. F. VITALE, M. J. GENTLES, H. L. HERZOG, E. B. HERSHBERG, J. Org. Chem., **24**, 695 (1959).

[4] CHANG, V. M., D. R. IDLER, Can. J. Biochem. and Physiol., **39**, 1127 (1961).

[5] DEGHENGHI, R., C. REVESZ, R. GAUDRY, J. Med. Chem., **6**, 301 (1963).

[6] KALCKAR, H. M., J. Biol. Chem., **167**, 461 (1947).

[7] KOGAN, L. M., E. A. YELIN, The II-nd International Congress on Hormonal Steroids, Milan, N 412 (1966).

[8] NORYMBERSKI, J. K., G. F. WOODS, J. Chem. Soc., 3426 (1955).

[9] RAKHIT, S., CH. R. ENGEL, Can. J. Chem., **40**, 2163 (1962).

[10] SUVOROV, N. N., L. V. SOKOLOVA, V. M. RYZHKOVA, G. G. DVORIANTSEVA, Dokl. Acad. Nauk USSR, **152**, 1130 (1963).

[11] TAKAHASHI, T., Y. UCHIBORI, Agr. Biol. Chem. (Tokyo), **26**, 89 (1962).

Studies of 3-Oxo-Δ^4-Steroid-Δ^1-Oxydoreductase of *Arthrobacter Simplex*

L. Penasse

Centre de Recherches Roussel-Uclaf, 93-Romainville, France

and E. E. Baulieu

Unité de Recherche sur le métabolisme moléculaire et la physiopathologie des stéroïdes (INSERM), Laboratoire de Chimie Biologique, Faculté de Médecine, Paris, France

Summary

An active cell-free preparation from *Arthrobacter simplex* has been obtained by a procedure involving the culture and induction by steroids of the bacteria, spheroplasts formation by lysozyme, nucleic acid elimination by streptomycin, and ammonium sulfate precipitation. The 3-oxo-Δ^4-steroid-Δ^1-oxydoreductase activity is found in the microsomal fraction.

Cell activity is measured by spectrophotometry using 19-nortestosterone as substrate, and acellular extracts activity is determined with dichlorophenol-indophenol as electron acceptor.

Examination of the enzyme biosynthesis process shows the independance of specific biosynthesis and general growth, and the high affinity between inducer and its receptor site.

The kinetic values of several steroids with the cell-free extract show an important contribution of the middle part of the α face in the enzyme-substrate interaction. Dicortinone (Velluz et al., 1966) is a powerful noncompetitive, and androstane(5α)dione a competitive inhibitor of Δ^1-oxydoreductase.

The dehydrogenation of the steroid nucleus, in position 1—2, has been described using different strains, mainly *Pseudomonas testosteroni* (Talalay et al., 1952; Levy and Talalay, 1959), *Nocardia restrictus* (Sih and Bennett, 1960), *Bacillus sphaericus* (Stoudt et al., 1955; Hayano et al., 1961; Ringold and Turner, 1962; Ringold et al., 1963), *Bacillus cyclooxydans* (Goodman et al., 1960; Iida et al., 1965), and *Arthrobacter* (syn. *Corynebacterium*) *simplex* (Nobile et al., 1955; Herzog et al., 1962; Charney et al., 1962; Kondo, 1961 and 1962).

The Δ^1-dehydrogenase (3-oxo-Δ^4-steroid-Δ^1-oxydoreductase, E.C. 1. 3. 99.4) of *B. sphaericus* works only on Δ^4-3-oxo-steroid substrates and favors the loss of trans-diaxial (1α, 2β) hydrogen atoms (Levy and Talalay, 1957; Hayano et al., 1961, Ringold et al., 1963) (fig. 1). There could be a flavinic co-substrate (Levy and Talalay, 1959; Jerussi and Ringold, 1965) and, in any case, the enzyme can transfer electrons to N-methyl-phenazonium sulfate (Levy and Talalay, 1957), different quinones and 2,6-dichlorophenol-indophenol (Sih and Bennett, 1960 and 1962; Stefanovic et al., 1963; Davidson and Talalay, 1964).

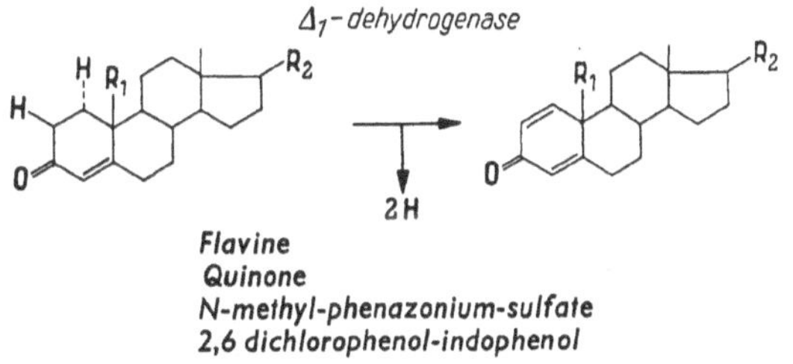

Fig. 1. 3-Oxo-steroid-Δ^1-Oxydoreductase

There is little information on the specificity and mechanism of action of *Arthrobacter* Δ^1-dehydrogenase and, in this communication, data are reported on two aspects of this enzyme:

enzyme biosynthesis and enzymatic specifity. Therefore methods have been devised for the preparation of an active cell-free extract and for the determination of activity both with intact bacteria and extracts. Some indications will be given first concerning these experimental procedures.

Experimental Procedures

1. *Enzymatic activity of acellular extracts* was determined by spectrophotometry at room temperature using cuvettes of 1 cm light path and a Beckman DU spectrophotometer. The 2,6-dichlorophenol-indophenol is used as an electron acceptor, and the decrease of the 600 mμ absorption is plotted automatically during 3 minutes.

One unit corresponds to the oxidation of 1 μmole of cortisol per minute per ml, on the basis of a stoechiometric relationship between steroid oxidation and acceptor reduction. The extinction coefficient used was 20.0×10^3 cm^{-1} mole^{-1} (ARMSTRONG, 1964) for 2,6-dichlorophenol-indophenol. From the initial velocity of the reaction, the activity is measured using a computer program for the validity of linearization (CLELAND, 1963) and for the calculation of K_m and K_i values, with an IBM 1401 computer.

2. *Enzymatic activity of intact bacteria* was determined by spectrophotometry of the transformation of nortestosterone into estradiol, directly on the suspension of the bacteria, in aerobiosis, without addition of foreign electron acceptor. At the absorption maximum of nortestosterone (249 mμ) the extinction of estradiol is almost nil; the measurement at this wavelength corresponds thus to the quantity of oxidized steroid (fig. 2). The bacteria are

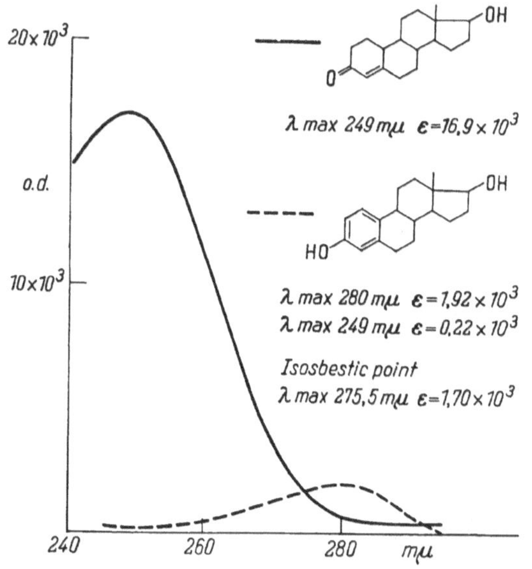

Fig. 2. UV-spectra of nortestosterone and estradiol

separated from the culture medium either by filtration with Millipore membrane or by 20 min × 10,000 G centrifugation and resuspended, with appropriate dilution, in both cases in 0.25 M sucrose, 0.003 M phosphate buffer, pH 7,0. Here also one unit corresponds to the oxidation of 1 μmole of nortestosterone per minute per ml, using an extinction coefficient of 16.9×10^3 for the substrate (this unit is obviously not identical with the first one). The specific activity is expressed as unit per mg of bacterial dry weight.

3. Steroids were identified with classical procedures including paper and thin layer chromatography, Δ^4- and $\Delta^{1,4}$-steroids were determined by circular dichroism (VELLUZ and

LEGRAND, 1961) or spectrophotometry (RÖSSLER and BRÜCKNER, 1961), proteins by the technique of LOWRY (LOWRY et al., 1951) or by UV-absorption at 260 and 280 mμ (KALCKAR, 1947) with a bovine albumin standard, and bacteria were counted by subculture, hemocytometer reading and optical density at 615 mμ.

4. *Culture and induction.* The culture medium contains per liter: peptone 5 g, Difco yeast extract 1.5 g, NaCl 3.5 g, glucose 2 g, and the pH is adjusted to 7.4. To 100 ml, 0.5 ml of a suspension of *Arthrobacter simplex* ATCC 6946 (5×10^6 bacteria per ml) are added and the culture left 24 h at 30 °C under agitation.

A concentration of approximately 1×10^{10} bacteria per ml is obtained. For studies of the induction process, these bacteria are separated by 1 h × 10,000 G centrifugation and resuspended, at a concentration of 1×10^9 bacteria per ml in 0.003 M phosphate buffer, pH 7.0. The steroid, usually 1×10^{-3} M cortisol, and if necessary 2 ‰ yeast extract, is added.

For the enzyme extraction studies, the culture is prepared in a 50 l stainless steel fermentor, under constant stirring, air is introduced at a rate of 5 l per min, and the transfer coefficient K is 0.3 $\mu M\ O_2 \times ml^{-1} \times min^{-1}$ (COOPER et al., 1944). After culture on the same medium, and induction, the bacteria are collected at 5° C with a Sharpless centrifuge and washed once in 0.05 M Tris buffer, pH 7.0.

Treatment of the bacteria by acetone (BOYER et al., 1965), or alcohol (TALALAY and BOYER, 1965), the use of the French press (SAGERS, 1962) or ultrasonic vibrations (22 KC, 40 min.) (KONDO, 1962) gave unsatisfactory results. Better results were obtained by the use of *lysozyme* (N-acetylmuramyl hydrolase, E.C. 3.2.1.17) which has been already used by WILSON and VESTLING (1965) for the preparation of the steroid 15-hydroxylase of *Bacillus megatherium.* A suspension of approximately $1-2 \times 10^{11}$ bacteria per ml, is made in 0.05 M Tris buffer pH 7.0, with 1.2 M sucrose, 0.01 M $MgCl_2$ and 250 μg/ml of lysozyme, and incubated at 37°C for 5—6 hours. When the initial opacity given by bacteria, measured on a 10 times diluted aliquot, is reduced by a factor of 3 or 4, spheroplasts are spun down at 4°C by 1 h × 10,000 G centrifugation and resuspended at the same concentration in 0.003 M phosphate buffer, 0.01 M $MgCl_2$, pH 7.0. After agitation of the flask for 3 minutes, lysis occurs and the supernatant after 1 h × 10,000 G centrifugation is collected. It has an enzymatic activity of approximately 500 mU/ml. The next steps involve nucleic acid precipitation by streptomycin, and then ammonium sulfate precipitation at a 2.3 M concentration. The precipitate is redissolved in phosphate buffer and this solution, which has an activity of 450 mU/mg protein, is used for enzymatic studies (table 1).

Table 1

Purification of Δ^1-Dehydrogenase

	Vol. (ml)	Proteins (mg/ml)	Activity		Yield
			mU/ml	mU/mg Protein	
Bacterial suspension	100	15–18 mg[1]	—	—	—
First extract	100	—	450	—	100
Supernatant after nucleic acid precipitation	200	3.6	240	70	105
$(NH_4)_2SO_4$ 2.3 M precipitate	10	9.6	4,000	450	80

[1] Baterial total dry weight.

Contrarily to the bacteria and spheroplasts which are stable in the cold for at least 10 days without loss of activity, the purified enzyme preparation is rapidly inactivated (50% in 4 hours at pH 7,0 and 5°C). The usual stabilizers, albumin, glycerol, mercaptoethanol, glutathion and tween do not help. However, in anaerobiosis, the presence of approximately 1×10^{-6} M cortisol, progesterone, estrenedione or other substrates (added in methanol to give a final concentration of 5% of this solvent) stabilizes the enzyme: the loss of activity is only about 25% in 20 hours at 5 °C.

Results and discussion

1. Enzyme biosynthesis

When the purified extract is obtained from bacteria grown without steroid, there is no activity. In the presence of cortisol the enzyme formation is progressive and the maximum (units per mg of protein) is obtained after approximately 20 hours, the mean generation time of *Arthrobacter simplex* being 1.5 hour (fig. 3).

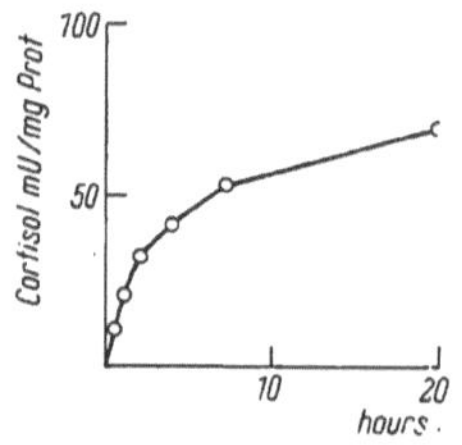

Fig. 3. Induction of Δ^1-dehydrogenase of *A. simplex* by cortisol

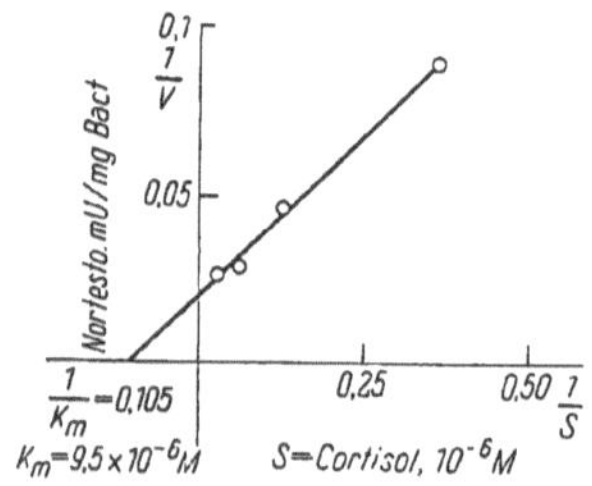

Fig. 4. Affinity of cortisol for induction receptor site.

Induction is performed in 0.003 M phosphate buffer, pH 7.0, at 1×10^9 bacteria per ml. After 2 hours at 30 °C activity is measured by direct spectrophotometry with nortestosterone as substrate

The influence of the steroid concentration on the induction kinetics, has been studied in the case of cortisol. For the very low concentrations ($3-30 \times 10^{-6}$ M), previous experiments having shown a large degradation of the steroid, it was necessary to introduce the bacterial suspension (10 ml $\times 1 \times 10^9$ bact. per ml) in a dialysis bag placed itself in 1000 ml 0.003 M

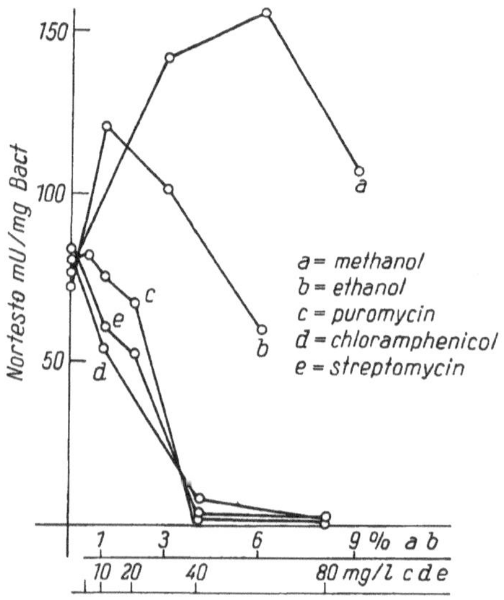

Fig. 5. Studies of some induction effectors.

Induction in 0.003 M phosphate buffer, pH 7.0, yeast extract 2 ‰, cortisol 1×10^{-3} M, aerobiosis, 30 °C, 1×10^9 bacteria per ml, with addition of: a) methanol; b) ethanol; c) puromycin; d) chloramphenicol; e) streptomycin. Specific activity is determined by direct spectrophotometry after 24 hours

phosphate buffer pH 7.0 containing cortisol and renewed every 30 min, so as to maintain the inducer concentration almost constant. Under these conditions, induction is linear for at least 4 hours and the use of a double reciprocal plot between induction velocity and steroid concentration shows the apparent affinity of cortisol for the induction receptor to be approximately:

$$\frac{1}{K_m} = \frac{1}{9.5} \times 10^6\,M^{-1} \simeq 1 \times 10^5\,M^{-1} \qquad \text{(fig. 4).}$$

Further, the use as medium during induction of 0.003 M phosphate buffer pH 7.0 or yeast extract at different concentrations, and the use of bacteria collected at different stages of growth do not modify the dehydrogenase activity.

Induction was performed in the presence of several antibiotics and alcohol (fig. 5). Puromycin, chloramphenicol and streptomycin inhibit induction more than growth, colimycin has no effect (up to 100 µg/ml), alcohols favor induction and inhibit growth. These facts seem to indicate a relative independence between induction and growth.

The induced enzyme of *Arthrobacter simplex* is attached to the subcellular fraction which sediments at 105,000 G as is the case also with *Pseudomonas testosteroni*, *Bacillus cyclooxydans* and *Nocardia restrictus*. Attempts of further solubilization were not successful.

2. Enzyme characteristics

The preparation which was used did not show any esterase, hydroxysteroid-oxydoreductase, nor $\Delta^{5,4}$-3-oxo-steroid-isomerase activity under the usual conditions of assays. However the presence of some 3β-, 17β-, and 20β-hydroxysteroid-oxydoreductase was detected by longer incubation. Attempts to obtain the reverse reaction (reduction of Δ^1-cortisol) have been made in anaerobiosis and have shown the formation of some cortisol, as was observed with *Bacillus cyclooxydans* (Iida et al., 1965).

Arthrobacter enzyme shares with the other microbial Δ^1-dehydrogenases the sensibility to several inhibitors (Levy and Talalay, 1959; Sih and Bennett, 1960), mainly the reagents of thiol and disulfide groups (Hg^{++}, Cu^{++}, iodoacetate, arsenite, ferrocyanide) (table 2).

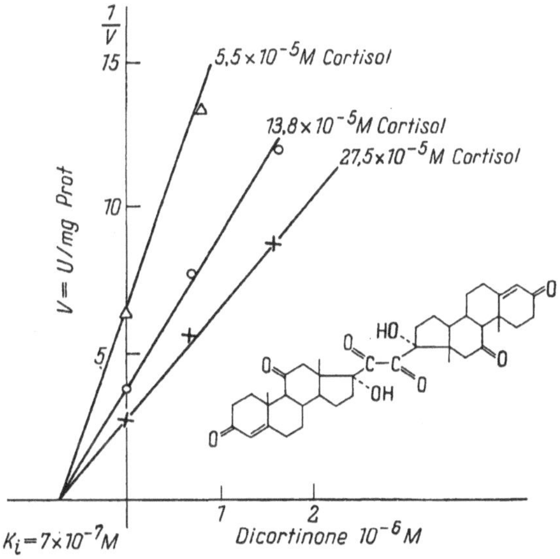

Fig. 6. Inhibition of Δ^1-dehydrogenase by dicortinone.

Determination of K_i. Substrate: cortisol at concentrations of $5{,}5 \times 10^{-5}$ M, 13.8×10^{-5} M, and 27.5×10^{-5} M. Extract is obtained from bacteria induced by cortisol, and used after precipitation of nucleic acids. Specific activity: 80 mU/mg. $K_i = 7 \times 10^{-7}$ M

Table 2

Δ^1-dehydrogenase inhibition

Inhibitors added to the enzymatic preparation 10 min before the dosage. Substrate: cortisol $3 \cdot 10^{-4}$ M, enzymatic extract is obtained after precipitation of nucleic acids of an extract of bacteria induced by cortisol, activity: 60 mU/mg protein

Effector	Concentration	% Inhibition
Potassium cyanide	$1 \cdot 10^{-2}$ M	20
Mercuric chloride	$1 \cdot 10^{-3}$ M	100
	$1 \cdot 10^{-4}$ M	100
	$1 \cdot 10^{-5}$ M	18
Sodium iodo-acetate	$3 \cdot 10^{-3}$ M	50
	$1 \cdot 10^{-3}$ M	25
p-Hydroxymercuribenzoate	$1 \cdot 10^{-4}$ M	25
Sodium arsenite	$3 \cdot 10^{-3}$ M	78
	$1 \cdot 10^{-3}$ M	25
Potassium ferrocyanide	$1 \cdot 10^{-3}$ M	100
	$1 \cdot 10^{-4}$ M	50
8-Hydroxyquinoline	$1 \cdot 10^{-3}$ M	15
o-Phenanthroline	$1 \cdot 10^{-3}$ M	0
EDTA	$1 \cdot 10^{-3}$ M	20
Diisopropyl fluorophosphate	$1 \cdot 10^{-3}$ M	35
N-Ethyl maleimide	$1 \cdot 10^{-3}$ M	35
Manganese sulfate	$1 \cdot 10^{-3}$ M	0
Copper sulfate	$1 \cdot 10^{-3}$ M	100
	$1 \cdot 10^{-4}$ M	100
	$1 \cdot 10^{-5}$ M	67
Zinc sulfate	$1 \cdot 10^{-3}$ M	10
Sodium azide	$1 \cdot 10^{-2}$ M	50
	$1 \cdot 10^{-3}$ M	0
Urea	1	63
Hydrogen peroxide	$1 \cdot 10^{-2}$ M	30
	$1 \cdot 10^{-3}$ M	20
Butanol	$3 \cdot 10^{-1}$ M	0
Chloramphenicol	$3 \cdot 10^{-4}$ M	0
Tween 20	5,000 μg/ml	0
Tween 80	5,000 μg/ml	0
Triton WR	10,000 μg/ml	0
Dicortinone	$1 \cdot 10^{-5}$ M	100
Bis-Δ^1-dicortinone	$1 \cdot 10^{-5}$ M	100

It is not inhibited by phenol steroids and $\Delta^{1,4}$-3-oxo-steroids as the *Pseudomonas* enzyme (LEVY and TALALAY, 1959). On the contrary, inhibition is observed with two groups of steroids.

First, the steroidal dimer structure dicortinone which was obtained by alcaline treatment of cortisone (VELLUZ et al., 1966), and bis- Δ^1-dicortinone are very effective noncompetitive (DIXON, 1953) inhibitors ($K_i = 7 \times 10^{-7}$ M and 7.5×10^{-7} M respectively) when tested under different conditions with different substrates and enzyme obtained after induction by different steroids (fig. 6).

Second, steroids with saturated A ring and A/B transjunction, 5α-androstanedione (XXVI) and 5α-androstanolone (XXV) which are inducers, are "pseudo competitive inhibitors" ($K_i = 2.5 \times 10^{-5}$ M) of the type described by FRIEDEN (1964) in which the substrate is less tightly bound in the presence of the inhibitor, while the stage of product formation and release takes place equally well. This inhibition is observed with intact bacteria as well as with the cell-free extracts; the inhibition characteristics are almost the same in both cases (fig. 7).

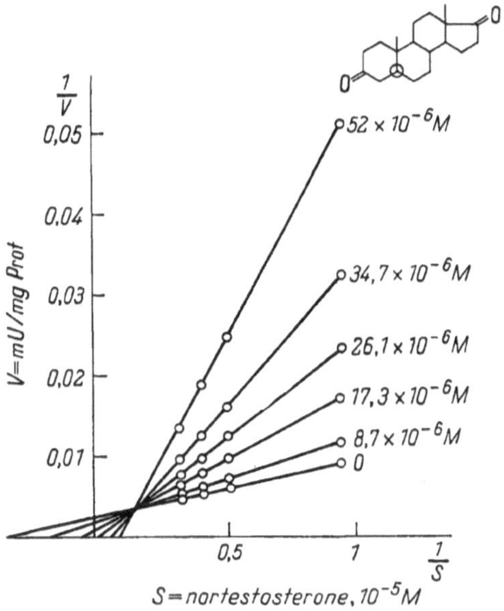

Fig. 7. Inhibition of Δ^1-dehydrogenase by 5α-androstanedione.
Substrate: nortestosterone at concentrations from 1×10^{-5} M to 3×10^{-5} M. Inhibitor: 5α-androstanedione at concentrations from 0.87×10^{-5} M to 5.2×10^{-5} M. Direct spectrophotometry with whole bacteria

Table 3

Studies on different substrates of Δ^1-dehydrogenase

K_m, V_m (μM/min) and V_3 (μM/min at 3×10^{-5} M substrate concentration). The enzymatic extract is obtained after precipitation of nucleic acids, using cortisol induced bacteria. The specific activity is 70 mU/mg protein

Substrate	$K \cdot 10^{-5}$	V_m	V_3	Substrate	$K_m \cdot 10^{-5}$	V_m	V_3
I 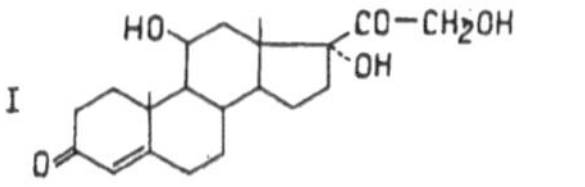	21.0	380	40	VIII 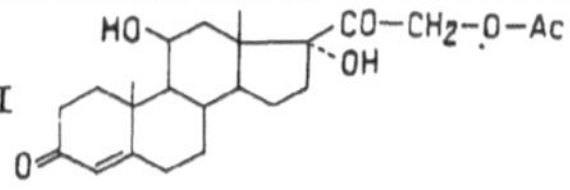	6.6	350	100
II 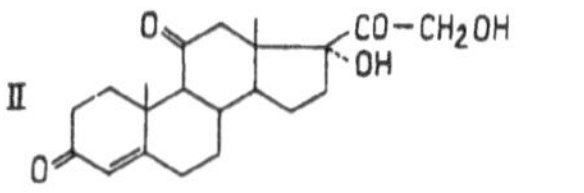	2.6	810	370	IX 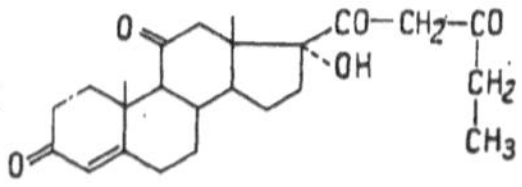	3.0	740	300
III 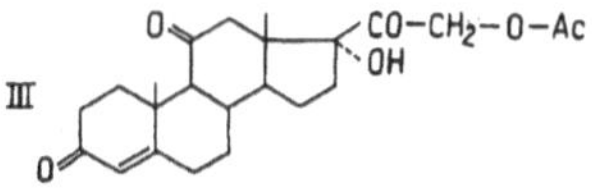	2.5	240	120	X 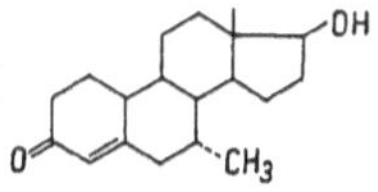	2.6	320	160
IV 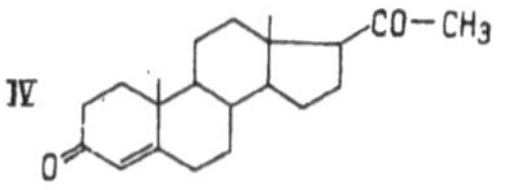	0.4	400	350	XI 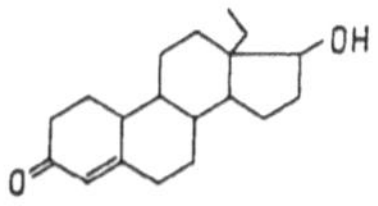	1.0	260	190
V 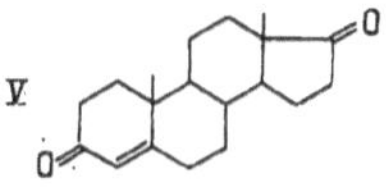	1.8	790	540	XII 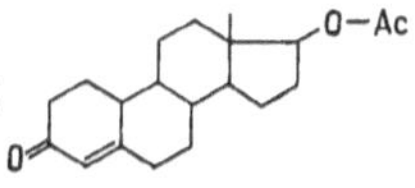	2.5	210	160
VI 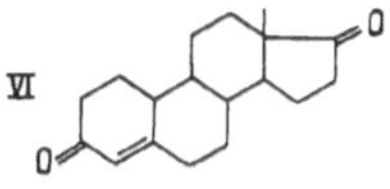	3.7	630	310	XIII 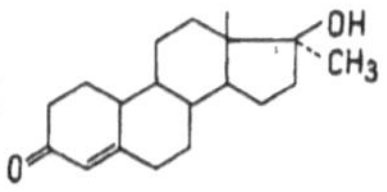	5.0	360	150
VII 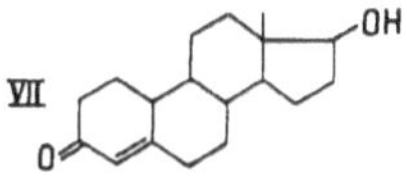	2.0	470	300				

Table 4

Studies on different substrates of Δ^1-dehydrogenase

Reaction rate (V_c: μM/min) at substrate concentration c. The enzymatic extract is obtained, using cortisol induced bacteria, after precipitation of nucleic acids. The specific activity is 70 mU/mg protein

Substrate	c 10^{-5}M	V_c	Substrate	V_c 10^{-5}M	V_c	Substrate	c 10^{-5}M	V_c
I 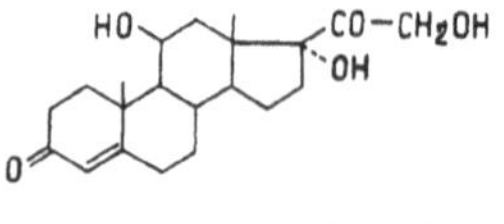	30	220	XXIV 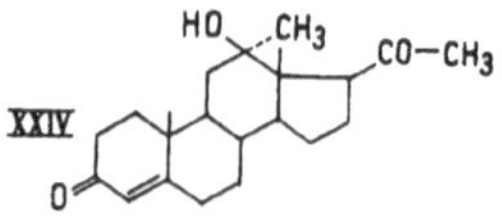	3	680	XXXV 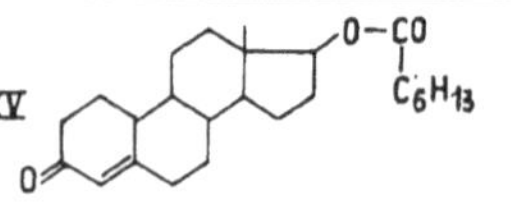	3	210
XIV 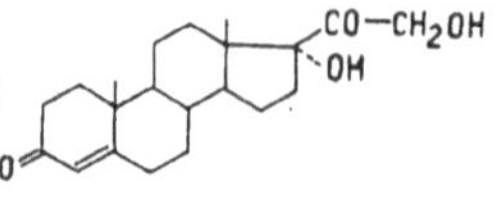	30	97	XXV 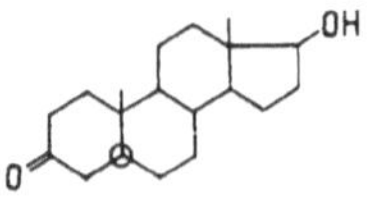	30	0	XXXVI 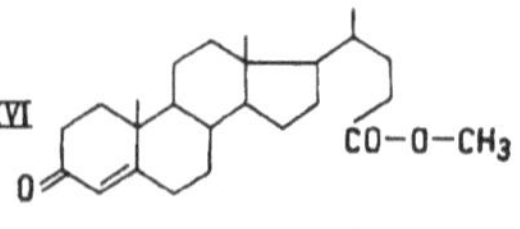	9	242
XV 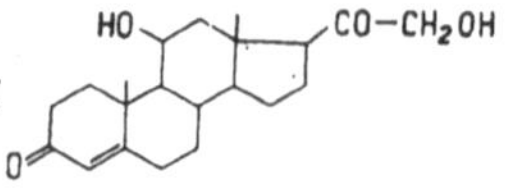	30	255	XXVI 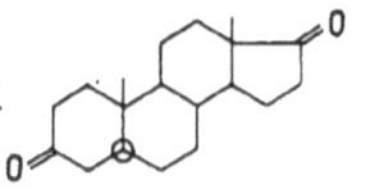	30	0	XXXVII 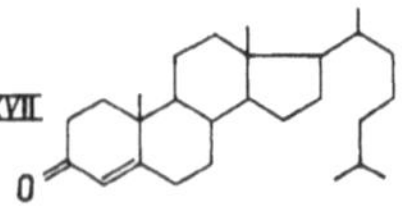	9	14
XVI 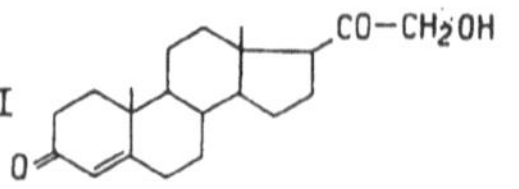	30	246	XXVII 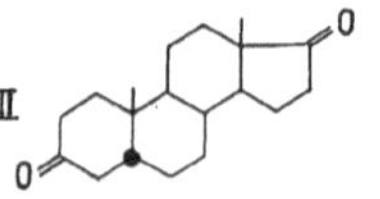	30	0	XXXVIII 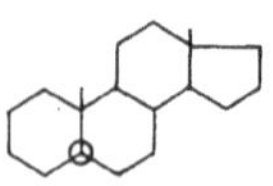	3	0
XVII 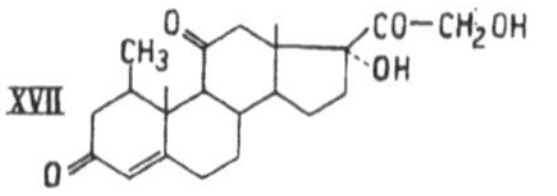	15	27	XXXVIII 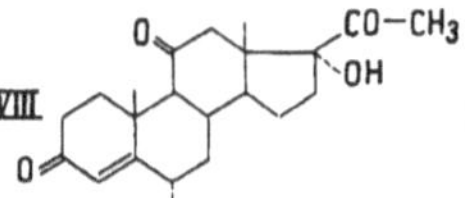	30	116	XXXIX 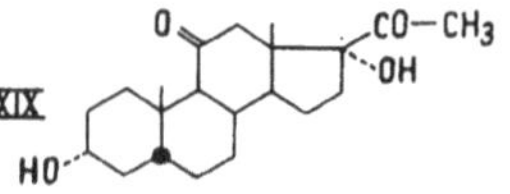	3 30	0 0
XVIII 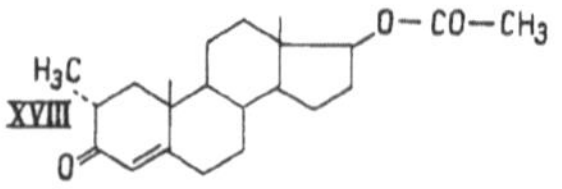	15	30	XXIX 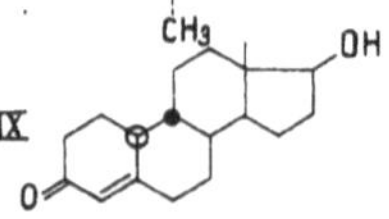	3 30	14 56	XL 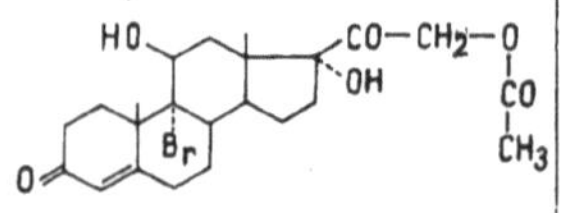	3	0
XIX 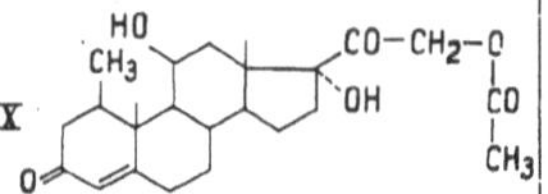	15	12	XXX 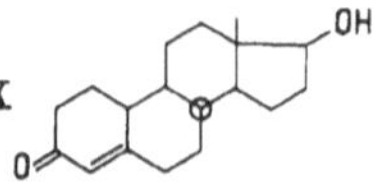	3	7	XLI 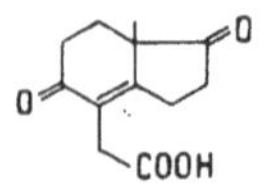	60	7
XX 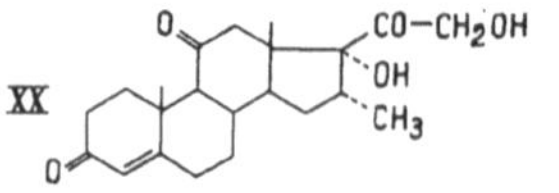	3	167	XXXI 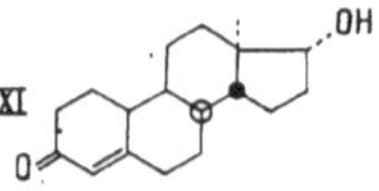	6 30	0 0	XLII 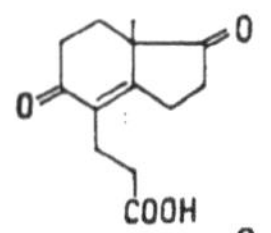	60	8
XXI 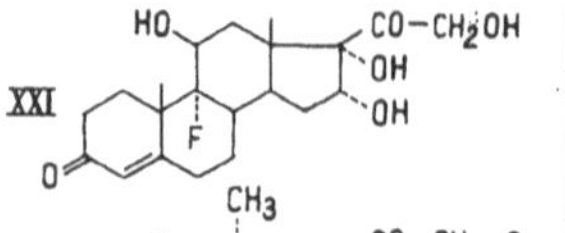	30	46	XXXII 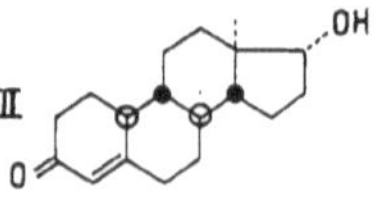	6 30	0 0	XLIII 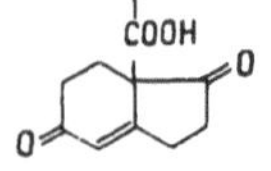	60	10
XXII 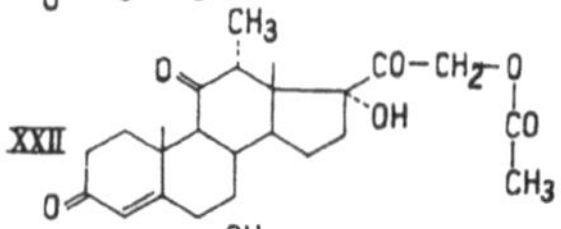	3	90	XXXIII 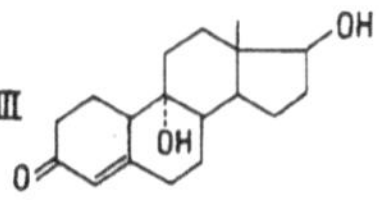	3	0	XLIV 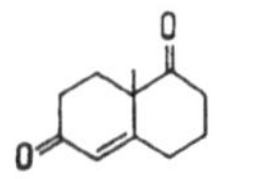	60	0
XXIII 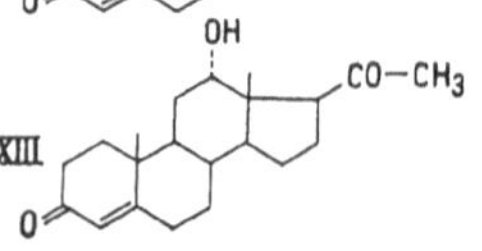	3	325	XXXIV 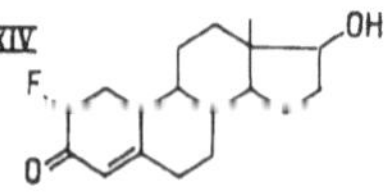	30	80			

In kinetic studies, reactions are possible only within 10^{-5} M and 10^{-4} or 10^{-3} M substrate concentration, since there is a solubility limit, and some steroids provoke an inhibition by excess of substrate. The double reciprocal plot (LINEWEAVER and BURK, 1934) has been used for the determination of K_m of different substrates. For a series of other substrates, the speed at only 1 or 2 concentrations has been measured (table 3, 4 and fig. 8).

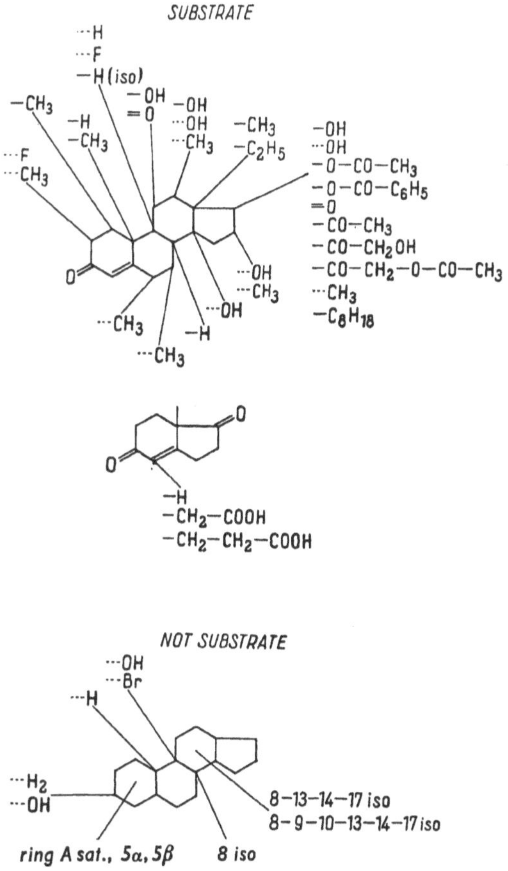

Fig. 8. Substrates of Δ^1-dehydrogenase

The kinetic data thus obtained with 44 substances show:

1. The Δ^4-3-oxo-group is necessary for the reaction to take place (XXV, XXVI, XXVII, XXXVIII, XXXIX) (CHARNEY et al., 1962), this group can be situated on nonsteroidal structures (indenone XLI, XLII, XLIII) and on C_{18}, C_{19}, C_{21}, C_{24} and C_{27} steroids.

2. Among the steroids, the equatorial (XVII, XVIII, XIX, XXIV, XXVIII, XXXIV), semi-equatorial (I, XIII, XV, XXI), β-axial (I, XI) as well as 7α (X), 12α (XXII, XXIII, XXIV) and 16α (XX, XXI) substituants modify slightly the K_m and V_m values, but do not prevent the reaction from taking place.

3. On the contrary, the presence of 9α-hydroxyl (axial, XXXIII) and 9α-bromine (XL) inhibits strongly the reaction, while 9α-fluorine (XXI) does not.

4. The C-8 isomer of 19-nortestosterone (XXX), in which 8-hydrogen is α-axial, reacts 40 times slower, and the C-9, C-10 isomer (XXIX) in which 10-hydrogen is α-axial, reacts 20 times slower than the natural derivative. The C-8, C-13, C-14, C17 isomer (XXXI), and the inverted steroid (XXXII) are no substrates, the Dreiding models show here that not only

the 8-hydrogen is α-axial, but also the 18-methyl group is almost in such an orientation. Recently (PAN, 1966) the dehydrogenation of several 14α-substituted steroids by *Corynebacterium simplex* has been described. Finally, the presence of substituants on the α face in position 8, 9, 10 and 13 inhibits considerably or prevents the reaction, whereas other substituants exert only limited influence on the reaction characteristics. These facts seem to indicate an important contribution of the middle part of the α face in the enzyme-substrate interaction (fig. 9).

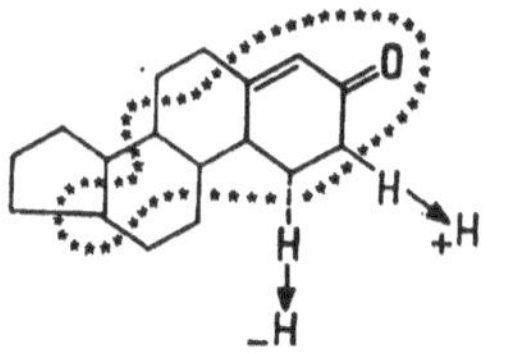

Fig. 9. Steroid molecule.
Part of α-face which could be implicated in enzyme-substrate interaction

On the other hand, the fact that the equatorial substitution of positions 1 and 2 by methyl group (XVII, XIX, XVIII) does not prevent the reaction, is in favor of a trans-diaxial loss of hydrogen atoms with *Arthrobacter simplex* enzyme as it is the case with *Bacillus sphaericus*. In this mechanism the 2β-hydrogen is supposed to be lost as a proton while the 1α-hydrogen should be fixed as a hydride ion by the enzyme (or coenzyme) thus located under this face.

In conclusion, the results obtained until now show that *Arthrobacter simplex* Δ^1-dehydrogenase constitutes a good model both for the study of enzyme biosynthesis and of enzyme-substrate interaction and deserves more work.

References

[1] ARMSTRONG, J. M., Biochim. Biophys. Acta, **86**, 194 (1964).
[2] BOYER, J., D. N. BARON, and P. TALALAY, Biochemistry, **4**, 1825 (1965).
[3] CHARNEY, W., A. NOBILE, C. FEDERBUSH, D. SUTTER, P. L. PERLMAN, H. L. HERZOG, C. C. PAYNE, M. E. TULLY, M. JEVNIK-GENTLES, and E. B. HERSHBERG, Tetrahedron, **18**, 591 (1962).
[4] CLELAND, W. W., Nature, **198**, 463 (1963).
[5] COOPER, C. M., G. A. FERNSTROM, and S. A. MILLER, Ind. Eng. Chem., **36**, 504 (1944).
[6] DAVIDSON, S. J., and P. TALALAY, Federation Proc., **23**, 275 (1964).
[7] DIXON, M., Biochem. J., **55**, 170 (1953).
[8] FRIEDEN, C., J. Biol. Chem., **239**, 3522 (1964).
[9] GOODMAN, J. J., M. MAY, L. and L. SMITH, J. Biol. Chem., **235**, 965 (1960).
[10] HAYANO, M., H. J. RINGOLD, V. STEFANOVIC, M. GUT, and R. I. DORFMAN, Biochem. Biophys. Res. Commun., **4**, 454 (1961).
[11] HERZOG, H. L., C. C. PAYNE, M. TULLY-HUGHES, M. JEVNIK-GENTLES, E. B. HERSHBERG, A. NOBILE, W. CHARNEY, C. FEDERBUSH, D. SUTTER, and P. L. PERLMAN, Tetrahedron, **18**, 581 (1962).
[12] IIDA, M., J. D. TOWNSLEY, M. HAYANO, and H. J. BRODIE, Steroids, Suppl. I, 159 (1965).
[13] JERUSSI, R., and H. J. RINGOLD, Biochemistry, **4**, 2113 (1965).
[14] KALCKAR, H. M., J. Biol. Chem., **167**, 461 (1947).
[15] KONDO, E., and E. MASUO, J. Gen. Appl. Microbiol. (Tokyo), **7**, 113 (1961).
[16] KONDO, E., Agr. Biol. Chem. (Tokyo), **27**, 69 (1962).
[17] LEVY, H. R., and P. TALALAY, J. Am. Chem. Soc., **79**, 2658 (1957).
[18] LEVY, H. R., and P. TALALAY, J. Biol. Chem., **234**, 2014 (1959).
[19] LINEWEAVER, H., and D. BURK, J. Am. Chem. Soc., **56**, 658 (1934).
[20] LOWRY, O. H., N. J. ROSEBROUGH, A. L. FARR, and R. J. RANDALL, J. Biol. Chem., **193**, 265 (1951).
[21] NOBILE, A., W. CHARNEY, P. L. PERLMAN, H. L. HERZOG, C. C. PAYNE, M. E. TULLY, M. A. JEVNIK, and E. B. HERSHBERG, J. Am. Chem. Soc., **77**, 4184 (1955).

[22] PAN, S. C., P. PRINCIPE, B. JUNTA, and R. C. ERICKSON, Abstracts of Papers, Am. Chem. Soc. 152th Meeting 16Q (1966).
[23] RINGOLD, H. J., M. HAYANO, and V. STEFANOVIC, J. Biol. Chem., **238**, 1960 (1963).
[24] RINGOLD, H. J., M. GUT, M. HAYANO, and A. TURNER, Tetrahedron letters, **18**, 835 (1962).
[25] RINGOLD, H. J., and A. TURNER, Chem. Ind. (London), 211 (1962).
[26] RÖSSLER, H., and K. BRÜCKNER, Naturwiss., **48**, 695 (1961).
[27] SAGERS, R. D., Appl. Microbiol., **10**, 37 (1962).
[28] SIH, C. J., and R. E. BENNETT, Biochim. Biophys. Acta., **38**, 378 (1960).
[29] SIH, C. J., and R. E. BENNETT, Biochim., Biophys. Acta., **56**, 584 (1962).
[30] STEFANOVIC, V., M. HAYANO, and R. I. DORFMAN, Biochim. Biophys. Acta, **71**, 429 (1963).
[31] STOUDT, T. H., W. J. MCALEER, J. M. CHEMERDA, M. A. KOZLOWSKI, R. F. HIRSHMANN, V. MARLATT, and R. MILLER, Arch. Biochem. Biophys., **59**, 304 (1955).
[32] TALALAY, P., M. M. DOBSON, and D. TAPLEY, Nature, **170**, 620 (1952).
[33] TALALAY, P., and J. BOYER, Biochim. Biophys. Acta, **105**, 389 (1965).
[34] VELLUZ, L., and M. LEGRAND, Angew. Chem., **73**, 603 (1961); Angew. Chem., **77**, 842 (1965); see also PESEZ, M., and J. BARTOS, Ann. Pharm. Franç., **22**, 675 (1964).
[35] VELLUZ, L., L. PENASSE, G. NOMINE, G. AMIARD, P. BARTHELEMY, and V. TORELLI, C. R. hebd. Séances Acad. Sci., Ser. C, **262**, 120 (1966).
[36] WILSON, J. E., and C. S. VESTLING, Arch. Biochem. Biophys., **110**, 401 (1965).

Enzymatic Mechanism of Microbiological Dehydrogenation and Hydrogenation of Steroid Ring A

N. N. Lestrovaya and G. K. Skryabin[1]

Institute of Microbiology, Academy of Sciences of USSR, Moscow, USSR

Summary

Evidence has been obtained that the oxidative-reductive process, which is accompanied by the introduction of 1,2-double bonds into steroids and its saturation is catalyzed in *Mycobacterium globiforme* by two separate enzymes. This conclusion was drawn basing on the character of induction, action of inhibitors, co-enzyme specificity and the protein behaviour during fractionation.

The culture of *Mycobacterium globiforme* 193 isolated in our laboratory carries out both Δ^1-dehydrogenation of the steroid molecule and the reduction of the corresponding 1,2-double bond, depending on the aeration conditions (Skryabin et al., 1964). It was proved possible to turn the process in the desired direction by means of changing the oxidative-reductive potential of the medium (Skryabin et al., 1965; Lestrovaya et al., 1965). It is evident that the above microorganism is useful for elucidating the enzymatic mechanism of the oxidative-reductive transformation.

It should be noted that in the recent years the steroid Δ^1-dehydrogenases of microorganisms have been studied in some laboratories (Levy et al., 1959; Sih et al., 1962; Stefanovic et al., 1963). The reverse process — the reduction of the 1,2-double bond — was also described for some microorganisms (Butenandt et al., 1940; Herzog et al., 1959; Goodman et al., 1960; Holmlund et al., 1962, Schubert et al., 1963). However, the enzymes responsible for the above transformation have not been studied.

Methods

The organism *Mycobacterium globiforme* was grown for 20 hours at 28 °C on a medium containing 1% glucose and 1% corn steep with the addition of progesterone to give a final steroid concentration of 200 μg per ml.

The cells were washed and suspended in 0.03 M phosphate buffer, pH 7.0, and treated in the ultrasonic desintegrator MSE 60 W for 15 min. After centrifugation of the disrupted cells a high speed supernatant fraction (105,000 × g) was used as the source of crude enzymes.

In our experiments cortisol was used as substrate in the oxidative process and prednisolone in the reductive process. The reaction mixture containing 150 μg of steroid, the cellfree extract (2—6 mg of protein) and 0.03 M phosphate buffer, pH 7.0, was incubated at 30 °C. The oxidative process was carried out for 1,5—2 hours under aerobic conditions and the reductive process for 15—16 hours under partially anaerobic conditions (a high column of liquid).

Reaction rates were determined by the increase or decrease of prednisolone, whose quantity was determined either by colorimetry at 550 μm (Mizsei et al., 1961) or by measuring the absorption at 240 μm after paper chromatography in the system: toluene — methanol — water — ethylacetate (9:5:5:1).

The specific enzymatic activity was estimated in μg of the product/mg of protein/hour. In some cases dehydrogenase activity was estimated by reduction of 2,6-dichlorphenol-

[1] Vortrag nicht gehalten.

indophenol followed up spectrophotometrically at 600 μm (SIH et al., 1962). The amount of enzyme which initiates the decrease of 0.01 absorbancy unit/min is defined as 1 indophenol unit, the specific activity being the units per mg of protein. The protein content was determined either by the Lowry method or light absorbance measurements at 280 μm and 260 μm.

Results and Discussion

Preliminary experiments were designed to determine conditions which would yield an enzyme extract with the highest specific activity. For this purpose Mycobacterium globiforme was grown on the above described medium with addition of some steroids (Table 1).

Table 1
The Influence of Steroid Inductors on Dehydrogenase and Reductase Activities

Inducer	Sp. Δ^1-Dehydrogenase activity	Sp. Δ^1-Reductase activity
—	0	0.39
progesterone	50.6	0.56
cortisol	4.7	0.65
prednisolone	9.7	0.56
Δ^4-androsten-3,17-dione	94.0	0.51
$\Delta^{1,4}$-androstadien-3,17-dione	37.9	0.39

The cell-free extracts proved active in Δ^1-dehydrogenation only if the cells were grown in the presence of an added inducer. Reductase activity was always present in the extracts regardless of the conditions of growth, the stimulating effect of the steroids being relatively small. It is of interest that the same steroids may display different activity in both cases. For example, cortisol, which has a somewhat stronger stimulating effect in the Δ^1-reductase activity as other steroids, was a very poor inducer of Δ^1-dehydrogenase. Experiments with $\Delta^{1,4}$-androstadiendione revealed the reverse relationship.

Table 2 shows the effect of some inhibitors on the above oxidative-reductive process. It was shown that dehydrogenation was inhibited only by p-chloromercuribenzoate and $HgCl_2$ while the reduction of the Δ^1-double bond proved extremely sensitive to several different inhibitors.

Table 2
Effect of Inhibitors on Δ^1-Dehydrogenase and Δ^1-Reductase

Compounds	Inhibitoin in %		Compounds	Inhibition in %	
	Δ^1-dehydrogenase	Δ^1-reductase		Δ^1-dehydrogenase	Δ^1-reductase
$HgCl_2$	48	100	jodacetamide	0	62
$CuSO_4$	0	100	jodacetate	0	43
			o-phenantroline	0	15
$ZnSO_4$	0	26	α,α'-dipyridil	0	58
$FeCl_3$	0	38	α-benzoinoxime	0	27
$K_3Fe(CN)_6$	0	75	EDTA	3	0
$MnSO_4$	0	35	acriflavin	0	23
$FeSO_4$	3	24			
p-chloromercuribenzoate	44	82			

The literature data (LEVY et al., 1959; SIH et al., 1962; STEFANOVIC et al., 1963) contain the evidence that NAD, NADP, FMN or FAD do not stimulate the dehydrogenation process

(Table 3). The natural electron acceptor could be imitated by means of phenazine methosulfate, methylene blue and to a lesser extent by 2,6-dichlorphenol-indophenol. The increase in the rate of Δ^1-dehydrogenation in the presence of cytochrome c was, evidently, due to the oxydation by cytochrome c of the natural co-enzyme of the steroid dehydrogenase.

The reductive process was stimulated by $NADH_2$ and to a lesser degree by $NADPH_2$ (Table 3). This phenomenon was especially well demonstrated with the extract from which the greater part of pyridine nucleotides was removed by "dialysis" with Sephadex G-50. In this case the rate of prednisolone transformation to cortisol increased in the presence of $NADH_2$ by 75 per cent as compared to the experiment with no reduced NAD added. Phenazine-methosulphate, an active electron acceptor in the oxidative process, entirely inhibited the reductive process, evidently, because of the nonenzymatic $NADH_2$ oxidation.

Table 3

Compounds Tested as Hydrogen Acceptors and Donators

acceptor	% conversion of F → Δ^1 F	donator	% conversion of Δ^1 F → F
—	56.3	—	19.0
NAD	56.3	$NADH_2$	49.5
NADP	33.6	$NADPH_2$	24.5
FMN	59.0	ascorbic acid	20.5
FAD	56.0	hydroquinone	12.5
cytochrom c	78.0	phenazinmethosulfate	0
2.6-dichlorphenol-indophenol	90.0		
methylenblue	100.0		
phenazinmethosulfate	100.0		

Using some methods of protein separation and purification an attempt was made to reveal the proteins responsible for catalyzing both processes. Ammonium sulphate fractionation of the crude extract revealed both enzymatic activities in all the fractions studied. But the

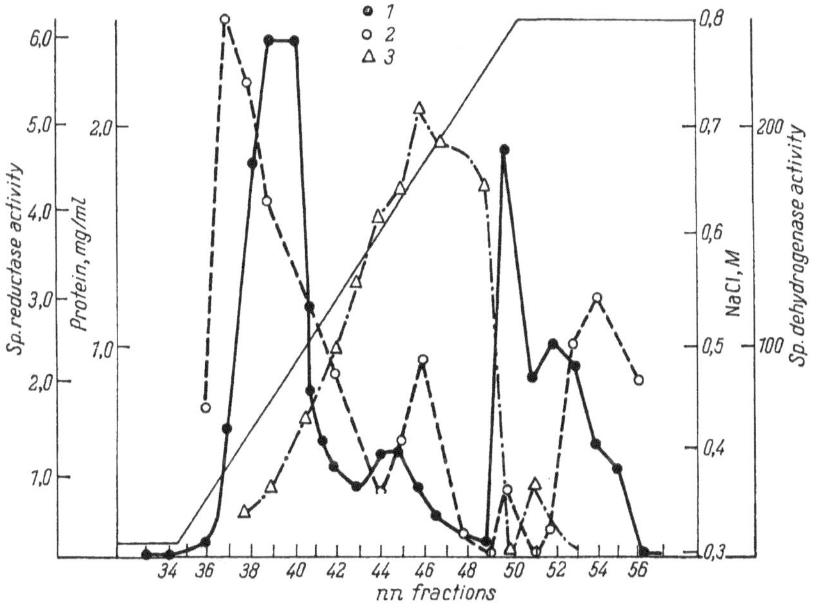

Fig. 1. Chromatography on DEAE-cellulose of the active fraction after gelfiltration on Sephadex G-75
1. protein; 2. reductase activity/μg cortisol/mg protein/hour; 3. dehydrogenase activity/indophenol units/mg protein

ratio of Δ^1-dehydrogenase and reductase activities in different fractions was 26.5 (0—30 per cent saturation), 53 (30—50 per cent saturation) and 77 (50—70 per cent saturation).

After desalting and partial purification of the 30—50% saturation fraction on Sephadex G-75 the sample was chromatographed on DEAE-cellulose which has been thoroughly equilibrated against 0.03 M phosphate buffer at pH 7.0.

Fig. 1 shows the distribution of proteins eluated from DEAE-cellulose by 0.3—1 M NaCl (linear gradient). The protein fractions eluated by 0.3—0.35 M NaCl has high reductase activity and contains practically no Δ^1-dehydrogenase.

References

[1] BUTENANDT, A., H. DANNENBERG, L. A. SURANY, Ber. **73**, 818, (1940).
[2] GOODMAN, J. J., M. MAY, L. L. SMITH, J. biol. Chem. **235**, 965, (1960).
[3] HERZOG, H. L., M. J. GENTLES, W. CHARNEY, D. SUTTER, E. TOWNLEY, M. JUDINS, P. KABASAKALIAN, E. B. HERSHBERG, J. Org. Chem., **24**, 691, (1959).
[4] HOLMLUND, C. E., L. I. FELDMAN, H. M. KISSMAN, M. J. WEISS, J. Org. Chem. **27**, 2122, (1962).
[5] LESTROVAYA, N. N., M. I. NASARUK, G. K. SKRYABIN, Docl. Acad. Nauc. SSSR, **163**, 763, (1965).
[6] LEVY, H. R., P. TALALAY, J. biol. Chem., **234**, 2014, (1959).
[7] MIZSEI, A., A. SCABO, J. Biochem. Microbiol. Techn., **3**, 119, (1961).
[8] SIH, C. J., R. E. BENNETT, Biochem. Biophys. Acta **56**, 584, (1962).
[9] SKRYABIN, G. K., I. S. ZVAYGINZEVA, L. V. SOKOLOVA, Isv. Acad. Nauc SSSR, ser. biol., N. 5, 715, (1964).
[10] SKRYABIN, G. K., I. S. ZVAYGINZEVA, M. I. NASARUK, L. V. SOKOLOVA, Docl. Acad. Nauc SSSR, **161**, 472, (1962).
[11] STEFANOVIC, V., M. HAYANO, R. I. DORFMAN, Biochim. Biophys. Acta, **71**, 429 (1963).

Closing Remarks

KURT SCHUBERT

In closing this symposium it is instructive to recall the advances made in the last few years.

The investigation of the biosynthesis of sterols has, after some time of lower activity again entered upon a stage of intensive treatment. Thereby the elucidation of certain stages of reaction is not only limited to the organism of mammalians, but it also refers to plants. Pregnenolone and progesterone were formed from cholesterol in digitalis plants. A new aspect is the biogenesis of sterols in bacteria.

Of special interest are the possibilities of inhibiting the biosynthesis, about which we have heard some new results, with the protozoae taken as a model. Now the problem centres round the function of sterols in the cells and, connected with this, their incorporation and alteration in the membranes. Recent objects of investigation were mycoplasma and mycobacteria.

Manifold are the possibilities of investigating the steroid metabolism with microorganisms. In this respect there is an especially close interconnection between the investigation of basic enzymatic reactions concerning the steroid molecule, the utilization of microbial reactions for the preparation of metabolites of naturally occurring steroids as well as of steroid drugs, and the use of microbial processes in the pharmaceutical industry.

Of basic importance are the interactions between steroids and microorganisms. The insights obtained from transforming the steroids and influencing the cells are essential for comparative biochemistry and physiology. Special information is to be expected from clearing up the role of steroid metabolites in cybernetic systems.

In this respect mention must be made of the advances achieved in the degradation of the side-chain of cholesterol, the differentiation of 2 enzymes in the dehydrogenation and hydrogenation of the ring A in 1-position, the observation of deviating reactions of hydroxylation in anomalous steroids and the microbial razemat separation, which is of special importance in connection with the total synthesis of steroids. Hitherto microbial transformations had chiefly been observed with the naturally occurring d-forms; now reports are available on the hydroxylation of l-forms, too. New aspects arise with regard to the elucidation of biologically active compounds of l-forms as well as metabolites of naturally occurring steroid hormones and steroid drugs. The microbial preparation of steroid metabolites is becoming more and more important with regard to the testing of these compounds for non-hormonal effects and investigations of their metabolism. Such processes are already being employed by the pharmaceutical industry.

A new microbiological reaction has been the introduction of an oxygen function in the 3-position. The introduction of such an oxygen function, which could be proved in some cases in the meantime, might be of importance in the metabolism of 3-desoxysteroid drugs in the organism of mammalians and in the totalsynthesis of steroids. Of special interest is the participation of intestinal microorganisms in the steroid metabolism, which has now been observed also with cardenolides by investigations in rats. Of increasing importance for the various investigations of the steroid metabolism are gas chromatography and mass spectrometry, whose advantages are high separation efficiency and high sensitivity.

The microbial degradation of the sterane skeleton is already elucidated in the main lines. It was shown that the end products are passing into the citric acid cycle. New intermediary products have been described and synthetized products have been investigated with regard to their further degradation.

On the basis of these results experiments can now be made in order to establish relations to the degradation of steroids in the organism of mammalians, which is still unknown. Another task is to investigate the functional significance of the intermediary products as well as to test the largely novel compounds and their analogues with regard to their biological effects. Some of the above mentioned biochemical investigations represent minor steps in the direction of the steroid mechanism of action or show new starting-points.

A good deal of progress has been made in the elucidation of control mechanisms. One feature of these is the regulation by feedback mechanisms in which metabolites modify the activity of enzymes. Investigations concerning the mechanism of action of steroids will be gaining in importance more and more. We have heard new results on the linkage of steroid hormones in the target organs, on steroid receptors, on the induction of enzymes, on the influencing of enzymatic mechanisms, on the steroid effect on the transport-ATP-ase and especially on the cell nucleus as a primary sphere of action of steroid hormones.

Investigations concerning the mechanism of action will cause even more specialists of different fields to cooperate in future.

Steroid research has a fascinating historic background. I believe that the aspects of biochemical research, as seen from the present state of development will be fascinating as well.

List of Participants

ČSSR

HANČ, O., Dr.	Institut für Pharmazie und Biochemie Prag 3 — Vinohrady, 17, Kouřimska
PROCHAZKA, Ž., Dr.	Institut für Organische Chemie und Biochemie der Akademie der Wissenschaften Prag
STARKA, L., Dr.	Forschungsinstitut für Endokrinologie Prag 1, Narodni 8

DBR

METZ, H., Dr.	E. Merck, 61 Darmstadt
SCHÜTZ, G., Dr.	Physiol.-Chemisches Institut der Philipps-Universität, Marburg/Lahn
SEKERIS, C. E., Dr.	Physiol.-Chemisches Institut der Philipps-Universität, Marburg/Lahn
TRÄGER, L., Dr.	Institut für Therapeutische Biochemie der Universität, 06 Frankfurt (Main) Ludwig-Rehn-Str. 14

DDR

BORCHERS, B., Dipl.-Chem.	Institut für Pflanzenchemie, Tharandt
BURCKHARDT, G., Dipl.-Chem.	Institut für Mikrobiologie und experimentelle Therapie der DAW, Jena
BÖHME, K.-H., Chem.-Ing.	Institut für Mikrobiologie und experimentelle Therapie der DAW, Jena
GIRBARDT, M., Prof. Dr.	Institut für Mikrobiologie und experimentelle Therapie der DAW, Jena
GROH, H., Dipl.-Chem.	Institut für Mikrobiologie und experimentelle Therapie der DAW, Jena
HELLER, K., Dr.	VEB Jenapharm, Jena
HERRMANN, I., Dr.	Institut für Biochemie der DAW, Berlin-Buch
HILLESHEIM, H.-G., Dr.	Institut für Mikrobiologie und experimentelle Therapie der DAW, Jena
HÖRHOLD, C., Dr.	Institut für Mikrobiologie und experimentelle Therapie der DAW, Jena
HÜBNER, M., Dr.	Institut für Mikrobiologie und experimentelle Therapie der DAW, Jena
KAUFMANN, G., Dipl.-Chem.	Institut für Mikrobiologie und experimentelle Therapie der DAW, Jena

Knöll, H., Prof. Dr.	Institut für Mikrobiologie und experimentelle Therapie der DAW, Jena
Kunstmann, F. W., Dipl.-Chem.	VEB Jenapharm, Jena
Lehmann, G., Dr.	Institut für Organische Chemie der DAW, Berlin-Adlershof
Müller, O., Dipl.-Biol.	VEB Jenapharm, Jena
Onken, D., Dr.	VEB Jenapharm, Jena
Ponsold, K., Dr.	Institut für Mikrobiologie und experimentelle Therapie der DAW, Jena
Portius, H. J., Dr.	Institut für Biochemie der DAW, Berlin-Buch
Rippberger, H., Dr.	Institut für Kulturpflanzenforschung der DAW, Gatersleben
Rose, G., Chem.-Ing.	Institut für Mikrobiologie und experimentelle Therapie der DAW, Jena
Schade, K., Dipl.-Chem.	Institut für Mikrobiologie und experimentelle Therapie der DAW, Jena
Schlegel, J., Dr.	Institut für Mikrobiologie und experimentelle Therapie der DAW, Jena
Schönegger, B., Dr.	Institut für Mikrobiologie und experimentelle Therapie der DAW, Jena
Schreiber, K., Prof. Dr.	Institut für Kulturpflanzenforschung der DAW, Gatersleben
Schubert, K., Prof. Dr.	Institut für Mikrobiologie und experimentelle Therapie der DAW, Jena
Schumann, W., Dipl.-Chem.	Institut für Mikrobiologie und experimentelle Therapie der DAW, Jena
Voigt, B., Dipl.-Chem.	Institut für Kulturpflanzenforschung der DAW, Gatersleben

England

Kellie, A. E., Dr.	The Courtauld Institute of Biochemistry. The Middlesex Hospital Med. School, London W. 1.

Finnland

Adlercreutz, A., Prof. Dr.	University Central Hospital Haartmaninkatu 4, Helsinki 29

Frankreich

Baulieu, E.-E., Prof. Dr.	Lab. Hormones, 94 Bicêtre 78. Avenue du Général Leclerc
Penasse, L., Dr.	Roussel UCLAF, Romainville

Holland

Doodewaard, L., Dipl.-Ing.	Res. Lab. of the Royal Netherland Fermentation Industries, Delft
Marx, A. F., Dipl.-Chem.	Res. Lab. of the Royal Netherland Fermentation Industries, Delft

WAARD, VAN DER, W. F., Dr.	Res. Lab. of the Royal Netherland Fermentation Industries, Delft

Indien

LAUMAS, K. R., Dr.	All India Institute of Medical Sciences. Reproductive Biology Res. Unit, New Delhi 16

Italien

GALLI, G., Dr.	University Degli Studi, Inst. di Farmacologia e di Terapia, Milano, Via Vanvitelli 32
WEISS, J. E., Dr.	University Degli Studi, Inst. di Farmacologia e di Terapia, Milano, Via Vanvitelli 32

Japan

HAYAKAWA, S., Dr.	Shionogi Research Laboratory Fukushima-Ko, Osaka

Schweden

GUSTAFSSON, I., Dr.	Karolinska Institutet, Kemiska Institutionen, Stockholm 60
LISBOA, B. P., Dr.	University of Stockholm, The Wenner-Gren Institute, Nortullgatan 16, Stockholm Va

UdSSR

KOGAN, L. M., Dr.	Institut für Naturstoffe der Akademie der Wissenschaften, Moskau, Ul. Vavilova 18
LESTROVAJA, N., Dr.	Institut für Mikrobiologie der Akademie der Wissenschaften, Moskau-B-133, Profsousnaja 7

Ungarn

BÜKI, K., Dr.	Institut für Arzneimittelforschung, Budapest IV, Ujpest
MATKOVICS, B., Dr.	Institut für Organische Chemie der Universität, Szeged, Dom-Tér 8
PETTKO, E., Dr.	Medizinische Universität Szeged, Chemisches Institut, Beloiannisz Ter 8 II.

USA

CASPI, E., Dr.	Worcester Foundation for Experimental Biology, Shrewsbury, Massachusetts 01545, USA
CONNER, R. L., Prof.	Bryn Mawr College, Dept. of Biology, Bryn Mawr (Pennsylv.) 19010
SMITH, P. F., Prof. Dr.	University of South Dakota, Dept. of Microbiology, Vermillion, South Dakota 57069 USA
TALALAY, P., Prof. Dr.	John Hopkins University, School of Medicine, Baltimore, Md.

Index

FSC
www.fsc.org
MIX
Papier | Fördert gute Waldnutzung
FSC® C083411

Kolibri 360 GmbH
Ferdinand-Jühlke-Straße 7
99095 Erfurt, Deutschland
produktsicherheit@kolibri360.de